PROTEINS INVOLVED IN DNA REPLICATION

ADVANCES IN EXPERIMENTAL MEDICINE AND BIOLOGY

Recent Volumes in this Series

Volume 172
EUKARYOTIC CELL CULTURES: Basics and Applications
Edited by Ronald T. Acton and J. Daniel Lynn

Volume 173
MOLECULAR BIOLOGY AND PATHOGENESIS OF CORONAVIRUSES
Edited by P. J. M. Rottier, B. A. M. van der Zeijst,
W. J. M. Spaan, and M. C. Horzinek

Volume 174
GANGLIOSIDE STRUCTURE, FUNCTION AND BIOMEDICAL POTENTIAL
Edited by Robert W. Ledeen, Robert K. Yu, Maurice M. Rapport,
and Kunihiko Suzuki

Volume 175
NEUROTRANSMITTER RECEPTORS: Mechanisms of Action and Regulation
Edited by Shozo Kito, Tomio Segawa, Kinya Kuriyama, Henry I. Yamamura,
and Richard W. Olsen

Volume 176
HUMAN TROPHOBLAST NEOPLASMS
Edited by Roland A. Pattillo and Robert O. Hussa

Volume 177
NUTRITIONAL AND TOXICOLOGICAL ASPECTS OF FOOD SAFETY
Edited by Mendal Friedman

Volume 178
PHOSPHATE AND MINERAL METABOLISM
Edited by Shaul G. Massry, Giuseppe Maschio, and Eberhard Ritz

Volume 179
PROTEINS INVOLVED IN DNA REPLICATION
Edited by Ulrich Hübscher and Silvio Spadari

A Continuation Order Plan is available for this series. A continuation order will bring delivery of each new volume immediately upon publication. Volumes are billed only upon actual shipment. For further information please contact the publisher.

PROTEINS INVOLVED IN DNA REPLICATION

Edited by

Ulrich Hübscher

University of Zurich
Zurich, Switzerland

and

Silvio Spadari

Institute of Biochemical and
Evolutionary Genetics, CNR
Pavia, Italy

PLENUM PRESS • NEW YORK AND LONDON

Library of Congress Cataloging in Publication Data

Main entry under title:

Proteins involved in DNA replication.

(Advances in experimental medicine and biology; v. 179)
"Proceedings of a workshop on proteins involved in DNA replication sponsored by
the European Molecular Biology Organization (EMBO) and held September 19-23,
1983 at Vitznau, Switzerland."—verso t.p.
Includes bibliographical references and index.
1. DNA replication—Congresses. 2. Protein—Congresses. 3. Enzymes—Congresses.
I. Hübscher, Ulrich. II. Spadari, Silvio. III. European Molecular Biology Organization.
IV. Series. [DNLM: 1. DNA Replication—congresses. 2. Proteins—biosynthesis—
congresses. W1 AD559/QU 55W927p 1983]
QP624.P76   1984   574.87'3282   84-13449
ISBN 0-306-41804-5

Proceedings of a workshop on Proteins Involved in DNA Replication
sponsored by the European Molecular Biology Organization (EMBO)
and held September 19-23, 1983, at Vitznau, Switzerland

© 1984 Plenum Press, New York
A Division of Plenum Publishing Corporation
233 Spring Street, New York, N.Y. 10013

Printed in the United States of America

PREFACE

    This book collects the Proceedings of a workshop sponsored by
the European Molecular Biology Organization (EMBO) entitled "Proteins Involved in DNA Replication" which was held September 19 to
23,1983 at Vitznau, near Lucerne, in Switzerland.

    The aim of this workshop was to review and discuss the status
of our knowledge on the intricate array of enzymes and proteins
that allow the replication of the DNA. Since the first discovery of
a DNA polymerase in Escherichia coli by Arthur Kornberg twenty eight
years ago, a great number of enzymes and other proteins were described that are essential for this process: different DNA polymerases, DNA primases, DNA dependent ATPases, helicases, DNA ligases, DNA topoisomerases, exo- and endonucleases, DNA binding proteins and others. They are required for the initiation of a round
of synthesis at each replication origin, for the progress of the
growing fork, for the disentanglement of the replication product,
or for assuring the fidelity of the replication process.

    The number, variety and ways in which these proteins inter-
act with DNA and with each other to the achievement of replication
and to the maintenance of the physiological structure of the chromosomes is the subject of the contributions collected in this volume.
The presentations and discussions during this workshop reinforced
the view that DNA replication in vivo can only be achieved through
the cooperation of a high number of enzymes, proteins and other
cofactors. The need for clean and refined enzymological work, coupled to the contribution of the genetic analysis and molecular cloning, is as pressing as ever in order to obtain a satisfactory
picture of the processes at the molecular level.

    The authors thank all participants for contributing to a
friendly and scientifically fruitful meeting. Fifty eight papers
were selected to cover some of the most relevant recent approaches
and efforts in molecular biology of DNA replication.

    We want to express our gratitude to all those who helped to
organize this meeting and to the European Molecular Biology

Organization for its generous financial support. Furthermore we are indebted to Ursula Hübscher-Faé for her secretarial assistance before, during and after the meeting and for carefully typing all the manuscripts.

        March 1984   Ulrich Hübscher
                                                       Silvio Spadari

CONTENTS

vii

VII. DNA BINDING PROTEINS, NUCLEASES
AND POLY ADP-RIBOSE POLYMERASE

VIII. FIDELITY OF DNA REPLICATION

IX. DNA METHYLATION AND DNA METHYLASES

I

In Vitro Prokaryotic DNA
Replication Systems

ENZYME STUDIES OF REPLICATION OF THE ESCHERICHIA COLI CHROMOSOME

Arthur Kornberg

Department of Biochemistry
Stanford University School of Medicine
Stanford, California 94305, USA

I want at the outset to express my gratitude to the organizers, Ulrich Hübscher and Silvio Spadari, for their wisdom and initiative in convening a conference on a subject that is important, timely and not adequately appreciated: The Proteins Involved in DNA Replication.

During the very days of this Workshop, a more highly publicized symposium is being held in Cambridge, Massachusetts. It is organized by Nature magazine to celebrate 30 years of DNA. The subjects include DNA structure, gene expression, developmental biology and biomedical applications, but none of the twenty-one contributions to the program deals with proteins in DNA replication, repair or recombination. What better way is there to chronicle and glorify the recent history of DNA than to celebrate the enzymes that create and maintain it! It is often forgotten that these enzymes are the reagents that gave rise to the recombinant DNA technology that made studies on DNA structure, gene expression and biomedical applications possible. Not only do the proteins involved in DNA replication, repair, recombination and transposition have an important place in the recent history of biologic science, but the elucidation of chromosome structure and function in the future will depend far more on understanding these and related proteins than it will on the sequence and organization of DNA itself.

After more than twenty years of studying the proteins of DNA replication and their mechanisms, a number of basic facts have become clear. These verities are: (i) $dNTPs \rightarrow (dNMP)_n + nPP_i$, (ii) $5' \rightarrow 3'$ elongation, (iii) Watson-Crick base pairing, (iv) auxiliary subunits for processivity and fidelity, and (v) generally, but not universally, chain initiation by RNA priming.

3

I will dwell here on predominant patterns, knowing that there
are exceptions or multiple variations, even within a single cell.
It is disadvantageous or even lethal for a cell to lack metabolic
alternatives; the metabolism of DNA is no exception. As we learn
more about DNA metabolism of a single cell, we discover auxiliary
and alternative enzymes and arrangements, cryptic origins of replication and many possibilities for suppressing otherwise lethal mutations. Among the variations known, or likely, in DNA replication are:
(i) sizes and subunits of polymerases, (ii) processivity and fidelity
of polymerases, (iii) primases and their mechanisms, (iv) chromosome
initiation, and (v) chromosome termination and segregation. The stages
and principal actors in DNA replication to be reviewed here will
include: (i) elongation of a DNA chain by DNA polymerase III holoenzyme, (ii) initiation of a chain by RNA priming dependent on a
primase or a primosome, (iii) organization of chain elongation and
initiations at the replication fork in a complex assembly (the putative "replisome"), and (iv) initiation of a cycle of chromosome replication at its unique origin, called oriC. Not included in this review
is the biochemistry of termination of replication and segregation of
the daughter chromosome. Too little is yet known about this subject.
It will surely be enriched by future studies of the partition functions of plasmids.

Studies of duplex DNA replication in Escherichia coli, including that of its viruses and plasmids, impinge on many features of
DNA repair, recombination and transposition. In all these aspects,
Escherichia coli DNA replication continues to be an experimentally
attractive subject and is proving to be prototypical for basic mechanisms in replication throughout nature.

The strategy of my experimental approach to understanding DNA
replication has been to resolve and reconstitute the responsible
proteins. Mutants, when available, are invaluable as the source of
assays for isolation of proteins by functional complementation and
as touchstones to verify the pathway under study. Even without mutants
as pillars and guides, fractionation and purification of the entities
needed for properly defined replication events can proceed. Proteins
purified without adequate functional criteria are likely to remain
chronically unemployed.

Elongation of a DNA chain: Escherichia coli DNA polymerase III holoenzyme

Replication of DNA duplexes is generally semidiscontinous. One
strand (leading) is synthesized continuously; the other strand
(lagging) is synthesized in small pieces (Okazaki fragments), that
is, discontinuously (1, 2). The principal synthetic enzyme for both
strands in Escherichia coli is the DNA pol III holoenzyme. Holoenzyme
has a core of 3 subunits with a polypeptide of 140 kdal (α subunit)
responsible for polymerization and proofreading (Table 1)(3). Four

Table 1: Components of DNA polymerase III holoenzyme

Subunit	Mass kdal	Genetic locus	Function
α	140	dnaE*	polymerase; 3'-5' exonuclease
ε	25	dnaQ (mutD)*	regulation of proofreading
θ	10	unknown*	unknown
γ	52	dnaZ	ATP activation
δ	32	dnaX (?)	ATP activation
τ	78	dnaZ,X	ATP activation
β	37	dnaN	holoenzyme properties

* The α, ε and θ subunits constitute the polymerase III core.

or more additional subunits endow the core with high catalytic
efficiency, fidelity and processivity. Animal cell polymerases, as
in the case of Drosophila, are also multisubunit holoenzymes with a
polypeptide of 180 kdal responsible for polymerization (4); primase
resides in two of the smaller Drosophila polymerase subunits.

     Which of the dozen polypeptide bands commonly seen in highly
purified preparations of Escherichia coli DNA polymerase III holoenzyme are genuine and which are adventitious? To make this judgment,
we rely on several criteria.

     First is the persistence of a polypeptide band throughout the
purification procedure in an abundance roughly equivalent to that
of an authenticated subunit (e.g., α, β, ε). A second criterion is
the functional contribution a polypeptide (e.g., τ or γ) makes to an
in vitro assay. Unfortunately, our assays do not generally impose
the stringency of intracellular environments, nor do they elicit the
special capacities of an enzyme that express its versatility. McHenry
(5) showed that the τ subunit improves the catalytic activity of
pol III core on single-stranded templates; it does so by increasing
processivity and endowing the core with some holoenzyme features.
Yet, separation of the τ subunit from holoenzyme by high performance
liquid chromatography (HPLC) caused no loss in activity in several
assay systems that measure holoenzyme function (S. Biswas and A.
Kornberg, unpublished work). The γ subunit, which remains in these
subassemblies lacking τ, is very likely processed from the τ subunit
(6) and may be substituting for τ in these particular assays.

Alternative and more probing assays of holoenzyme function are always of prime importance. Beyond catalytic efficiency measured by DNA synthesis, are association and dissociation rates with various forms of templates and primers, tests of fidelity, and interactions with other proteins responsible for template binding, priming, repair and recombination.

An important criterion of subunit legitimacy is sometimes supplied by a specific inhibitor, such as rifampicin for the β subunit of RNA polymerase. Unfortunately, none is yet in hand for pol III holoenzyme. An antibody specific for a polypeptide would likely be useful, and C.S. McHenry (personal communication) has made a significant start with antibodies to the α, β and τ subunits. A powerful alternative to specific inhibitors is alteration of the polypeptide by mutations in the gene that encodes it. On this basis, holoenzyme has been judged to have at least 5 subunits (α, τ, γ, β and ε); it also seems likely that one of the δ bands is a product of the dnaX-Z locus. Several other polypeptides in the holoenzyme still lack an assigned genetic locus.

Initiation of a DNA chain: RNA priming by a primase or a primosome

With respect to initiation of chains, no DNA polymerase has this capacity. An oligoribonucleotide is transcribed from the template by a primase or RNA polymerase. With only a few notable exceptions (e.g., adenoviruses, phage φ29 and parvoviruses), starts of DNA chains are primed by RNA. Primases differ greatly from each other as well as from RNA polymerases. In some instances, the primase expresses a demand or preference for a template sequence; in others the primase counts the number of template residues to be transcribed but is indifferent to their sequence. Some primases also accept a deoxynucleotide in place of a ribonucleotide in extending the initial rNTP residue (generally ATP).

Of considerable mechanistic interest regarding Escherichia coli primase is how the enzyme is organized within a very complex primosome of some 20 polypeptides with a total mass near 10^6 daltons (Figure 1) (1-3). The primosome is translocated on the template in a locomotive-like structure with protein n' as the "engine" using ATP energy; dnaB protein serves as the "engineer" using ATP for conformational changes that stabilize or generate secondary structures in the DNA suitable for primase transcription. We need to know how these several components of the primosome are related to one another and how the primosome remains processive on the template strand, moving with a polarity that keeps it at the growing fork. With the primosome moving opposite to the direction of elongation, how does the primase within it add residues to elongate the oligonucleotide primers?

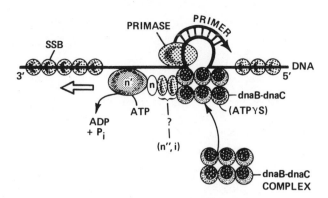

*Figure 1:   Chain initiation by the primosome. ATP and its analog
(ATPγS) support assembly of the primosome on SSB-coated DNA. For
primosome movement, ATP (or dATP) hydrolysis is essential and presumably the function of protein n'. The stoichiometry of the primosome subunits is known only for the dnaB-dnaC complex, protein n'
and protein n.*

Recent studies have examined Escherichia coli primase in its
synthesis of a 28-residue at the unique origin of phage G4 DNA (7).
These studies show that more than one primase molecule, very likely
two, are essential. Possibly, a primase molecule, unlike polymerases,
cannot translocate. Being able to add only one nucleotide within its
domain, it may rely on its partner primase molecule to add the next
nucleotide, and so on.

We have been concerned also with how a holoenzyme molecule
moves from one completed Okazaki fragment to the most nascent primer
at the replication fork. Recent results (8) show that transfer of a
holoenzyme molecule to another primer on the same template is far
more rapid than an intermolecular transfer (Figure 2). The holoenzyme
diffuses rapidly along DNA as proposed for repressors and other polymerases (9).

Organization of chain elongation and initiations at the replication fork: the "Replisome"

Very likely, the elongation of DNA chains and their initiations
on the lagging strand at the replication fork are organized in a
mechanism more elegantly designed than is described on paper (Figure
3) (1, 2). Although never taken alive, faith abides in a "replisome",
an assembly of enzymes responsible for advancing the replication
fork (1, 2). Such a structure operating at each replicating fork of
an Escherichia coli chromosome might contain a dimeric polymerase,
a primosome, and one or more helicases. An even more complex repli-

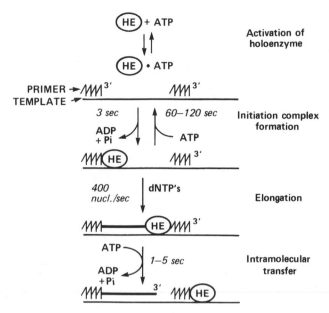

Figure 2: Scheme for the actions of holoenzyme dependent on ATP, including: (i) activation for formation of an initiation complex, and (ii) dissociation for cycling to another primer terminus.

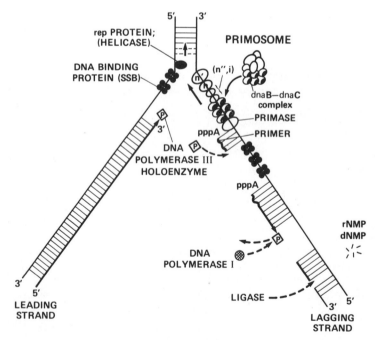

Figure 3: A scheme for DNA chain elongation and initiation in the semidiscontinuous growth at one of the forks of a bidirectionally replicating Escherichia coli chromosome.

some could be envisioned should both forks of the bidirectionally
replicating chromosome remain contiguous because of an attachment
to the cell membrane by specific proteins.

   This proposal of Escherichia coli DNA polymerase III holoenzyme
functioning as a pair of identical oligomers, each with an active
site, replicating both strands simultaneously has the attractive
feature of providing a model for concurrent replication of both
strands. Physical and functional linkage of the polymerase to the
primosome in a replisome structure would coordinate their actions.
Evidence in favor of a dimeric polymerase is still fragmentary: (i)
stoichiometry of two β subunits, two γ subunits, two primases, and
two molecules of ATP per replicating center; (ii) anomalously large
size of pol III' (core + τ) suggestive of a dimeric pol III core
and τ subunits (2); and (iii) looped DNA structures observed in association with the primosome (Figure 4) in view of the requirement
for looping of DNA in concurrent replication and the topological
feasibility of looping.

   In the scheme for concurrent replication (Figure 5) (2), the
leading strand is always ahead by the length of one nascent fragment;
the regions of the parental template strands undergoing simultaneous
synthesis are therefore not complementary. By looping the lagging
strand template 180° (perhaps halfway around the polymerase), the

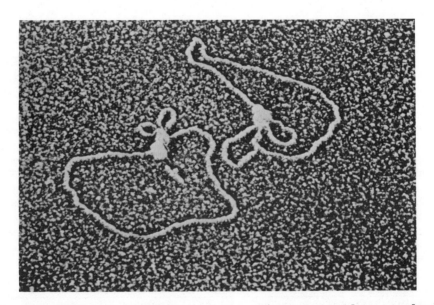

*Figure 4:   Electron micrograph of the primosome bound to covalently
closed ϕX174 duplex replicative form. These enzymatically synthesized
duplexes invariably contain a single primosome with one or two associated small DNA loops.*

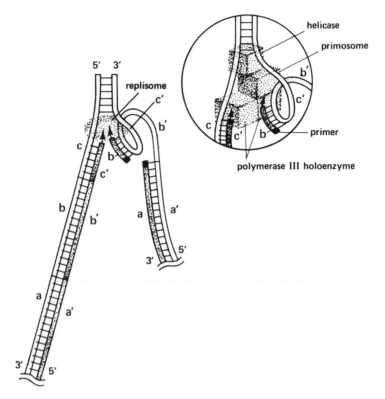

Figure 5: *Hypothetical scheme for concurrent replication of leading and lagging strands by a dimeric polymerase associated with a primosome and one or more helicases in a "replisome".*

strand has the same 3'→5' orientation at the fork as the leading strand. A primer generated by the primosome is extended by polymerase as the lagging strand template is drawn through it. When synthesis approaches the 5' end of the previous nascent fragment, the lagging strand template is released and becomes unlooped. Synthesis of the leading strand has, in the meanwhile, generated a fresh, unmatched length of template for synthesis of the next nascent fragment for the lagging strand. By this mechanism, one polymerase molecule not only copies both templates concurrently but also remains linked to the primosome whose movement is in the direction opposite the elongation of the lagging strand.

Initiation of a cycle of replication of the Escherichia coli chromosome at its unique origin (oriC)

Having considered the enzymology responsible for progress of the replication fork, we come now to the basic question of control and regulation of chromosome replication. In Escherichia coli, the major control is at the commitment to initiate a cycle of replication at the unique chromosomal origin (oriC) (Figure 6). What is the molecular anatomy of the complex switch that determines whether replication will be active or quiescent? The numerous sequences and spacer regions in oriC conserved through hundreds of millions of years in the Enterobacteriaceae imply a comparable conservation of proteins that recognize these sequences and their distinctive topological arrangements.

Several criteria were viewed as important in supporting the validity of a cell-free system responsible for starting a cycle of chromosome replication: (i) RNA polymerase dependence (rifampicin sensitivity), (ii) oriC sequence specificity, (iii) bidirectional replication from oriC, (iv) dependence on dnaA, and (v) dependence on key replication proteins (e.g., pol III holoenzyme, primase, dnaB protein, dnaC protein).

Availability of the Escherichia coli chromosomal origin (oriC) in a functional form as part of a small plasmid, first prepared by

```
                        BglII          BglII        50                              BamHI    100
CONSENSUS SEQUENCE: AGAAGATCTnTTTATTTAGAGATCTGTTnTATTGTGATCTCTTATTAGGATCGnnntnnnnTGTGGATAAGnnnGGATCCnnn
                                                                                ACACCTATT

    Escherichia coli: ..        A        .    . C    ..        CACTGCCC        CAAG.....GGC
Salmonella typhimurium: ..      T        .    . C    ..        CGCCAGGC        CCCG.....TGT
Enterobacter aerogenes: ..      -        .    . C    ..        ACTCTCTA        GTCG.....ACG
  Klebsiella pneumoniae: ..     -        .    . T    ..        GCTTGTCT        GTCA.....GCG
   Erwinia carotovora: GA       T      A      C  -   A-        TCGTGTTG        GTGATTATTCAT

101                                          150   AvaII                                      200
nTTtAAGATCAAnnnnnTggnAAGGATCncTAnCTGTGAATGATCGGTGATCcTGGnCcGTATAAGCTGGGATCAnAATGnGGnTTATACACAgcCCAA

T. T A        CAACC GGA...      AT..A              . ..A .        .G  . AG..G    .   A.T
AA A A        TGCGT GGA...      AC..G              G ..T .        .A  C GGTAC    .   A.T
A. T A        ACGCT AAG...      ACA.T              . .TT .        .G  . AA..G    G   G.T
G. T G        CCGTT AAG...      GC.TT              . ..T .        .A  . AA..G    .   G.A
A. A A        GAGAA GGCGTT      CT..C              . A.C A        TT  . TG..T    .   GGA

201                                  HindIII 250
AAAncgnACaaCGGTTaTTcTTTGGATAACTACCGGcTTGATCcCAAGCTTTtnAnCAgAGTTATCCACAntnGAnnGcnn-GAT :CONSENSUS SEQUENCE
                     AACCCTATT

.CTGA. AA.A    G .. .        . .. ... .......CCTGA  G..       GTA. TC.CAC-.   :Escherichia coli
.GTGA. AA..    A .. .        . . .. .........CC.C  G.T        ATG. TC.CAC-.   :Salmonella typhimurium
.GCAT. TC..    A .. .        . . .. .........TG.G  GG.        GAA. AGCTGCG.   :Enterobacter aerogenes
TTCAGG AA..    A A. .        . . ... ........TA.G  T...       GAA. AA.TAT-.   :Klebsiella pneumoniae
.ACGC. TCG.    G .A  G       G C  TTA ACCAGAA.TA.G  T..        TTCA CT.CCG-A   :Erwinia carotovora
```

Figure 6: Consensus sequences of the minimal origin of enteric bacterial chromosomes (10). The consensus sequence is derived from five bacterial oriC sequences. A large capital letter signifies that the same nucleotide is found in all five origins and a small capital letter in four of the five; lower case letters signify near randomness at that site. The minimal Escherichia coli oriC is enclosed in the box. The upper left end is the 5'end. The 9-basepair repeats are indicated by arrows with the 5'→3' sequence of the complementary strand given below the arrows. (Courtesy of Dr.J. Zyskind.)

Yasuda and Hirota (11), or as small phage chimera, prepared by
Kaguni, et al. (12), was crucial to discovery of the oriC enzyme
system (13, 14). It took nearly ten man-years of effort before an
enzyme system for oriC replication was achieved. Two things mattered
most. One was proceeding with an inactive lysate to prepare an ammonium sulfate fraction in which the numerous required proteins were
highly concentrated and from which inhibitors were excluded. The
second was the inclusion of a high concentration of polyethylene
glycol, a serendipitous vestige from trials with reagents that would
sustain transcription and translation concurrent with assays of
replication.

      In the resolution and reconstitution of the oriC system, two
standard approaches have been taken (15). In one approach, the assays
are based on complementation of a crude extract from mutant cells.
This has been successful only in the identification and isolation
of the dnaA protein. To date, no other gene previously implicated or
suspected of being involved specifically in chromosome cycle initiation (e.g., dnaI, dnaP, dnaJ, dnaK, dnaL) has been validated or proved
useful in pursuing the oriC problem. From a strain (constructed in
vitro) which overproduces the dnaA protein more than 200-fold, the
52-kdal polypeptide was purified to near homogeneity (16). Although
the protein tends to aggregate, monomer-sized protein purified by
HPLC is fully active for replication and participates at an early
stage in initiating bidirectional replication from oriC.

      Specific binding of dnaA protein to four 9-basepair (bp) se-
quences, highly conserved within oriC of Enterobacteriaceae (Figure
6) (10), provides the basis for interaction with supercoiled oriC
DNA. Plasmids with sites of moderate affinity for dnaA protein contain a copy of at least an 8/9 match of the consensus sequence
TTAT^C_ACACA. From ^{32}P-end-labeled TaqI digests of a variety of plasmid
DNAs, dnaA protein retains on Millipore filters only those fragments
which contain the 9-bp sequence. Fragments containing this 9-bp sequence within or near the following sequences are specifically bound:
(i) oriC, (ii) the promoter of a 15.5-kdal protein coding sequence
adjacent to oriC (found in both M13oriC26 and M13oriC26Δ221), (iii)
the region between the two promoters for the dnaA gene (17), (iv) the
IR$_L$ sequence of Tn5 (18), (v) the region between the origin of DNA
synthesis and the protein n' ATPase site on the L strand of pBR322
(19), and (vi) the origin of replication of pSC101. Fragments containing oriC with its four 9-bp sequences are bound more tightly
than other fragments containing only one. Thus, the dnaA protein-
binding may have a positive role at certain sites (e.g., oriC and
ori-pSC101) and a negative role at others (e.g., ori-pBR322, ori-
ColE1 and the control region for the dnaA gene).

      The other approach to clarifying the oriC system relies on
assembling the replication proteins known to be needed and on fractionating the crude extract for the additional proteins required

for optimal activity and specificity. The replication proteins include RNA polymerase, DNA gyrase, dnaA, dnaB, and dnaC proteins,
primase, pol III holoenzyme, and single-strand binding protein (SSB).
The novel factors identified by fractionation include those that act
in a positive sense in contrast to "specificity proteins" that suppress replication of plasmids that lack oriC and elicit the dnaA
protein dependence of oriC plasmids. The specificity role bears some
analogy to that of SSB in single-strand phage DNA replication, where
SSB maintains specificity of priming at complementary strand origins
and dependence on specific priming mechanisms.

One of the positive factors has been purified to homogeneity
(20) and has proven to be identical to protein HU, the well-known
and abundant, small basic protein that binds both single- and double-
stranded DNA (21, 22) and is associated with the bacterial nucleoid
(23). Forty dimers of HU per template circle stimulate replication
of a 12.2 kb oriC plasmid 3- to 5-fold. This represents one molecule
of the protein per 300 basepairs, or about one-tenth the level required for activation of transcription or extensive condensation of
the DNA. The more stringent requirement for HU in replication, and
its apparent lack of positive effect on dnaA-independent reactions
suggest oriC-specificity in its action. Its binding to double-stranded DNA, production of nucleosome-like structures in the presence
of a type I topoisomerase (24), and stimulation of transcription of
λ DNA (21), indicate two possible functions in replication. One is
preparation of the template, where HU, in concert with gyrase and
topoisomerase I (see below), could introduce and preserve some favorable structure in the DNA in the vicinity of oriC, a process perhaps
directed by the dnaA protein. The second possibility is that HU specifically binds and stabilizes the annealed nascent RNA transcript
destined for use as primer for leading strand synthesis.

Further study of the mechanism of the actions of HU will become
accessible with more complete reconstitution of the replication of
several oriC-containing DNAs. Assays in hand also suggest that additional positively acting factors are required in their enzymatic replication. One of the "specificity proteins" purified to homogeneity
has been identified as topoisomerase I. In its absence, the reconstitution reaction is not specific for oriC-containing templates, nor
is it dependent on dnaA protein. Specificity and dnaA dependence are
restored by topoisomerase I (J.M. Kaguni and A. Kornberg, unpublished
observation). An alternate reconstitution reaction which differs in
being largely dnaA-independent in the presence of topoisomerase I
indicates a requirement for still more specificity proteins. Recent
studies indicate that one of these activities is ribonuclease H present as minor impurity in some of the partially purified preparations
(T. Ogawa, N.E. Dixon and A. Kornberg, unpublished oberservation).

Escherichia coli rnh mutants defective in RNase H activity
display the features of previously described sdrA (stable DNA repli-

cation) and dasF (dnaA suppressor) mutants: (i) sustained DNA replication in the absence of protein synthesis, (ii) lack of requirement
for dnaA protein and the origin of replication (oriC), and (iii)
sensitivity of growth to a rich medium (25). Both the sdrA mutants
(selected for continued DNA replication in the absence of protein
synthesis) and the dasF mutants (selected as dnaA suppressors) are
defective in RNase H activity measured in vitro. Furthermore, a
760-base pair fragment containing the rnh[+] structural gene complements
the phenotype of each of the rnh, sdrA and dasF mutants, indicative
of a single gene. Thus, one function of RNase H in vivo is in the
initiation of a cycle of DNA replication at oriC dependent on dnaA[+].
In keeping with these results, RNase H contributes to the specificity of dnaA protein-dependent replication initiated at oriC (25) in
the partially purified enzyme system (15).

SUMMARY AND CONCLUSIONS

     Progress of the replication forks of the Escherichia coli
chromosome depends on a multisubunit DNA polymerase (for chain elongation) and a primosome (for chain initiations), together comprising
about 30 polypeptides with a mass in excess of 10^6 daltons. Integration of their actions with those of helicases and DNA binding proteins suggest a more complex and integrated replisome assembly with
novel possibilities for concurrent replication of both parental
strands. Initiation of a new cycle of chromosome replication at its
unique 245-bp (oriC) is being studied in a soluble enzyme system
with plasmids, autonomous replication of which depends on the oriC
sequence. Required proteins include RNA polymerase, DNA gyrase, dnaA
protein (with 4 strong binding sites in oriC), HU protein, and additional proteins (e.g., topoisomerase I and ribonuclease H) that confer
oriC specificity by suppressing initiation of replication elsewhere
on the duplex DNA.

     Clarification of the biochemical mechanisms of replication is
fundamental for understanding cell growth and development. Knowledge
of the biochemistry of initiating a cycle of chromosome replication
opens the way toward exploring the regulation of the cell cycle.

     I remain faithful to the conviction that anything a cell can
do, a biochemist should be able to do. He should do it even better,
being freed from the constraints of substrate and enzyme concentrations, pH, ionic strength, and temperature, and by having the license
to introduce novel reagents to drive or restrain a reaction. Put
another way, one can be creative more easily with a reconstituted
system. One can grapple directly with the molecules instead of trying
by remote means to manipulate their structures or levels in the intact cell.

Enzyme purification carries many dangers beyond the well-known exposure of the fragile enzyme to the hostilities of an unfamiliar environment, high dilution, glass containers and a denaturable investigator. But the rewards of enzyme purification have justified the effort. The polymerases, nucleases, ligases purified out of curiosity about the mechanisms of replication, repair and recombination have supplied the cast of actors responsible for the current drama of genetic engineering. Beyond the uses of these enzymes as reagents, understanding the mechanisms of DNA metabolism will have practical value in manipulating the replication of plasmids and viruses and the expression of their genes and, beyond that, in obtaining a more secure grasp of chromosome structure and function.

ACKNOWLEDGEMENTS

Understanding of DNA replication has relied on studies of several systems described in this volume. In particular, these would include the findings with phages T7 and T4, the small phages, plasmids, animal viruses and eukaryotic as well as prokaryotic chromosomes. With regard to contributions from my laboratory, I want to express my indebtedness to the many students and postdoctoral fellows who did so much to make this work possible. I also wish to state that this work depended on grants from the National Institutes of Health and the National Science Foundation.

REFERENCES

1.   Kornberg,A. (1980) DNA Replication, San Francisco:
           W.H. Freeman.
2.   Kornberg,A. (1982) Supplement to DNA Replication, San Francisco:
           W.H. Freeman.
3.   Dixon,N.E., Bertsch,L.L., Biswas,S.B., Burgers,P.M.J.,
           Flynn,J.E., Fuller,R.S., Kaguni,J.M., Kodaira,M.,
           Stayton,M.M. and Kornberg,A. (1983) in: Mechanisms
           of DNA Replication and Recombination, Cozzarelli,N.R.,
           ed., UCLA Symposia on Molecular and Cellular Biology,
           News Series, Vol. 10, Alan R. Liss, Inc. New York, 93.
4.   Kaguni,L.S., Rossignol,J.-M., Conaway,R.D. and Lehman,I.R.
           (1983) in: Mechanisms of DNA Replication and Recombina-
           tion, Cozzarelli,N.R. ed., UCLA Symposia on Molecular
           and Cellular Biology, New Series, Vol. 10, Alan R. Liss,
           Inc., New York, 495.
5.   McHenry,C.S. (1982) J. Biol. Chem. 257, 2657.
6.   Kodaira,M., Biswas,S.B. and Kornberg,A. (1983) Molec. Gen.
           Genet. 192, 80.
7.   Stayton,M. and Kornberg,A. (1983) J. Biol. Chem. 258, 13205.
8.   Burgers,P.M.J. and Kornberg,A. (1983) J. Biol. Chem. 258, 7669.
9.   Berg,O.G., Winter,R.B. and von Hippel,P.H. (1981) Biochemistry
           20, 6929.

10. Zyskind,J.W., Cleary,J.M., Bruslow,W.S.A., Harding,N.E.
        and Smith,D.W. (1983) Proc. Natl. Acad. Sci. USA
        80, 1164.
11. Yasuda,S. and Hirota,Y. (1977) Proc. Natl. Acad. Sci. USA
        74, 5458.
12. Kaguni,J.M., LaVerne,L.S. and Ray,D.S. (1981) Proc. Natl.
        Acad. Sci. USA 76, 6250.
13. Fuller,R.S., Kaguni,J.M. and Kornberg,A. (1981) Proc. Natl.
        Acad. Sci. USA 78, 7370.
14. Kaguni,J.M., Fuller,R.S. and Kornberg,A. (1982)
        Nature 296, 623.
15. Fuller,R.S., Bertsch,L.L., Dixon,N.E., Flynn,J.E., Kaguni,J.M.,
        Low,R.L., Ogawa,T. and Kornberg,A. (1983) in:
        Mechanisms of DNA Replication and Recombination,
        Cozzarelli,N.R., ed., UCLA Symposia on Molecular and
        Cellular Biology, New Series, Vol. 10, Alan R. Liss,
        Inc., New York, 275.
16. Fuller,R.S. and Kornberg,A. (1983) Proc. Natl. Acad. Sci. USA
        80, 5817.
17. Hansen,E.B., Hansen,F.G. and vonMeyenberg,K. (1982)
        Nucleic Acids Res. 10, 7373.
18. Auserwald,E.-A., Ludwig,G. and Schaller,H. (1981)
        Cold Spring Harbor Symp. Quant. Biol. 45, 107.
19. Marions,K.J., Soeller,W. and Zipursky,S.L. (1982)
        J. Biol. Chem. 257, 5656.
20. Dixon,N.E. and Kornberg,A. (1984) Proc. Natl. Acad. Sci. USA,
        in press.
21. Rouvière-Yaniv,J. and Gros,F. (1975) Proc. Natl. Acad. Sci.
        USA 73, 3428.
22. Geider,K. and Hoffmann-Berling,H. (1981) Ann. Rev. Biochem.
        50, 233.
23. Rouvière-Yaniv,J. (1977) Cold Spring Harbor Symp. Quant. Biol.
        42, 439.
24. Rouvière-Yaniv,J., Yaniv,M. and Germond,J.-E. (1979) Cell
        17, 265.
25. Ogawa,T., Pickett,G.G., Kogoma,T. and Kornberg,A. (1984)
        Proc. Natl. Acad. Sci. USA, in press.

ENZYMOLOGICAL STUDIES OF THE T4 REPLICATION PROTEINS

C.Victor Jongeneel[1], Timothy Formosa,
Maureen Munn and Bruce M. Alberts

Department of Biochemistry and Biophysics
University of California, San Francisco
San Francisco, California 94143, USA

[1]Swiss Institute for Experimental Cancer Research
CH-1026 Epalinges, Switzerland

INTRODUCTION

Ever since the pioneering efforts of Delbruck and the early
phage workers, bacteriophage T4 has been a favorite model for un-
raveling the fundamental processes of molecular genetics. The replication of double-stranded DNA is no exception, even though it was
not until the last ten years or so that substantial progress was
made in purifying the gene products involved in this process. Extensive genetic analysis of bacteriophage T4 has identified eleven genes
whose products appear to be directly involved in the formation and
subsequent moving of DNA replication forks: genes 32, 39, 41, 43,
44, 45, 52, 60, 61, 62, and dda (1-5). The proteins specified by
these genes have all been isolated and purified to near homogeneity
(6-10, Jongeneel, Formosa and Alberts, submitted).

The functions of the individual gene products have been found
to be the following: DNA polymerase (the product of gene 43), polymerase accessory proteins (genes 44, 62, and 45), helix-destabilizing
protein (gene 32), DNA helicases (genes 41 and dda), RNA primase
(genes 41 and 61), and type II DNA topoisomerase (genes 39, 52 and
60). We shall briefly review our current understanding of the contribution of each protein to the function of the bacteriophage T4 DNA
replication multienzyme. A more comprehensive review of the properties of the bacteriophage T4 replication proteins has been recently
published (11).

17

By itself, the T4 DNA polymerase can elongate pre-existing primers on single-stranded DNA templates (12, 13). Short regions of helical structure in the template strand, or the long helix created by a complementary strand annealed to this template strand, act as barriers to continued polymerization by the DNA polymerase (14). Even on templates without regions of helical structure, the processivity of the polymerase is very low at physiological salt concentrations (15). Two different types of proteins affect the rate and the processivity of the DNA polymerase on single-stranded DNA templates. When added in amounts sufficient to cover the template strand, the gene 32 protein can melt out its short helical regions and present the strand to the polymerase in a conformation that increases both the rate and the processivity of DNA synthesis (16, 17). This effect is specific and reflects direct physical interactions between the DNA polymerase and the 32 protein (18). The polymerase accessory proteins stimulate the DNA polymerase in a very different manner. Together, the 44/62 and 45 proteins have a DNA-dependent ATPase activity that is activated by single-stranded DNA ends (19). Frequent hydrolysis of ATP is required for their stimulatory effect on the DNA polymerase (20). They are thought to act by holding the polymerase in place at its primer-template junction, thereby greatly increasing the processivity of its synthesis (11, 17).

The mixture of the gene 32, 43, 44/62 and 45 proteins constitutes a "core" replication system, being the minimum number of proteins required for efficient DNA synthesis by strand displacement starting from a nick on a double-stranded DNA template (21, 22). When any one of these proteins or ATP is omitted, almost no DNA synthesis is detected. The rate of fork movement observed with the core replication system is highly dependent on the 32 protein concentration, and more than 500 µg/ml of this protein is required to approach the fork rates observed in vivo.

The gene 41 protein has a dual function as a DNA helicase and as a component of the RNA primase. It has a DNA-dependent GTPase and ATPase activity, and it uses the energy of nucleotide hydrolysis to move rapidly and processively in the 5' to 3' direction along a single-stranded DNA molecule (23, 24). At high GTP concentrations, the 41 protein can displace short DNA fragments paired to the strand along which it is moving, thus justifying its classification as a DNA helicase (24). When added to the core reactions described above, the 41 protein greatly enhances the rate of replication fork movement (20, 24). This effect is particularly strong at 32 protein concentrations below 50 µg/ml, where reactions in the core system are very slow.

In order to reveal the RNA primase activity of the 41 protein, the product of gene 61 is required. Together these two proteins produce the pentaribonucleotides that serve as primers for the synthesis of Okazaki fragments on the lagging side of the replication fork.

It is not known which of these two proteins contains the active site at which ribonucleoside triphosphates are polymerized (25-27).

A mixture of the seven proteins described so far constitutes the "core-41-61" system, which produces replication forks that closely resemble those that function in vivo (26, 28-30): 1) the template DNA is replicated semi-conservatively at about the in vivo rate and with high fidelity; 2) the complex that moves a replication fork is very stable once it is assembled, proceeding for more than 50,000 nucleotides on the same DNA template; 3) Okazaki fragments with an average length of about 1500 nucleotides are synthesized on the lagging strand; 4) the RNA primers that start the Okazaki fragments have the same length and sequence (pppApCpNpNpN) as in vivo.

Bacteriophage T4 codes for another DNA helicase, the product of the dda gene (31-33). This protein is not essential for viral replication, presumably because it can be complemented in vivo by an Escherichia coli host protein with a similar function. In vitro studies have revealed that the dda protein can stimulate strand displacement synthesis in the T4 core replication system and that it does so by acting in a distributive fashion (Jongeneel, Bedinger and Alberts, submitted). Clues to an important in vivo function for the dda protein came from experiments designed to study the collisions of DNA replication forks with RNA polymerase molecules bound to the DNA template. While core or core-41 replication complexes are unable to bypass the obstacle created by a bound RNA polymerase molecule, addition of small amounts of the dda protein removes the inhibition (34). The dda protein may thus be required for replication forks to traverse regions in the DNA template that are occupied by RNA polymerase molecules and other proteins that normally function by binding tightly to double-stranded DNA.

The role of the T4-specific type II topoisomerase in viral DNA replication is still unclear. Mutants in the three "DNA-delay" genes that code for the topoisomerase subunits (35, 36) are deficient in the number of replication forks that form rather than their rate of movement (4), suggesting that the topoisomerase is involved in the initiation of replication forks. Recent work in our laboratory has shown that the topoisomerase preferentially recognizes a limited number of sites in the T4 genome, reinforcing this hypothesis (37, Kreuzer and Alberts, submitted). However, definitive evidence for the direct involvement of the topoisomerase in the initiation of T4 DNA replication forks is still lacking.

In the infected cell, two different pathways or mechanisms seem to be used for the initiation of new replication forks (38-40). The primary mode of initiation, which is used only early in infection, generates replication bubbles at one or a few specific origins on the parental DNA molecule, and directly requires RNA synthesis by the host RNA polymerase (40, 41). Later in infection, initiation of

new replication forks becomes independent of the host RNA polymerase
(40). This secondary mode of initiation requires an active T4 genetic
recombination system, and it has been suggested that recombination
intermediates directly prime the initiation of these replication
forks (39, 40). The most abundant component of the T4 recombination
system is the product of the uvsX gene, which has a function analogous
to the Escherichia coli recA protein (T. Minagawa, personal communication). In this sense, the uvsX protein, as well as other proteins
involved in T4 genetic recombination, can also be thought of as DNA
replication proteins.

       A long-standing goal of our laboratory has been to gain a de-
tailed understanding of all of the steps involved in the replication
of double-stranded DNA, and of the role of the various replication
proteins in each of those steps. In this report, we address three
specific questions related to that overall goal: 1) We had demonstrated previously that the dda protein is required for a replication
fork to proceed past a RNA polymerase molecule that is moving in the
same direction as the leading strand of the replication fork (34).
We have now extended these studies to show that this is also true
for RNA polymerase molecules moving in the opposite direction. 2)
The spatial arrangement of the DNA polymerase and the other components
of the replication complex around the primer:template junction is
still unknown. We report some recent experiments that explore the
recognition of primer:template junctions by the polymerase accessory
proteins. 3) As the first step toward the reconstitution in vitro of
the mechanism that uses genetic recombination to initiate replication
forks, we have purified and begun a study of the T4 uvsX protein,
comparing it to the closely-related Escherichia coli recA protein.

RESULTS AND DISCUSSION

Effects of the dda protein on replication forks blocked by RNA
polymerase

       In order to study the effects of RNA polymerase on the movement
of DNA replication forks, we devised the following assay (34, 42).
A covalently closed circular DNA containing the filamentous phage
(fd or M13) replication origin is nicked at a specific site in the
viral strand by the action of purified fd gene 2 protein (43). The
nick provides a unique site on the DNA template at which the T4
replication proteins can assemble to initiate a replication fork.
This DNA template is incubated with the Escherichia coli RNA polymerase to allow the polymerase to bind at promoter sites on the DNA.
DNA synthesis is then started in a synchronous fashion in the presence of labeled deoxyribonucleoside triphosphates, and the size of
the radioactively labeled DNA made during the resulting rolling
circle DNA replication is determined by gel electrophoresis under
denaturing conditions (Figure 1). The size distribution of the pro-

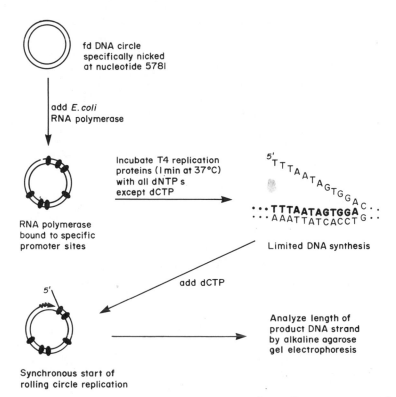

Figure 1: *Schematic outline of the method used to measure the effect of template-bound RNA polymerase molecules on DNA replication fork movement.*

ducts, as determined by scanning an autoradiogram of the dried gel, allows one to determine the rate of replication fork movement, as well as the location of any site-specific blocks to continued DNA synthesis.

Using this approach, we have demonstrated previously that a single RNA polymerase molecule bound at a promoter site can completely block the progress of a replication fork beyond that site (34). This holds true for the replication forks formed by both core and core-41 reactions, and for both stationary and transcribing RNA polymerase molecules. The addition of as little as 2 µg/ml of the dda protein completely reverses this block, as measurements of fork rates no longer show any effect of the RNA polymerase, and no replication complexes blocked at promoter sites are found.

For the DNA templates used in our previous experiments, the initial block to replication fork movement was caused by an RNA polymerase molecule bound to the gene 2 promoter of filamentous phage, which was located 180 nucleotides downstream from the specific nicking site where replication forks began (the same configuration found in the phage). Since RNA transcription from this promoter proceeds in the same direction as the replication fork, the RNA polymerase uses the same template strand as does the DNA polymerase molecule on the leading strand at the fork. In order to test whether the blocking of the fork by RNA polymerase occurs irrespective of promoter orientation, we have constructed a plasmid in which the same gene 2 promoter faces in the opposite direction relative to the specific nicking site. To construct such a DNA molecule, we took advantage of the fact that there is an AvaI restriction enzyme site between the specific nicking site and the gene 2 promoter in the genome of the filamentous phage M13. The pJMC110 plasmid, which contains 393 nucleotides from the M13 origin region inserted into the BamHl site of pBR322 (44), also contains a second AvaI cutting site located in the pBR322 sequence, 1049 nucleotides from the end of the insert that is close to the gene 2 promoter. (Therefore, the gene 2 promoter and the site at which the gene 2 protein nicks are on different DNA fragments after cutting the pJMC110 plasmid DNA with AvaI.) A new plasmid, designated as pCVJ2, was constructed by cutting the pJMC110 plasmid with AvaI, filling in the ends of the resulting two fragments with DNA polymerase to create blunt ends, treating with DNA ligase, and screening for transformants containing plasmids with the desired inversion of the fragments. In this new plasmid, the gene 2 protein nicking site and the gene 2 promoter are separated by 1094 nucleotides, and the promoter faces in a direction opposite to the direction of movement of a replication fork that starts at the specific nick.

Figure 2 illustrates the results of an experiment in which either the pJMC110 or the pCVJ2 plasmid, after nicking with gene 2 protein, was used as the template for DNA replication with and without RNA polymerase present, but in the absence of RNA synthesis. In the left panel, pJMC110 DNA was used as the template. The result is the same as reported previously (34): in the absence of the dda protein, the addition of RNA polymerase to the core replication system blocks the fork at the gene 2 promoter site, but the addition of the dda protein completely releases this block. In the right panel, the same experiment was carried out with the pCVJ2 DNA template. Three discretely-sized DNA bands appear upon addition of RNA polymerase, corresponding to forks stopped at the three promoters known to exist on this plasmid. The shortest DNA fragment is produced when the fork stops at the M13 gene 2 promoter, since the RNA polymerase bound here is the first one to be encountered by the moving fork. However, some forks proceed past this point and become arrested instead at either the promoter for the RNA that primes DNA synthesis at the origin of plasmid DNA replication, or the promoter cluster for the transcription of the β-lactamase and tetracycline resistance genes (45). Thus, the blocking effect of an RNA polymerase molecule that faces in the direction opposite to fork movement does not seem quite

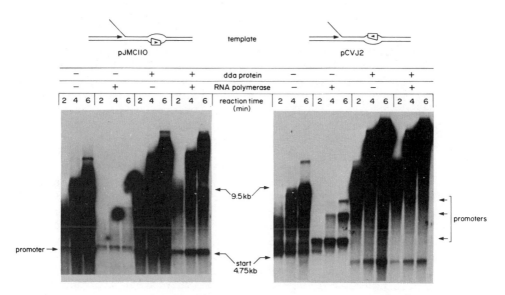

*Figure 2: Effects of a stationary RNA polymerase molecule on core
replication reactions with and without the dda protein. In vitro DNA
synthesis was carried out in the core replication system exactly as
described in Bedinger et al.(34). In the indicated reactions, Escherichia coli RNA polymerase was added at 4 µg/ml, and the dda protein
was added at 2 µg/ml. At various times after the addition of dCTP
(which begins DNA synthesis), 12 µl aliquots were removed and further
reaction stopped by adding an equal volume of 20 mM Na₃EDTA and 2%
SDS. After a rapid gel filtration step to remove unincorporated
nucleotides, the reaction products were analyzed by electrophoresis
through a 0.5% agarose gel run in 30 mM NaOH, 0.1 mM Na₃EDTA, followed
by autoradiography of the dried gel.*
*The positions of the primer strand before elongation (4.75 kb) and
after one round of rolling circle replication (9.5 kb) were determined
from the migration rate of appropriate DNA size markers. Other arrows
mark the positions expected for strands terminating at known promoter
positions on the DNA template. For the pJMC110 DNA, the position of
the gene 2 promoter is shown, while the promoters for gene 2 (bottom
arrow), the plasmid origin (middle arrow), and the β-lactamase gene
(top arrow) are marked for the pCVJ2 plasmid.*

as absolute as that seen with an RNA polymerase molecule facing in
the same direction, inasmuch as more forks eventually make it past
the first promoter that they encounter (the gene 2 promoter) in the
pCVJ2 DNA than in the pJMC110 DNA. Nevertheless, RNA polymerase molecules bound to the DNA template in either orientation can be seen to
interfere significantly with the progress of the replication fork.

The effect of adding the dda protein to the core replication reaction is the same for either promoter orientation: the blocking effect of the RNA polymerase molecule completely disappears (Figure 2). As found previously (34), the polymerase still inhibits net DNA synthesis slightly in the presence of the dda protein. As discussed in our earlier communication, this inhibition seems to be mostly due to interference with assembly of the replication complex at the nick in the template DNA, and it is attributed to the known strong binding of RNA polymerase to nicks in double-helical DNAs.

The experiment depicted in Figure 3 is identical to the one in Figure 2, except that all four ribonucleoside triphosphates were added, thereby allowing the RNA polymerase present to synthesize RNA. As expected, the discrete bands on the gel corresponding to forks blocked at promoter sites are no longer seen. Instead, the reactions in which RNA polymerase was added without the dda protein show smears on the gel autoradiograms, each smear corresponding to a population of forks blocked at the variety of sites expected to be occupied by the transcribing RNA polymerase molecules that start from a particular promoter site. In the reaction where pJMC110 DNA was used as the

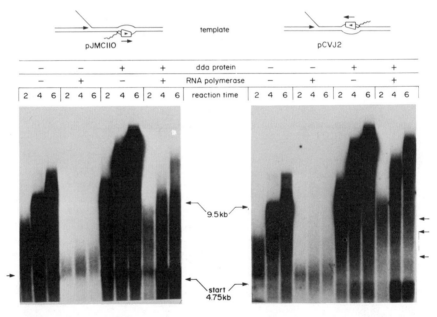

Figure 3: Effects of a transcribing RNA polymerase molecule on core replication reactions with and without the dda protein. The experimental conditions and analysis of the reaction products were identical to those in Figure 2, except that, in addition to ATP, the ribonucleoside triphosphates CTP, GTP and UTP were present in the reaction at 0.1 mM each. The arrows on the outside of each panel mark the same positions as in Figure 2.

template, the size distribution of the DNA made at blocked forks
reveals that DNA synthesis has continued beyond the promoter site,
suggesting that the forks follow a transcribing RNA polymerase molecule at the slow rate of its transcription. In the reaction with a
pCVJ2 DNA template, however, all of the labeled DNA strands are
shorter than would have been expected if the fork had reached the
first promoter, suggesting that these DNA strands were terminated
by a "head-on" collision between a replication fork and a transcribing RNA polymerase molecule. In either case, the replication fork
clearly cannot proceed past a RNA polymerase molecule when only the
core replication system is present. However, the addition of the dda
protein removes this block to fork movement, regardless of the relative orientation of the promoter (Figure 3).

    These observations extend our previous results. A template-
bound RNA polymerase molecule can block the progress of the T4 bacteriophage DNA replication complex irrespective of the relative
orientation of the RNA polymerase and replication fork movements,
and whether or not the RNA polymerase is synthesizing RNA. In every
case, addition of small amounts of the dda protein reverses this
block. This result lends further credence to our hypothesis that the
dda protein (and presumably at least one homologous Escherichia coli
host protein) plays an important role in DNA replication by allowing
the replication fork to proceed past RNA polymerase molecules bound
to the DNA template. We plan to investigate whether the same type
of results are observed for replication fork encounters with other
proteins that are known to bind tightly to double-stranded DNA, such
as gene regulatory proteins and histones.

Recognition of primer-template junctions by the polymerase accessory
proteins

    In the absence of the T4 DNA polymerase, the 44/62 protein com-
plex exhibits a DNA-dependent ATPase (and dATPase) activity that is
stimulated by the 45 protein. Studies of the DNA cofactor requirements in the ATPase reaction have been useful for detecting the DNA
binding site for these proteins. It was previously shown that ATP
hydrolysis is maximally stimulated by single-stranded DNA ends (19).
In recent experiments, we have compared the stimulatory effect of
double-stranded DNAs cut with various restriction enzymes in order
to determine whether the short single-stranded overhangs that protrude
from the ends are sufficient to stimulate the ATP hydrolysis event
and whether any difference is seen between 3' versus 5' single-stranded overhangs in this regard. The results of this study are summarized in Table 1.

    The data reveal several interesting features. First, the V_{max}
obtained in the presence of restriction cut DNA is 5-10 times higher
than that observed with the uncut circular DNA, confirming the requirement for DNA ends reported before by Piperno et al. (19).

C. V. JONGENEEL ET AL.

Table 1: Stimulation of ATPase activity of the T4 DNA polymerase accessory proteins by DNAs cut with various restriction endonucleases *

Enzyme	Type of end produced	K_m (nM DNA ends)	V_{max} (nmol min^{-1}ml^{-1})
1. none	none	-	0.25
2. AluI	blunt	34	2.0
3. HaeIII	4 base 3' overhang	32	1.5
4. HgiaI	4 base 3' overhang	13	1.4
5. HgaI	5 base 5' overhang	9.1	2.3
6. Sau3A	4 base 5' overhang	7.4	2.5

* Reaction mixtures of 10 μl containing 40 mM Tris-acetate, pH 7.8, 80 mM potassium acetate, 12 mM magnesium acetate, 120 μg/ml bovine serum albumin, 0.6 mM dithiothreitol, 250 μM ATP, 10⁶ cpm (γ-³²P)-ATP, 5 μg/ml 44/62 protein, 15 μg/ml 45 protein, and either cut or uncut pJMC110 DNA (see ref. 44) were incubated for 15 minutes at 37ºC. Reactions were stopped by the addition of 10 μl of cold 40 mM Na₃EDTA and placed on ice. To separate inorganic phosphate from ATP, 5 μl aliquots were spotted on a polyethyleneimine cellulose chromatography plate which was developed using 1 M formic acid, 0.5 M LiCl. Radioactive phosphate and ATP spots were located by autoradiography, cut out and quantitated by measuring their Cerenkov emmission. Stimulation of ATPase activity was calculated by subtracting the basal level of activity obtained without DNA from the raw data. The K_m values are expressed in terms of nM DNA ends to account for the fact that the number of restriction sites for each enzyme differed by as much as a factor of two. Values for K_m and V_{max} were estimated graphically from plots of activity versus DNA concentration.

Furthermore, there is a distinct difference in the degree of stimulation depending on the type of end present. Linear DNA molecules
with protruding 5' ends stimulate the reaction most efficiently,
being equivalent to single-stranded DNA. Next best are DNA molecules
with blunt-ends, while DNA molecules with 3' overhangs are the least
effective. The type of double-stranded end affects both the apparent
K_m and the V_{max}. The effect on V_{max} suggests that the accessory proteins undergo a conformational change that increases their ATP hydrolysis activity when they bind to a site with a 5' overhang.

The preference of these polymerase accessory proteins for
double-stranded DNA ends with 5' overhangs (and corresponding recessed
3' hydroxyl groups) as cofactors for their ATPase activity suggests
that these proteins preferentially recognize and bind to a primer-
template junction. Clearly, single-stranded ends are not in themselves
sufficient for this recognition. Moreover, because a higher V_{max} is
achieved in the presence of blunt-ended DNA fragments than with those
with 3' overhangs, we suggest that the blunt-ended DNA molecules may
partially fulfill the requirement for a primer-template junction by
providing a base-paired 3' hydroxyl terminus.

The precise locations of the polymerase accessory proteins and
other replication proteins at the primer-template junction are presently under study using DNA footprinting techniques.

Studies of the mechanism of initiation of T4 DNA replication forks by a genetic recombination pathway

Genetic studies demonstrate that the initiation of DNA replication in T4 infected cells requires a set of genetic recombination
proteins at late times of infection. Because of this dependence on
genetic recombination, Luder and Mosig (40) have proposed that intermediates formed during recombination events are used directly to generate new replication forks. A likely pathway for such an event is
shown schematically in Figure 4. The ability of similar recombination
intermediates to prime DNA synthesis in this way is also an important
feature of many models designed to explain the general phenomenon of
"post-replication" DNA repair (46). We are therefore interested in
reconstituting this form of replication fork initiation in vitro in
order to determine how such intermediates are converted to replication
forks and what protein factors are required to mediate this conversion.

The products of T4 genes uvsX and uvsY are known to be involved
in a major genetic recombination pathway in T4-infected cells (47).
T. Minagawa and coworkers have found that the purified uvsX protein
shares many properties in common with the Escherichia coli recA
(personal communication). In order to carry out in vitro studies of
the role played by genetic recombination intermediates in the initiation of T4 DNA replication, we have purified both the uvsX and

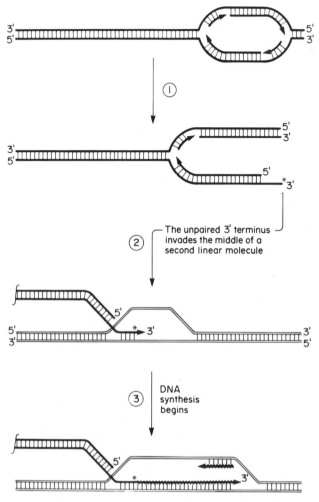

*Figure 4:   A possible scheme for initiating DNA replications forks
in a recombination-dependent manner. 1) The primary mechanism for
replication fork initiation produces a bidirectional replication
bubble as shown at the top. Because of the RNA-primed mechanism by
which the lagging strand is synthesized as a series of short Okazaki
fragments, whenever a fork reaches the end of the linear DNA molecule,
the lagging strand cannot be filled in and replication is incomplete,
leaving a single-stranded 3' tail (designated with an asterisk).
2) Since the T4 genome is organized as a terminally redundant linear
DNA molecule, its ends will be homologous to an internal sequence on
other copies of the T4 genome. (This is true even in singly infected
cells, where several rounds of primary replication should produce
many identical copies of the original genome.) If the single-stranded
3' tail produced in step one invades the homologous internal sequence
in another linear molecule, the "D-loop" shown on the third line will*

be produced. 3) The inserted 3' end is proposed to serve as a primer
for deoxyribonucleoside triphosphate polymerization, leading to DNA
synthesis that displaces a large single-stranded DNA loop. Once this
loop serves as a template for lagging strand synthesis, a normal
replication fork is produced. Depending on how this structure is re-
solved, this could either lead to simple initiation of DNA replication
using the second DNA molecule as the template, or to DNA synthesis
that is accompanied by recombination between the two interacting
DNA molecules (see ref. 40).

uvsY proteins to apparent homogeneity. This was accomplished initially
by taking advantage of the ability of a T4 gene 32 protein-agarose
affinity matrix to bind both the uvsX and uvsY proteins (18 and
Hosoda,J., personal communication). After this identification and
partial purification, both proteins were then further purified using
standard chromatographic procedures.

The uvsX protein has an apparent molecular weight on SDS polyacrylamide gels of 40,000. It is a single-stranded DNA-dependent
ATPase with a turnover of about 350 moles of ATP hydrolyzed per min
per mole of protein at 37°C, a specific activity 10-20 times greater
than that of the Escherichia coli recA protein. Interestingly, the
products of this reaction are a mixture of ADP + Pi and AMP + pyrophosphate. The ATPase activity was measured by incubating the protein
with (α-^{32}P)-ATP, Mg^{++} and single-stranded DNA in a Tris-HCl buffered
solution (pH 7.4) at 37°C. At various times, samples were spotted
onto a PEI-cellulose thin layer plate, dried, and then developed with
0.75 M KH$_2$PO$_4$ to resolve ATP, ADP and AMP from one another. Alternatively, (γ-^{32}P)-ATP was used and the plates developed with 0.5 M
LiCl plus 1.0 M formic acid to separate ATP, phosphate and pyrophosphate. We find that hydrolysis to ADP and to AMP are similarly, but
not identically, DNA-dependent and produce about a 2:1 ratio of ADP:
AMP at saturating levels of ATP. Both activities appear to be intrinsic to the uvsX protein (data not shown).

We have also assayed the uvsX protein for its ability to catalyze homologous pairing of DNA strands in a system analogous to those
used to characterize the Escherichia coli recA protein (48-50). The
single-stranded circular genome of fd and the homologous linear
double-stranded DNA molecule, RFIII (the latter labeled with ^{3}H or
^{32}P), were mixed and incubated with the uvsX protein plus various
other additions, and the products were then separated by electrophoresis through an agarose gel. A limited pairing results in the formation of "D-loops", whereas each complete strand exchange that occurs
will result in the formation of one labeled linear single-stranded
DNA molecule plus one nicked double-stranded DNA circle.

The main product of the reaction under our standard conditions
is some form of DNA network that will not enter the agarose gel. The

production of this network requires (1) ATP hydrolysis, (2) the addition of gene 32 protein and (3) homologous single- and double-stranded
DNAs. The molar ratios of 32 to uvsX proteins can vary between about
1:8 to 2:1 without severely affecting the reaction, and a rather high
level of ATP (greater than 1 mM) must be added for optimal network
formation. We have not determined the structure of the product, but
we believe it to reflect the joining of many double-stranded DNA molecules together via multiple D-loops formed by each single-stranded
DNA circle. The products expected for a complete strand exchange can
be obtained by either adding small amounts of dda protein to the
reaction (1:30:30 molar ratio of dda:uvsX:32), by heating the products
to 50°C for 10 minutes, or by treating the product with phenol.

We conclude that the uvsX and 32 proteins are together required
to catalyze homologous pairing between single-stranded DNA circles
and double-stranded linear DNA molecules. While this initial pairing
event is very efficient, the extensive branch migration needed to
complete the strand exchange process requires additional factors.
This is different from the results obtained with the analogous recA
and Escherichia coli SSB protein-catalyzed reaction, where complete
strand exchange is readily obtained (49, 51, 52).

The T4 uvsY protein acts in the same pathway as the uvsX protein.
It was first identified by Junko Hosoda as a protein that binds to an
agarose matrix containing immobilized 32 protein (personal communication). It has an apparent molecular weight of 16,000 and stimulates
the uvsX ATPase by about 1.2-fold in the presence of a single-stranded
DNA cofactor. The uvsY protein has a greater, although complex,
effect when double-stranded DNA is present. Addition of large amounts
of the uvsY protein to the uvsX protein and 32 proteins (2:4:1 molar
ratio of uvsY:uvsX:32) allows the formation of a network between two
homologous double-stranded linear DNA molecules that does not enter
an agarose gel. This reaction exhibits a long lag time and could be
due to a trace of exonuclease activity in our preparation of uvsY
protein that creates a short single-stranded DNA region at the terminus of each double-stranded linear DNA molecule. We have not yet
characterized the effects of the uvsY protein adequately to draw any
firm conclusions at this time.

We have examined the protein-protein interactions that involve
the uvsX protein by the method employed previously with gene 32 protein (18) and other replication proteins (37). Briefly, purified
uvsX protein was immobilized on an agarose matrix by an N-hydroxy-
succinimide displacement reaction that leaves the protein bound
through a peptide bond to the matrix. This matrix was then used to
chromatograph a crude lysate of T4-infected cells in which the T4
encoded proteins were selectively labeled with (^{35}S)-methionine.
One- and two-dimensional polyacrylamide gel analysis revealed that
the uvsX protein-agarose matrix selectively binds the uvsX, uvsY,
32, 42, and dda proteins, relative to a control column containing

immobilized bovine serum albumin. One <u>Eschrichia coli</u> protein with
an apparent molecular weight of 32,000 was also selectively bound
to the uvsX-agarose matrix.

After further characterizing the uvsX and uvsY proteins, we
will use them to construct recombinational intermediates in vitro.
The resulting DNA intermediates should allow us to assay for further
protein factors that are suspected to be involved in the conversion
of recombination intermediates to replication forks in vivo.

REFERENCES

1.   Epstein,R.H., Bolle,A., Steinberg,E., Kellenberger,E.,
         Boy de la Tour,E., Chevalley,R., Edgar,R.S., Susman,M.,
         Denhardt,G.H. and Lielausis,A. (1963) Cold Spring
         Harbor Symp. Quant. Biol. 28, 375.
2.   Warner,H.R. and Hobbs,M.D. (1967) Anal. Biochem. 93, 158.
3.   Curtis,M.J. and Alberts,B.M. (1976) J. Mol. Biol. 102, 793.
4.   McCarthy,D., Minner,C., Bernstein,H. and Bernstein,C. (1976)
         J. Mol. Biol. 106, 963.
5.   Gauss,P., Doherty,D.H. and Gold,L. (1983) Proc. Natl. Acad.
         Sci. USA 80, 1669.
6.   Morris,C.F., Hama-Inaba,H., Mace,D., Sinha,N.K. and Alberts,
         B.M. (1979) J. Biol. Chem. 254, 6787.
7.   Morris,C.F., Moran,L.A. and Alberts,B.M. (1979) J. Biol. Chem.
         254, 6797.
8.   Bittner,M., Burke,R.L. and Alberts,B.M. (1979) J. Biol. Chem.
         254, 9565.
9.   Kreuzer,K.N. and Jongeneel,C.V. (1983) Methods Enzymol.
         100, 144.
10. Silver,L.L. and Nossal,N.G. (1982) J. Biol. Chem. 257, 11696.
11. Nossal,N.G. and Alberts,B.M. (1983) in: "Bacteriophage T4"
         (C.K. Mathews, E.M. Kutter, G. Mosig, P.B. Berget, eds.),
         Academic Society for Microbiology, Washington, D.C.
         pp. 71.
12. Aposhian,H.V. and Kornberg,A. (1962) J. Biol. Chem. 237, 519.
13. Goulian,M., Lucas,Z.J. and Kornberg,A. (1968) J. Biol. Chem.
         243, 627.
14. Huang,C.-C. and Hearst,J.E. (1980) Anal. Biochem. 103, 127.
15. Newport,J. (1980) Ph. D. Thesis, University of Oregon.
16. Huberman,J.S., Kornberg,A. and Alberts,B.M. (1971) J. Mol.
         Biol. 62, 39.
17. Huang,C.-C., Hearst,J.E. and Alberts,B.M. (1981) J. Biol. Chem.
         256, 4087.
18. Formosa,T., Burke,R.L. and Alberts,B.M. (1983) Proc. Natl.
         Acad. Sci. USA 80, 2442.
19. Piperno,J., Kallen,R. and Alberts,B.M. (1978) J. Bio. Chem.
         253, 5180.

20.   Alberts,B.M., Barry,J., Bedinger,P., Burke,R.L., Hibner,U.,
         Liu,C.-C. and Sheridan,R. (1980) in: "Mechanistic Studies
         of DNA Replication and Genetic Recombination" (ICN-UCLA
         Symposia on Molecular and Cellular Biology, Vol. 19)
         (B. Alberts, ed.) Academic Press, New York, pp. 449.
21.   Liu,C.-C., Burke,R.L., Hibner,U., Barry,J. and Alberts,B.M.
         (1979) Cold Spring Harbor Symp. Quant. Biol. 43, 469.
22.   Nossal,N.G. and Peterlin,B.M. (1979) J. Biol. Chem. 254, 6032.
23.   Liu,C.-C. and Alberts,B.M. (1981) J. Biol. Chem. 256, 2813.
24.   Venkatesan,M., Silver,L.L. and Nossal,N. (1982) J. Biol. Chem.
         257, 12426.
25.   Nossal,N.G. (1980) J. Biol. Chem. 255, 2176.
26.   Liu,C.-C. and Alberts,B.M. (1980) Proc. Natl. Acad. Sci. USA
         77, 5698.
27.   Liu,C.-C. and Alberts,B.M. (1981) J. Biol. Chem. 256, 2821.
28.   Sinha,N.K., Morris,C.F. and Alberts,B.M. (1980) J. Biol. Chem.
         255, 3290.
29.   Hibner,U. and Alberts,B.M. (1980) Nature 285, 300.
30.   Sinha,N.K. and Haimes,M.D. (1981) J. Biol. Chem. 256, 10671.
31.   Behme,M.T. and Ebisuzaki,K. (1975) J. Virol. 15, 50.
32.   Purkey,R.M. and Ebisuzaki,K. (1977) Eur. J. Biochem. 75, 303.
33.   Krell,H., Durwald,H. and Hoffmann-Berling,H. (1979) Eur. J.
         Biochem. 93, 387.
34.   Bedinger,P., Hochstrasser,M., Jongeneel,C.V. and Alberts,B.M.
         (1983) Cell 34, 115.
35.   Liu,L.F., Liu,C.-C. and Alberts,B.M. (1979) Nature 281, 456.
36.   Stetler,G.L., King,G.J. and Huang,W.M. (1979) Proc. Natl.
         Acad. Sci. USA 76, 3737.
37.   Alberts,B.M., Barry,J., Bedinger,P., Formosa,T., Jongeneel,C.V.
         and Kreuzer,K. (1983) Cold Spring Harbor Symp. Quant.
         Biol. 47, 655.
38.   Mosig,G., Luder,A., Garcia,G., Dannenberg,R. and Bock,S.
         (1979) Cold Spring Harbor Symp. Quant. Biol. 43, 501.
39.   Mosig,G., Luder,A., Rowen,L., McDonald,P. and Bock,S. (1981)
         in: "The Initiation of DNA Replication" (ICN-UCLA Symp.
         on Molecular and Cellular Biology, Vol. 22) (D.S. Ray,
         ed.) Academic Press, New York, pp. 277.
40.   Luder,A. and Mosig,G. (1982) Proc. Natl. Acad. Sci. USA
         79, 1101.
41.   Snyder,L.R. and Montgomery,D.L. (1974) Virology 62, 184.
42.   Meyer,T.F., Baumel,I., Geider,K. and Bedinger,P. (1981)
         J. Biol. Chem. 256, 5810.
43.   Meyer,T.R. and Geider,K. (1979) J. Biol. Chem. 254, 12642.
44.   Cleary,J.M. and Ray,D.S. (1980) Proc. Natl. Acad. Sci. USA
         77, 4638.
45.   Stuber,D. and Bujard,I. (1981) Proc. Natl. Acad. Sci. USA
         78, 167.
46.   Howard-Flanders,P. (1981) Scientific American 245, 72.
47.   Bernstein,C. and Wallace,S. (1983) in:"Bacteriophage T4" (C.K.
         Mathews, E.M. Kutter, G. Mosig and P.B. Berget, eds.),
         American Society for Microbiology, pp. 138.

48.   Kahn,R., Cunningham,R.P., DasGupta,C. and Radding,C.M.
          (1980) Proc. Natl. Acad. Sci. USA 78, 4786.
49.   West,S.C., Cassuto,E. and Howard-Flanders,P. (1981) Proc.
          Natl. Acad. Sci. USA 78, 6149.
50.   Cox,M.M. and Lehman,I.R. (1981) Proc. Natl. Acad. Sci. USA
          78, 3433.
51.   Cox,M.M., Soltis,D.A., Livneh,Z. and Lehman,I.R. (1983)
          Cold Spring Harbor Symp. Quant. Biol. 47, 803.
52.   Radding,C.M., Flory,J., Wu,A., Kahn,R., DasGupta,C., Gonda,D.,
          Bianchi,M. and Tang,S.S. (1983) Cold Spring Harbor Symp.
          Quant. Biol. 47, 821.

IN VITRO REPLICATION OF BACTERIOPHAGE φ29

Margarita Salas, Luis Blanco, Ignacio Prieto,
Juan A. García, Rafael P. Mellado, José M. Lázaro
and José M. Hermoso

Centro de Biología Molecular (CSIC-UAM)
Universidad Autónoma
Canto Blanco
Madrid-34, Spain

INTRODUCTION

The Bacillus subtilis phage φ29 has as genetic material a linear, double-stranded DNA of about 18,000 base pairs (1) with a protein covalently linked to the 5' ends by a phosphoester bond between serine and 5' dAMP, the terminal nucleotide at both 5' ends (2). The protein, product of the viral gene 3, p3 (3), is required for the initiation of replication (4). An inverted terminal repetition, six nucleotides long, exists at the DNA ends (5, 6). φ29 replication starts at either DNA end and proceeds by strand displacement (7, 8).

The terminal protein, p3, primes φ29 DNA replication by reaction with dATP and formation of a protein p3-dAMP covalent complex that provides the 3' OH group needed for elongation. This reaction, which has been shown to take place in vitro, requires φ29 DNA-protein p3 as template and it does not occur when proteinase K-treated φ29 DNA is used (9-11). In addition, the initiation reaction in vitro requires the gene 2 product (11, 12), in agreement with its in vivo requirement (4). The viral genes 5, 6 and 17 are not required for the in vitro initiation of φ29 DNA replication (12), in agreement with in vivo results (4).

To further study the initiation reaction in vitro we have cloned genes 2 and 3 in pBR322 derivative plasmids under the control of the P_L promoter of phage lambda. After heat induction, the two proteins

are overproduced in <u>Escherichia coli</u> (13, 14). Protein p3 has been
obtained in a highly purified form (15), and protein p2 is being
purified (12) from <u>Escherichia coli</u> transformed with the recombinant
plasmids containing gene 3 and gene 2, respectively. A DNA polymerase
activity is associated with protein p2. The initiation and elongation
reactions using proteins p3 and p2 will be described as well as the
template requirements for the in vitro formation of the initiation
complex.

By using cloning techniques two carboxy-terminal mutants of
protein p3 have been obtained and the effect of the mutations on the
in vitro initiation reaction has been studied.

RESULTS AND DISCUSSION

<u>Template requirements for the formation of the initiation complex</u>
<u>between the terminal protein p3 and dAMP</u>

A covalent complex between the terminal protein p3 and 5'dAMP
is formed when extracts from <u>Bacillus subtilis</u> infected with ϕ29 are
incubated in vitro with dATP (9). This reaction requires ϕ29 DNA-
protein p3 complex as a template and it does not occur when proteinase
K-treated ϕ29 DNA is used, indicating that the parental protein p3
is an essential requirement for the initiation reaction. This requirement for ϕ29 DNA-protein p3 complex is also observed when ^{3}H-TTP incorporation is followed. As shown in Figure 1, ^{3}H-TMP is specifically
incorporated when ϕ29 DNA-protein p3 is used as a template and very
little incorporation occurs with proteinase K-treated ϕ29 DNA.

To study whether or not an intact ϕ29 DNA molecule is needed
as template, the ϕ29 DNA-protein p3 complex was treated with several
restriction nucleases. Treatment with Bst EII, Cla I, Bcl I, Eco RI
or Hind III which produce different number of cuts on ϕ29 DNA essentially did not affect the amount of protein p3-dAMP complex formed
relative to a control undigested or treated with Bam HI which does
not produce any cut on ϕ29 DNA (Figure 2). Taking into account the
above results, isolated terminal fragments, containing protein p3,
as well as internal fragments were used as template. The terminal
fragments Eco RI A (9000 bp) and C (1800 bp) were quite active in
the initiation reaction, being the internal fragment Eco RI B completely inactive (13). To study the size effect, the terminal fragments Eco RI A or C were treated with different restriction nucleases
to produce p3-containing terminal fragments of decreasing length.
Fragment Eco RI A, from the left end, was treated with Hha I, Bcl I,
Taq I or Mnl I to produce terminal fragments 175, 158, 73, 42 and
26 bp long, respectively. The formation of the initiation complex
was 74%, 47%, 51%, 32% and 27%, respectively, of the value obtained
with fragment Eco RI A. The right terminal fragment Eco RI C was
treated with Hind III, Acc I, Hinf I, Taq I or Rsa I which produce

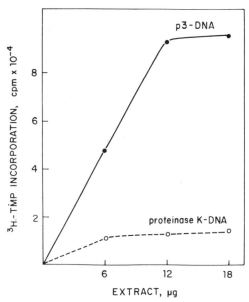

*Figure 1: Incorporation of (³H)-TTP dependent on φ29 DNA-protein
p3 as template. DNA-free extracts from* Bacillus subtilis *infected
with φ29 were prepared as described (10). The incubation mixture contained, in a final volume of 0.05 ml, 50 mM Tris-HCl, pH 7.5, 10 mM
MgCl₂, 1 mM ATP, 1 mM spermidine, 1 mM DTT, 5% (vol/vol) glycerol,
40 μM dATP, dGTP and dCTP, 0.3 μM ³H-dTTP (2 μCi), 0.5 μg of either
φ29 DNA-protein p3 or proteinase K-treated φ29 DNA and the indicated
amounts of the cell extract. After 15 min at 30ºC the trichloroacetic
acid insoluble radioactivity was determined.*

protein-containing fragments of 269, 125, 59, 52 and 10 bp, respectively, the initiation complex formed being 181%, 157%, 36%, 29% and
14% of that obtained with fragment Eco RI C. Taking into account that
the background value in the absence of DNA was 9% of the one obtained
with fragment Eco RI C, it can be concluded that the 10 bp long terminal Rsa I fragment is very little active. To determine whether this
low activity was due to the lack of specific DNA sequences or to the
small size of the fragment, an excess of Rsa I-digested lambda DNA
was ligated to the latter 10 bp long fragment. A marked stimulation
in the initiation reaction was obtained suggesting that the low
activity of the 10 bp long fragment was due to its small size rather
to the lack of specific DNA sequences (13).

Cloning and expression of gene 3 in Escherichia coli. Purification
of protein p3

A fragment of φ29 DNA containing genes 3, 4 and 5 and most of
gene 6 has been cloned in plasmid pKC30 (16) under the control of
the P_L promoter of phage lambda giving rise to the recombinant plasmid

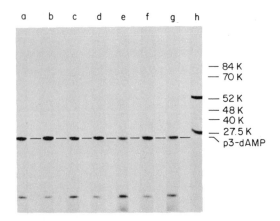

*Figure 2: Formation of the protein p3-dAMP initiation complex with
φ29 DNA-protein p3 treated with different restriction nucleases.
φ29 DNA-protein p3 complex (0.6 ug) was treated with the restriction
endonuclease Bam HI (b), Bst E II (c), Cla I (d), Bcl I (e), Eco RI
(f) or Hind III (g) and used as template for the formation of the
initiation complex protein p3-dAMP as described by Peñalva and
Salas (9). Undigested DNA was used as a control (a). After treatment
with micrococcal nuclease the samples were subjected to SDS-polyacrylamide gel electrophoresis. Lane h shows the φ29 structural
protein labelled with (^{35}S)-sulfate.*

pKC30 Al (17). After heat inactivation of the thermosensitive lambda
repressor carried in the <u>Escherichia coli</u> lysogen host four specific
polypeptides of Mr 27,000, 18,500, 17,500 and 12,500 were labelled
with ^{35}S-methionine. The 27,000 polypeptide, accounting for about
3% of the de novo synthesized protein, was identified as protein
p3 by radioimmunoassay and by its activity in the formation in vitro
of the initiation complex when supplemented with extracts from
<u>Bacillus subtilis</u> infected with a <u>sus</u>3 mutant (17). Protein p3 has
been highly purified from <u>Escherichia coli</u> cells harbouring the
gene 3-containing recombinant plasmid by DEAE-cellulose chromatography (protein p3 elutes at 0.7 M salt), precipitation of the
nucleic acids with polyethyleneimine in the presence of 1 M NaCl
and chromatography on phosphocellulose. The purified protein (see
Figure 3) was active in the formation of the p3-dAMP initiation
complex when supplemented with extracts from <u>sus</u>3-infected <u>Bacillus
subtilis</u> (15).

Effect of mutations at the carboxyl end of protein p3 on the in vitro formation of the initiation complex

The φ29 DNA fragment Hind III G, containing all but the last
15 nucleotides coding for protein p3 (18), has been cloned in the
Hind III site of plasmid pBR322 (19) or in that of plasmid pKTH601

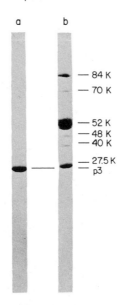

Figure 3: SDS-polyacrylamide gel electrophoresis of purified protein p3. Protein p3 was purified from Escherichia coli cells transformed with the gene 3-containing recombinant plasmid pKC30 A1 (17) as described (15) and subjected to SDS-polyacrylamide gel electrophoresis. a: Purified protein p3, 10 μg. b: φ29 structural proteins as molecular weight marker.

upstream of an inserted oligonucleotide with translation stop signals in the three possible reading frames (20). In the first case the normal protein p3 which has the sequence -ser-leu-lys-gly-phe at the carboxyl end is mutated to another protein, p3', with the sequence -ser-phe-asn-ala-val-val-tyr-his-ser, having four additional amino acids (21). In the second case the normal sequence of protein p3 is changed to -ser-leu-leu-ile-asp (protein p3") (22). After a further cloning of each of the p3-containing sequences in plasmid pPLc28 (23) under the control of the P_L promoter of phage lambda, both proteins p3' and p3" were made in 6% and 9%, respectively, of the total de novo synthesized protein after heat inactivation of the thermosensitive lambda repressor contained in the Escherichia coli lysogen host. The activity of proteins p3' and p3", assayed by the formation of the initiation complex in vitro, was about 15% and 12%, respectively, relative to that of the normal protein p3, suggesting that the carboxyl end of the protein is an important requirement for the normal function of protein p3 in the initiation of replication.

Cloning and expression of gene 2 in Escherichia coli: Partial
purification of protein p2

     The φ29 DNA Hind III B fragment, containing gene 2, was cloned
first in plasmid pBR322 and then, to achieve a high level of synthesis of protein p2, in plasmid pPLc28 to put protein p2 under the
control of the lambda P$_L$ promoter. After heat inactivation of the
lambda repressor a protein of Mr 68,000, the size expected for protein
p2 (24), was synthesized in Escherichia coli, accounting for about
2% of the de novo synthesized protein (see Blanco et al., this
volume). The protein p2 synthesized in Escherichia coli was active
in the in vitro formation of the p3-dAMP initiation complex when
supplemented with extracts from sus2-infected Bacillus subtilis,
with extracts from Escherichia coli transformed with the gene 3-con-
taining recombinant plasmid pKC30 Al or with purified protein p3.

     Protein p2 has been partially purified from the Escherichia
coli cells transformed with the gene 2-containing recombinant plasmid.
The protein eluted from a blue dextran-agarose column at 0.5 M NaCl
and it was active in the formation of the initiation complex in vitro
when supplemented with extracts from sus2-infected Bacillus subtilis
or with extracts from Escherichia coli transformed with the gene 3-
containing recombinant plasmid described above (14).

DNA polymerase activity associated with protein p2

     The protein p2 fraction eluted from the blue dextran-agarose
column with 0.5 M NaCl had DNA polymerase activity when assayed with
poly dA-(dT)$_{12-18}$ as template (14 and Figure 4). Moreover, the DNA
polymerase present in the protein p2-containing fraction could also
use the φ29 DNA-protein p3 complex as template (Figure 4). In this
case, however, the presence of protein p3 was also required since
the initiation complex p3-dAMP has to be formed to initiate replication, previous to the elongation step. To show that the product elongated under these conditions was specific a pulse-chase experiment
was carried out as described (9). After a pulse in the presence of
0.25 μM (α-^{32}P)dATP, 40 μM dATP, dGTP and dTTP and 100 μM ddCTP were
added to stop elongation at nucleotides 9 and 12 from the left and
right DNA ends, respectively. Figure 5 shows, in addition to the
protein p3-dAMP band, the appearance after the chase of two slower
moving bands at the expected positions.

Study of the requirement of other proteins for the initiation and
elongation of φ29 DNA replication

     Figure 6 shows that formation of the protein p3-dAMP initiation
complex takes place when highly purified protein p3 and partially
purified protein p2 as described above are used as the only viral
protein factors, in the presence of φ29 DNA-protein p3 as template.
However, addition of extracts from uninfected Bacillus subtilis or

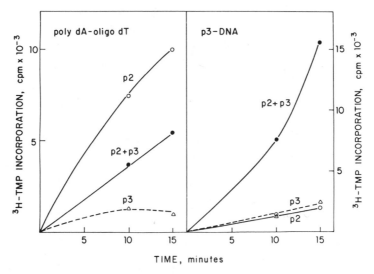

*Figure 4:   DNA polymerase activity in the partially purified prepara-
tion of protein p2. The incubation mixture was as described in Fig. 1
with 0.5 μg of either poly dA-(dT)$_{12-18}$ or φ29 DNA-protein p3 as
template. This partially purified fraction of protein p2 (0.4 μg) and
highly purified protein p3, (20 ng), prepared as described by Blanco
et al. (this volume) and Prieto et al. (15), respectively, were used.*

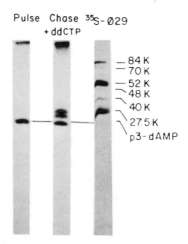

*Figure 5:   Elongation of the protein p3-dAMP complex by the partially
purified preparation of protein p2. Partially purified protein p2
(2 μg) and highly purified protein p3 (20 ng) were incubated with
φ29 DNA-protein p3 as template and 0.25 μM (α-^{32}P)dATP (5 μCi) as
described (9). After 5 min at 30°C (pulse) 40 μM dATP, dGTP and dTTP
and 100 μM ddCTP were added and the incubation continued for 15 min
(chase).*

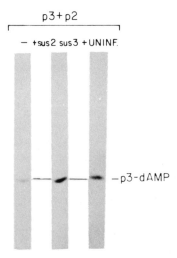

Figure 6: Requirement of a host factor for the formation of the protein p3-dAMP initiation complex. Partially purified protein p2 (0.4 μg) and highly purified protein p3 (30 ng) were incubated with φ29 DNA-protein p3 and 0.25 μM (α-³²P)dATP (5 μCi) as described (9) in the absence or presence of extracts from Bacillus subtilis infected with a double sus mutant in genes 2 and 3 (3 μg) or uninfected (3 μg).

infected with a sus2sus3 double-mutant, clearly stimulated the initiation reaction (Figure 6). Since a similar stimulation was obtained by addition of either extract, a bacterial protein factor(s) seems to be required, besides protein p2 and p3, for an efficient formation of the p3-dAMP initiation complex.

Figure 7 shows a summary of our present knowledge of the initiation of φ29 DNA replication. A free molecule of the terminal protein p3, after interaction with the parental terminal protein and possibly also with the inverted terminal repetition reacts with dATP in a reaction probably catalyzed by the DNA polymerase activity of protein p2 and stimulated by some host factor(s) and forms a protein p3-dAMP covalent complex with a free 3'OH group which is used for elongation. The DNA polymerase activity of protein p2 can elongate the p3-dAMP complex formed in vitro. Whether this is the only DNA polymerase involved in elongation of φ29 DNA or there is some other DNA polymerase activity remains to be determined.

The product of genes 5 and 6 and, probably also 17, are involved in vivo in φ29 replication in elongation steps (4). The specific function of these proteins remains to be determined in the in vitro system. With the aim of purifying the corresponding proteins to study

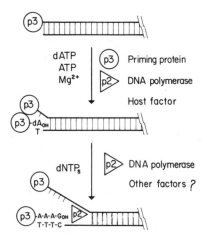

Figure 7:   Model for the initiation of ϕ29 DNA replication.

its function, genes 5 and 6 have been cloned under the control of
the lambda P_L promoter (17, 25) and we intend to clone also gene 17.
Purification of the overproduced proteins will allow us to study their
function in the process of ϕ29 DNA replication.

ACKNOWLEDGEMENTS

        This investigation has been aided by Grant 2 R01 GM27242-04
from the National Institutes of Health and by Grants from the Comisión
Asesora para el Desarrollo de la Investigación Científica y Técnica
and Fondo de Investigaciones Sanitarias. L.B., J.A.G., R.P.M. and
I.P. were Fellows of Plan de Formación de Personal Investigador,
Spanish Research Council, Ministerio de Educación y Ciencia (senior
Fellowship) and Spanish Research Council, respectively.

REFERENCES

1.   Sogo,J.M., Inciarte,M.R., Corral,J., Viñuela,E. and Salas,M.
        (1979) J. Mol. Biol. 127, 411.
2.   Hermoso,J.M. and Salas,M. (1980) Proc. Natl. Acad. Sci. USA
        77, 6425.
3.   Salas,M., Mellado,R.P., Viñuela,E. and Sogo,J.M. (1978)
        J. Mol. Biol. 119, 269.
4.   Mellado,R.P., Peñalva,M.A., Inciarte,M.R. and Salas,M. (1980)
        Virology 104, 84.
5.   Escarmís,C. and Salas,M. (1981) Proc. Natl. Acad. Sci. USA
        78, 1446.

6. Yoshikawa,H., Friedman,T. and Ito,J. (1981) Proc. Natl. Acad.
       Sci. USA 78, 1336.
7. Harding,N.E. and Ito,J. (1980) Virology 104, 323.
8. Inciarte,M.R., Salas,M. and Sogo,J.M. (1980) J. Virol. 34, 187.
9. Peñalva,M.A. and Salas,M. (1982) Proc. Natl. Acad. Sci. USA
       79, 5522.
10. Shih,M., Watabe,K. and Ito,J. (1982) Biochem. Biophys. Res.
       Commun. 105, 1031.
11. Matsumoto,K., Saito,R. and Hirokawa,H. (1983) Mol. Gen.
       Genetics, in press.
12. Blanco,L., García,J.A., Peñalva,M.A. and Salas,M. (1983)
       Nucl. Acids Res. 11, 1309.
13. García,J.A., Peñalva,M.A., Blanco,L. and Salas,M. (1983)
       Proc. Natl. Acad. Sci. USA, in press.
14. Blanco,L., García,J.A., Lázaro,J.M. and Salas,M. (1984),
       this volume.
15. Prieto,I., Lázaro,J.M., García,J.A., Hermoso,J.M. and Salas,M.
       (1983), submitted for publication.
16. Shimatake,H. and Rosenberg,M. (1981) Nature 292, 128.
17. García,J.A., Pastrana,R., Prieto,I. and Salas,M. (1983)
       Gene 21, 65.
18. Escarmís,C. and Salas,M. (1982) Nucl. Acids Res. 10, 5785.
19. Bolivar,F., Rodriguez,R.L., Greene,P.J., Betlach,M.C.,
       Heynecker,H.L., Boyer,H.W., Crossa,J.H. and Falkow,S.
       (1977) Gene 2, 95.
20. Pettersson,R.F., Lundstrom,K., Chattopadhyasa,J.B.,
       Josephson,S., Philipson,L., Kaariainen,L. and Palva,I.
       (1983) Gene, in press.
21. Mellado,R.P. and Salas,M. (1982) Nucl. Acids Res. 10, 5773.
22. Mellado,R.P. and Salas,M. (1983) Nucl. Acids Res. 11, 7397.
23. Remaut,E., Stanssens,P. and Fiers,W. (1981) Gene 15, 81.
24. Yoshikawa,H. and Ito,J. (1982) Gene 17, 323.
25. Salas,M., García,J.A., Peñalva,M.A., Blanco,L., Prieto,I.,
       Mellado,R.P., Lázaro,J.M., Pastrana,R., Escarmís,E. and
       Hermoso,J.M. (1983) UCLA Symposia on Molecular and
       Cellular Biology, New Series, ed. Cozzarelli,N.R. (Alan
       R. Liss, Inc., New York) vol. 10, in press.

PROTEINS AND NUCLEOTIDE SEQUENCES INVOLVED IN DNA REPLICATION OF FILAMENTOUS BACTERIOPHAGE

K. Geider, T.F. Meyer, I. Bäumel and A. Reimann

Max-Planck-Institut für medizinische Forschung
Abteilung Molekulare Biologie
Jahnstrasse 29
D-6900 Heidelberg, Federal Republic of Germany

INTRODUCTION

Filamentous bacteriophages (fd, M13, fl) contain single-stranded circular DNA of about 6400 bases (1). They penetrate the host cell via pili induced by the F-episome of an <u>Escherichia coli</u> cell. During the penetration process their single-stranded genome is converted into double-stranded DNA which is the main viral component in the first minutes after infection. Later in the life cycle, viral single strands are formed and complexed with gene 5 protein. They are assembled in the host membrane into phage particles, which penetrate the cell wall without severe damage of the host. Filamentous phages are quite flexible on the size of the DNA to be packaged. They spontaneously generate miniphages of about 1 kb (2, 3), but they can also comprise artificial DNA sequences up to a length of 15 kb, if the phage packaging signal is provided (4).

The present aspects of molecular interest in these phages are directed on signals and proteins involved in replication of their DNA, assembly steps of phage particles in the bacterial cell membrane and their use as cloning vectors including the convenient isolation of single-stranded DNA from purified phage particles.

Synthesis of the DNA strand complementary to the infecting viral genome

Special interest on replication of single-stranded bacteriophages came up with the question, how to get a circular single strand initiated in order to allow a DNA polymerase to synthesize the

45

complementary strand. The addition of rifampicin, a specific inhibitor of bacterial RNA polymerase, to bacteria briefly after infection
with filamentous phages prevented the conversion of their single-
stranded genome into the double-stranded form (5). Chloramphenicol,
an inhibitor of protein synthesis, did not influence this step in
DNA synthesis. The conclusion that bacterial RNA polymerase initiates
the DNA chain by synthesis of an RNA primer was later on substantiated by several arguments: (i) The conversion of single-stranded
circles from filamentous phages to double-stranded DNA in soluble
cell extracts can also be inhibited by rifampicin (6). (ii) The
extracts from Escherichia coli cells deficient in a gene coding for
an RNA polymerase subunit were also temperature-sensitive for SS→RF
conversion. This defect could be supplemented by the addition of wild
type RNA polymerase (7). (iii) RNA polymerase in the presence of DNA
binding protein I protects a hairpin region on fd DNA, suggesting
the formation of an initiation complex (8). (iv) With purified
Escherichia coli RNA polymerase and DNA binding protein I a 26 to
30 nucleotides long RNA primer could be isolated and localized on
the fd genome (9).

     Sequence and terminus of the initiator RNA suggest that chain
start occurs at the 3'-end of a hairpin forming sequence localized
in the intergenic region of the phage genome (Figure 1 and 2). The
RNA primer runs up to the loop region. It is assumed that the DNA/
RNA hybrid is more stable than the hairpin itself considering the
complete base pairing of the hybrid and incomplete pairing of the
DNA sequence. This hybrid formation causes opening of the hairpin
structure and exposition of the DNA sequence opposite the RNA. The
consequence is complex formation of DNA binding protein I with the
opened single-stranded region. This complex interferes with the
formation of the RNA primer by the polymerase, causing termination
of the RNA chain. Chain elongation by DNA polymerase III holoenzyme
suggests a free 3'-end of the RNA. The promoter property of this
hairpin could be derived from less base pairing than realized in
the other hairpin structures. The non-pairing sequences in the hairpin
might strongly attract RNA polymerase holoenzyme and allow the polymerase to start the RNA chain just at the beginning of the pairing
sequence. It is still an open question how DNA polymerase III holoenzyme manages to pass the secondary structure of the hairpins upstream the initiation site.

     It was shown by us in electron microscopical studies that in-
sertion of a transposon with its loop-stem structure (Figure 1)
prevents proceeding of DNA polymerase III holoenzyme in vitro (10).
This is in contrast to Escherichia coli DNA polymerase I, which can
easily perform strand switching (11). The block of DNA polymerase
III holoenzyme does not only shut off DNA synthesis initiated at the
c-strand origin, it also allows primer formation at secondary sites.
This event is seen in the electron microscope as a large section of
double-stranded DNA in the fd loop or in the transposon loop together

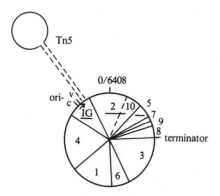

Figure 1: *Genetic map of phage fd and insertion site of transposon Tn5. The elements involved in phage DNA replication are underlined (origins, gene 2, gene 5). The localization of the transposon in fd::Tn5 (fd-73; (4)) is indicated by dashed lines. The map is drawn in the same polarity as used in Figure 2, which is opposite to the conventional direction. Terminator, stop for transcription of polycistronic messenger.*

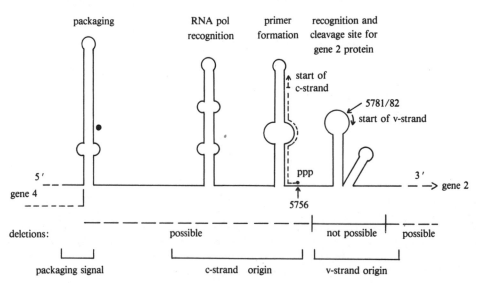

Figure 2: *Replication origins and base pairing in the intergenic region. The function of the hairpin structures is derived from published results: mutation analysis (transposition and deletion) (9, 12, 13); gene 2 protein cleavage site (14), insertion site of Tn5 in fd-73.*

with the intact stem structure of the transposon. Addition of rep-helicase to the enzyme system performing SS→RF conversion (Table 1) of single-stranded fd::Tn5 results in unwinding of the transposon stem structure and conversion of the whole fd::Tn5 DNA into the double-stranded form.

This result agrees with the in vivo finding of Kim et al. (13) that deletion of the c-strand origin does not abolish the viability of filamentous phages. Faint plaques formed by the mutated phages suggest that deletion of the primary origin induces the use of other sites for initiation of c-strand synthesis.

The DNA synthesized as complementary sequence of the infecting viral strand is further processed in several steps. DNA polymerase III holoenzyme can apparently not remove the primer and easily bring the 5'- and 3'-end of the complementary strand in the adjacent position for ligation into the doubly-closed form (RFIV and RFI). It was shown that Escherichia coli DNA polymerase I has a supporting function for this step (7). The bacterial DNA ligase is used for strand joining. The next step in replication of the viral genome is the removal of turns from the helix of the double-stranded circle. The formation of RFI is a prerequisite of replication of double-stranded phage DNA. It was shown in vitro that purified DNA gyrase can perform the task to convert the relaxed DNA into the replication active superhelical form (15). Nalidixic acid, an inhibitor of gyrase, shuts off double strand replication of filamentous phage in vivo which indicates the involvement of supercoiled viral DNA as replication intermediate (16).

Replication of double-stranded phage fd DNA

Early genetical experiments showed the requirement of gene 2 for replication of viral DNA, because mutations in the gene prevent the formation of double-stranded parental RFII DNA under non-permissive conditions (17). The gene product was later isolated and characterized (18). Gene 2 protein has a molecular weight of 46,000 and its sequence can be deduced from the nucleotide sequence (19). The question arose, if the initiation codon close to the ribosome binding site (distance four nucleotides) or the next initiation codon (16 nucleotides from the ribosom binding site) is used for the translational start. By selective labelling of the protein with radioactive amino acids we showed that 90% of the gene 2 protein molecules start the first initiation codon and 10% use the next start codon (20). There is no hint if the two forms of gene 2 protein synthesized are distinct in their enzymatic properties and if they are differentially expressed in the phage life cycle. A protein coded at the end of gene 2 (gene 10 protein) might also influence gene 2 protein action.

Gene 2 protein is an endonuclease which is specific for a nucleotide sequence and for the DNA conformation. It cleaves supercoiled DNA of filamentous phages at position 5781/5782 (in the stand-

Table 1:   Proteins for DNA replication of filamentous phages and their
           function in the sequential steps of DNA synthesis

SS→RF conversion: (Escherichia coli proteins only)

DNA binding protein I:
    (i) complex formation with single-stranded genome, leaving a
    hairpin structure open for the initiation process
    (ii) competition with initiating RNA polymerase
    (iii) helper function in DNA synthesis by DNA polymerase III
    holoenzyme

RNA polymerase:
    (i) complex formation at hairpin structure in intergenic region
    (ii) synthesis of RNA primer:pppAGGGCGAAAAACCGUCUAUCAGGGCG(AUGG)

DNA polymerase III holoenzyme:
    (i) complex at the 3'-terminus of the RNA primer (26)
    (ii) chain elongation as DNA (fast compared to DNA polymerase I)

DNA polymerase I:
    (i) degradation of primer
    (ii) gap filling

DNA ligase:
    Strand sealing at adjacent 3'-5' ends

DNA gyrase:
    removal of helical turns, formation of supercoil

Replication of fd RF into viral strands

viral gene 2 protein:
    (i) complex at v-strand origin
    (ii) cleavage of viral strand at nucleotides 5781/82
    (iii) helper function in start of strand unwinding
    (iv) cleavage of single-stranded tail at the origin sequence
    after one round of replication in the rolling circle mode
    (v) circularization of viral strands subsequent to the cleavage
    reaction in replicating DNA

Escherichia coli rep-helicase:
    strand unwinding, starting at the cleaved v-strand origin sup-
    ported by gene 2 protein

Escherichia coli DNA binding protein I:
    (i) complex formation with single strands generated by strand
    unwinding
    (ii) support of DNA polymerase III holoenzyme in DNA synthesis

DNA polymerase III holoenzyme:
    chain elongation starting at nick in origin succeeding strand
    unwinding
viral gene 5 protein: (inhibition of c-strand synthesis) competition
    with Escherichia coli DNA binding protein I

ard fd map) (14). Single-stranded fd DNA or relaxed double-stranded
DNA are not cleaved (21). In the presence of Mn^{2+}-ions the enzyme
is less selective and can also cleave fd RFI in both strands producing
linear fd DNA.

Cleavage of supercoiled RF by purified gene 2 protein is accompanied with re-sealing of strands, which gives rise to 2/3 of the
molecules to be open in one strand (RFII) and 1/3 of the molecules
to be relaxed and doubly-closed (RFIV). This property of the enzyme
is required for circularization of viral single strands during replication, but the direct conversion of RFI into RFIV might possibly
not occur in the phage infected cell.

The isolation of fd gene 2 protein (18) is facilitated using
phage fd mutants in gene 5 and even more by the use of cloned viral
gene 2 inserted into a pBR-plasmid (22). Enzyme production can be
increased 20 or 200 fold, respectively, compared to the synthesis
in cells infected with wild type phages.

After cleavage of supercoiled fd DNA by gene 2 protein, replication is initiated by strand unwinding. The enzyme involved in this
step is Escherichia coli rep-helicase. The rep-gene was earlier shown
to be required for the propagation of small bacteriophages (23). The
gene product, now called rep-helicase, was isolated and shown to
react in unwinding of φX174 RF (24) and also for unwinding of fd RF
(15). Although the enzyme cannot completely unwind long DNA duplices
(10), in the presence of φX174 gene A protein or fd gene 2 protein
it can separate phage double strands. We assume that for fd RF DNA
the start of unwinding by rep-helicase is facilitated by complex formation of gene 2 protein at the replication origin (25). The unwinding
reaction requires, besides rep-helicase, the presence of fd gene 2
protein and Escherichia coli DNA binding protein I and ATP or dATP
as energy source. The binding protein may prevent the reannealing of
the separated strands. If Escherichia coli DNA polymerase III holoenzyme is also present in the incubation mixture, the cleavage site
of the nick produced by gene 2 protein is used for chain elongation.
DNA unwinding and replication give rise to a rolling circle structure
of the phage DNA. After one round of replication gene 2 protein
cleaves off the single-stranded tail and immediately circularizes
the viral strand. For Escherichia coli DNA polymerase III holoenzyme
strand unwinding is a prerequisite to replication. To study the
cleavage reaction of gene 2 protein in rolling circle structures we
have used two model systems which allowed unwinding and DNA replication in the absence of fd gene 2 protein: The substrate fd RFI was
cleaved with gene 2 protein and the enzyme inactivated. DNA synthesis
was started by addition of proteins isolated for phage T4 DNA replication (27). The replication products were mainly rolling circles
with long tails. Inactivation of the proteins and subsequent addition
of gene 2 protein did not alter the rolling circle structures.
However, the presence of gene 2 protein during replication produced

single-stranded fd DNA in linear and circular form. Similar results
were obtained by using replication enzymes induced by phage T7 (28).
This indicates that gene 2 protein can cleave in the replication
fork around its recognition sequence if the replication machinery
proceeds. We assume that the DNA conformation in the fork allows the
enzyme to transfer the cleavage energy immediately to the sealing
reaction. We have never found an indication that gene 2 protein is
attached to the 5'-end of the rolling circle. Cleavage and sealing
seems to be almost synchronous, otherwise a loss of energy should
impair strand circularization. - Functional aspects of gene 2 protein
in double strand replication of filamentous phage have also been
summerized elsewhere (25).

Simultaneous replication of phage fd viral and complementary strands

We have also combined the enzymes required for the conversion
of fd single strands to double-stranded DNA (Table 1) which are
Escherichia coli RNA polymerase holoenzyme, DNA binding protein I
and DNA polymerase III holoenzyme with the proteins additionally
required for double strand replication (Table 1) which are Escherichia
coli rep-helicase and fd gene 2 protein (29). For density labelling
we used dBUTP. The main reaction products observed in a CsCl-sensity
gradient and on agarose gels were rolling circle structures of intermediate density, and we could not find free molecules with a heavy
complementary strand or a heavy viral strand. The distance of the
replication origins for complementary and viral strand synthesis is
only 24 nucleotides. The c-strand origin is exposed at the very end
of a round of replication. We assume that RNA polymerase can rapidly
form a complex at this origin thereby preventing the cleavage and
circularization step done by gene 2 protein. If RNA polymerase holoenzyme saturated with intact δ-factor (30) is more suitable for
simultanous strand replication of filamentous phage has not been investigated.

The involvement of viral gene 5 protein in phage replication

Filamentous phages with a mutation in gene 5 accumulate double-
stranded DNA in large amounts within infected cells (18, 31). Gene
5 protein forms complexes with single-stranded phage DNA (32), thereby
preventing the conversion of circular viral strands into the double-
stranded form (7). It was shown that gene 5 protein can displace
Escherichia coli DNA binding protein I associated with phage single
strands (33, 34). In single strand to RF conversion it was inhibitory
for DNA synthesis, although this replication system could tolerate
small amounts of gene 5 protein without an effect on the rate of
DNA synthesis (7). For the replication of double-stranded fd DNA
gene 5 protein was not inhibitory and it could partially substitute
Escheriachia coli binding protein, although small amounts of the
latter protein were still required for DNA synthesis (Meyer, Reimann
and Geider, unpublished observations). The establishment of a

replication system simultanously synthesizing both phage strands
should allow better understanding, how gene 5 protein acts discrimi-
natory in strand synthesis.

The use of replication functions for DNA cloning

The relatively simple replication functions, used by filamentous
phages appeared to be attractive for construction of new cloning
vehicles.

A DNA fragment with gene 2 cloned in a pBR-plasmid was used
for synthesis of gene 2 protein in order to propagate an artificial
plasmid bearing the replication origins of phage fd connected to a
gene for kanamycin resistance (22). Furthermore, we transferred the
DNA fragment with fd gene 2 from the plasmid DNA to phage λ (35).
Escherichia coli cells were made lysogenic for this phage. They could
propagate artificial plasmids with the fd replication origins. Some
of the vectors constructed are not homologous to pBR-plasmids. The
pfd vectors are synthesized in large copy numbers in competent cells.
Since fd gene 2 protein is temperature-sensitive in vitro and in vivo,
these vectors can be easily removed by incubation of the cells at
elevated temperature.

CONCLUSIONS

The studies of enzymatic replication of filamentous phages
have revealed properties of replication mechanisms in general. Events
during the formation of an RNA primer were characterized. Strand unwinding could be studied with and without concomitant replication.
The circularization of fd viral strands is only dependent on fd gene
2 protein, which is not attached to the 5'-end of rolling circles.
These findings reveal some differences to replication of bacteriophage
φX174, especially in the behaviour of fd gene 2 protein and φX174
gene A protein (25, 36). The replication functions of filamentous
phages could be exploited for the construction of a plasmid DNA
cloning system.

Open questions are the requirements for simultanous replication
of both strands, the replication of pfd plasmids in other cells than
Escherichia coli, and intermediate steps in strand circularization.

REFERENCES

1.   Beck,E. and Zink,B. (1981) Gene 16, 35.
2.   Griffith,J. and Kornberg,A. (1974) Virology 59, 139.
3.   Enea,V. and Zinder,N.D. (1975) Virology 68, 105.
4.   Herrmann,R., Neubebauer,K., Pirkl,E., Zentgraf,H. and
        Schaller,H. (1980) Mol. Gen. Genetics 177, 231.

5. Brutlag,D., Schekman,R. and Kornberg,A. (1971)
        Proc. Natl. Acad. Sci. USA 68, 2826.
6. Wickner,W., Brutlag,D., Schekman,R. and Kornberg,A. (1972)
        Proc. Natl. Acad. Sci. USA 69, 965.
7. Geider,K. and Kornberg,A. (1974) J. Biol. Chem. 249, 3999.
8. Schaller,H., Uhlmann,A. and Geider,K. (1976)
        Proc. Natl. Acad. Sci. USA 73, 49.
9. Geider,K., Beck,E. and Schaller,H. (1978) Proc. Natl. Acad.
        Sci. USA 75, 645.
10. Bäumel,I., Meyer,T.F. and Geider,K. (1983) submitted for
        publication.
11. Schildkraut,C.L., Richardson,C.C. and Kornberg,A. (1964)
        J. Mol. Biol. 9, 24.
12. Schaller,H. (1978) Cold Spring Harbor Symp. Quant. Biol.
        43, 410.
13. Kim,M.H., Hines,J.C. and Ray,D.S. (1981) Proc. Natl. Acad.
        Sci. USA 78, 6784.
14. Meyer,T.F., Geider,K., Kurz,C. and Schaller,H. (1979)
        Nature 278, 365.
15. Geider,K., Bäumel,I. and Meyer,T.F. (1982) J. Biol. Chem.
        257, 6488.
16. Schneck,P.K., Staudenbauer,W.L. and Hofschneider,P.H. (1973)
        Eur. J. Biochem. 38, 130.
17. Fidanián,H.M. and Ray,D.S. (1972) J. Mol. Biol. 72, 51.
18. Meyer,T.F. and Geider,K. (1979) J. Biol. Chem. 254, 12636.
19. Schaller,H., Beck,E. and Takanami,M. (1978) in: "The Single-
        Stranded DNA phages" Cold Spring Harbor Laboratory,
        Cold Spring Harobor pp. 139.
20. Meyer,T.F., Beyreuther,K. and Geider,K. (1980) Mol. Gen.
        Genetics 180, 489.
21. Meyer,T.F. and Geider,K. (1979) J. Biol. Chem. 254, 12642.
22. Meyer,T.F. and Geider,K. (1981) Proc. Natl. Acad. Sci. USA
        78, 5416.
23. Denhardt,D.T., Dressler,D.H. and Hathaway,A. (1967)
        Proc. Natl. Acad. Sci. USA 57, 813.
24. Scott,J.F., Eisenberg,S., Bertsch,L.L. and Kornberg,A. (1977)
        Proc. Natl. Acad. Sci. USA 74, 193.
25. Meyer,T.F. and Geider,K. (1982) Nature 296, 828.
26. Wickner,W. and Kornberg,A. (1973) Proc. Natl. Acad. Sci. USA
        70, 3679.
27. Meyer,T.F., Bäumel,I., Geider,K. and Bedinger,P. (1981)
        J. Biol. Chem. 256, 5810.
28. Harth,G., Bäumel,I., Meyer,T.F. and Geider,K. (1981)
        Eur. J. Biochem. 119, 663.
29. Reimann,A. (1981) Diploma work, Heidelberg.
30. Kaguni,J.M. and Kornberg,A. (1982) J. Biol. Chem. 257, 5437.
31. Salstrom,J.S. and Pratt,D. (1971) J. Mol. Biol. 61, 489.
32. Alberts,B., Frey,L. and Delius,H. (1972) J. Mol. Biol. 68, 139.
33. Zentgraf,H., Berthold,V. and Geider,K. (1977) Biochem. Biophys.
        Acta 474, 629.

34.   Geider,K. (1978) Eur. J. Biochem. 87, 617.
35.   Geider,K., Hohmeyer,C., Haas,R. and Meyer,T.F. (1983)
          submitted for publication.
36.   Meyer,T.F. and Geider,K. (1980) in: "Mechanistic studies of
          DNA replication and genetic recombination", ICN-UCLA
          Symposium on Molecular and Cellular Biology (Alberts,B.M.
          ed.) pp. 579, Academic Press, New York.

IN VITRO REPLICATION OF Rl MINIPLASMID DNA

Ramon Diaz[1] and Walter L. Staudenbauer[2]

Max-Planck-Institut für Molekulare Genetik
Abteilung Schuster
D-1000 Berlin 33, Federal Republic of Germany

INTRODUCTION

The antibiotic resistance plasmid Rl is composed of a cluster of drug resistance genes (r-determinant) and the resistance transfer factor (RTF). These two segments are flanked by directly repeated ISl elements (ISla and ISlb). The RTF carries all the genetic information necessary for autonomous plasmid replication and conjugal transfer. The basic replicon of Rl (designated RepA) is situated adjacent to the ISlb element and includes the replication origin oriV and genes involved in replication and copy number control (see Figure 1). Rl replication is controlled by the 90 nucleotide copA RNA and the 11 kdal copB protein which act by inhibiting the expression of the 33 kdal repA protein (1). The repA protein is considered as the limiting factor for the initiation of Rl replication making plasmid DNA synthesis strictly dependent on de novo protein synthesis (2).

In order to develop an in vitro replication system for the Rl plasmid, we employed the mini-Rl plasmids pKN177 and pKN182, which had been spontaneously generated from the Rl copy mutant pKN104 (3). The 6.5 kb plasmid pKN182 carries only the replication region of Rl, whereas the 12.8 kb plasmid pKN177 also contains the Tn3 transposon of the parent plasmid (4). Due to their small size and increased copy number these miniplasmids offer experimental advantages for replica-

---

[1]Present address: C.S.I.C., Instituto de Biologia Celular, Velazquez 144, Madrid-6, Spain.
[2]Present address: Technische Universität München, Lehrstuhl für Mikrobiologie, Arcisstr. 21, D-8000 München 2, Federal Republic of Germany

tion studies. It should be noted, however, that the in vitro system
described in this paper is not restricted to a particular type of Rl
miniplasmid or copy mutant.

MATERIALS AND METHODS

Preparation of cell extracts

Cultures of Escherichia coli C600(pKN177) were grown with
shaking at $37^{\circ}C$ in 2 l of L broth supplemented with 0.2% glucose. At
an OD_{600} of 1.0 cells were poured onto 600 g of crushed ice, sedimented, washed once with 30 ml of buffer A (25 mM HEPES-KOH pH 8.0,
100 mM KCl, 1 mM dithiothreitol, 1 mM p-aminobenzamidine), and suspended in buffer A (1 ml per g of wet cells). Lysis was carried out by
the freeze-thaw method (5), in the presence of 1 mM EDTA. Extracts
were divided into small aliquots and stored in liquid nitrogen.

Assay of DNA synthesis

Standard reaction mixtures contained final concentrations of
50 mM HEPES-KOH (pH 8.0), 100 mM KCl, 11 mM magnesium acetate, 15 mM
creatine phosphate, 0.1 mg of creatine kinase per ml, 2 mM ATP, 0.4
mM each CTP, GTP and UTP, 0.05 mM NAD, 0.05 mM cyclic AMP, 0.025 mM
each dATP, dCTP, dGTP and (^{3}H)-dTTP (500 cpm per pmol), 0.5 mM each
of the 20 amino acids, and 2.5% polyethylene glycol 6000. The optimal
amounts of extract and exogenous plasmid DNA had to be determined for
each extract preparation.

RESULTS

Properties of the system

Plasmid DNA synthesis was studied in extracts from the plasmid-
carrying Escherichia coli strain C600(pKN177) and in extracts from
plasmid-free Escherichia coli cells. With both systems a significant
amount of DNA synthesis was observed (Table 1). However, the replicative capacity of extracts from plasmid-free cells was consistently
by an order of magnitude lower than that from plasmid-carrying cells.
Moreover, plasmid DNA synthesis in extracts from plasmid-free cells
was strictly dependent upon the addition of exogenous template DNA
whereas in extracts from plasmid-carrying cells some endogenous DNA
synthesis was observed. For both systems the magnesium ion concentration was of critical importance with a rather sharp optimum around
10 mM. DNA synthesis required not only deoxynucleoside triphosphates
but also all four ribonucleoside triphosphates. A strict dependence
on an ATP-regenerating system was observed even with ATP at an initial concentration of 2 mM. Addition of polyethylene glycol stimulated DNA synthesis 2-3 fold. A variable degree of stimulation, de-

Table 1: Properties of the Rl miniplasmid DNA replication system[*]

Omission / addition	Source of extract	
	C600(pKN177)	C600
None	94.9	13.8
- DNA	15.7	0.2
- Mg-acetate	0.8	0.3
- ATP	2.1	0.3
- CTP, GTP, UTP	3.5	1.1
- Creatine phosphate	1.0	0.5
- Polyethylene glycol	42.0	5.0
- Cyclic AMP	36.7	4.0
- Amino acids	40.3	4.0
+ Novobiocin	1.6	0.2
+ Oxolinic acid	2.4	0.5
+ Chloramphenicol	1.1	0.2
+ Puromycin	1.5	0.3
+ Rifampicin	14.9	0.1

[*] Extracts were prepared from the Escherichia coli strains indicated
and assayed for plasmid DNA synthesis. Reaction mixtures (25 µl)
were supplemented with pKN177 DNA (30 µg per ml) and incubated for
60 min at 30°C. Antibiotics were added at concentrations of 50 µg
per ml. Values shown are dTMP incorporated (pmol).

pending on the extract concentration, was noted on adding cyclic AMP
and/or an amino acid mixture.

In vitro DNA synthesis was highly sensitive to inhibitors of
DNA gyrase (novobiocin and oxolinic acid). Furthermore in both types
of extracts the incorporation was completely blocked by various antibiotics affecting protein synthesis. Plasmid DNA synthesis was also
drastically reduced upon inhibiting transcription with rifampicin.
Some residual incorporation was observed in the presence of rifampicin with extracts from plasmid-carrying (but not plasmid-free) cells.
Under these conditions only the endogenous plasmid DNA could function
as template for DNA synthesis (6).

The reaction product of in vitro replication consisted almost
exclusively of monomeric supercoiled miniplasmid DNA. However, in
the presence of the chain terminator arabinosylcytosine triphosphate
accumulation of theta-shaped replicative intermediates was observed
(4). Analysis of these intermediates indicated that replication initiated at a unique origin located in the RepA region at 82.4 kb and
proceeded unidirectionally towards the RTF-proximal edge of IS1b
(see Figure 1). Thus both origin usage and directionality of in vitro
replication were in full agreement with the pattern observed for in
vivo replication (7).

Requirement for protein synthesis

The effect of antibiotics on mini-R1 DNA synthesis indicate
that plasmid replication is tightly coupled to transcription and
translation. The kinetics of these three processes in extracts prepared from plasmid-free cells are shown in Figure 2. It can be seen
that DNA synthesis commences after a lag period of 30 min at a time
when protein synthesis has nearly reached a plateau. Therefore protein synthesis appears to be a prerequisite for the initiation of
plasmid replication.

The proteins synthesized in the in vitro system were labeled
with (^{14}C)-amino acids and analysed by polyacrylamide gel electrophoresis. As shown in Figure 3, three distinct bands of plasmid-encoded
polypeptides with molecular weights of approximately 10, 30 and 50
kdal were observed, whereas a faster moving band of 8 kdal was also

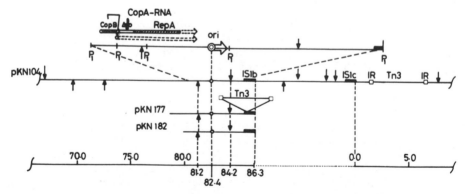

Figure 1: *Physical maps of pKN177 and pKN182 and partial physical
map of pKN104 (1, 4). Circular plasmid DNA is depicted in a linear
presentation. The replication region of R1 is shown in an expanded
scale. Indicated are the target sites for the restriction enzymes
EcoRl (downward arrow), SalI (upward arrow, and Pstl (P1). The direction of replication is denoted by an open arrow. The numbers indicate the R1 coordinates in kb.*

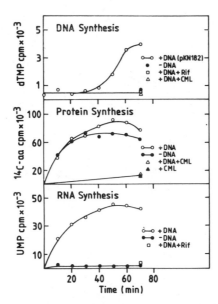

Figure 2: *Kinetics of DNA, protein, and RNA synthesis in extracts of Escherichia coli C600. Standard reaction mixtures were supplemented with pKN182 DNA (30 μg per ml) without addition of amino acids.* (^3H)*-dTTP (40 μCi per ml, 50 mCi per mmol),* (^{14}C)*-protein hydrolysate (80 μCi per ml, 55 mCi per mmol), and* (^3H)*-dUTP (80 μCi per ml, 45 mCi per mmol) were used as radioactive label respectively. Antibiotics were added at concentrations of 50 μg per ml. Incubations were carried out at 30°C and aliquots (25 μl) removed at the times indicated and assayed for incorporation of radioactive label.*

labeled in the absence of miniRl DNA. It should be noted that the molecular weights of the 10-kdal and 30-kdal polypeptides correspond closely to that of the copB and repA gene products respectively. Confirming this interpretation it could be shown by lacZ gene fusions that both the copB and the repA promoter are functional in the cell-free system. The nature of the 50-kdal protein is unknown. Addition of novobiocin had little or no effect on the synthesis of the plasmid-encoded proteins and no degradation was noticed during a 60-min chase in the presence of chloramphenicol (data not shown).

DISCUSSION

In vitro replication of Rl miniplasmid DNA closely resembles in vivo replication with respect to dependence on de novo protein synthesis, which is presumeably required to supply adequate amounts of repA protein. A polypeptide of corresponding molecular weight was found to be one of the major protein species synthesized in the

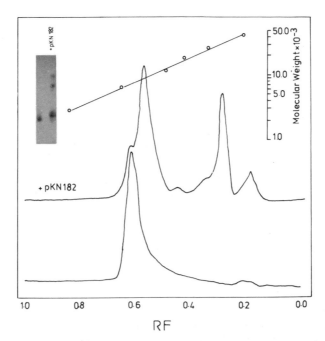

Figure 3: *Polyacrylamide gel electrophoresis of proteins synthesized in* Escherichia coli *C600 extracts. Standard reaction mixtures (25 μl) were incubated at 30°C without addition of amino acids in the presence or absence of pKN182 DNA (20 μg per ml). After 10 min 2 μCi (^{14}C)-protein hydrolysate (55 mCi per mmol) was added and the incubation continued for 40 min at 30°C. Labeled proteins were separated by electrophoresis in a 15% polyacrylamide gel and identified by auto-radiography. Densitometer tracings of the autoradiograms are shown. Standards for molecular weight determination were from BRL.*

cell-free system. The role of this protein in the initiation of plas-mid replication is unclear. Since the repA function is poorly comple-mented in trans (8) one may assume that this protein binds tightly to a replication control site on the template DNA. Recent cloning experiments show that removal of R1 sequences downstream from the repA termination codon eliminates the titration site for the repA protein (S. Ortega and R. Diaz, unpublished results). Insertion of the Sau3A fragment containing most of the repA sequence into the BamH1 site of the expression vector pAS1 (obtained from M. Rosenberg) resulted in a fused protein, which specifically stimulated R1 mini-plasmid replication in trans.

The rifampicin sensitivity of R1 replication might simply re-flect the requirement for the synthesis of the repA messenger RNA. It is conceiveable, however, that these transcripts also function as primer for the initiation of DNA synthesis at oriV (1). An invol-

vement of RNA polymerase in the synthesis of a cis-acting RNA associated with the template DNA is consistent with the differential
effect of rifampicin on endogenous and exogenous R1 miniplasmid DNA
synthesis (6). Supercoiling of the plasmid DNA by DNA gyrase while
having little effect on the expression of the repA gene could be
essential for melting the primer RNA into the double-stranded DNA
template (9, 10). One might speculate that the function of the repA
protein consists in facilitating the formation and/or utilization of
such a template-primer complex at the origin of replication.

REFERENCES

1.   Light,J. and Molin,S. (1983) EMBO J. 2, 93.
2.   Uhlin,B.E. and Nordström,K. (1978) Mol. Gen. Genet. 165, 167.
3.   Molin,S. and Nordström,K. (1980) J. Bacteriol. 141, 111.
4.   Diaz,R. and Staudenbauer,W.L. (1982) J. Bacteriol. 150, 1077.
5.   Staudenbauer,W.L. (1976) Mol. Gen. Genet. 145, 273.
6.   Diaz,R., Nordström,K. and Staudenbauer,W.L. (1981)
           Nature 289, 326.
7.   Timmis,K.N., Danbara,H., Brady,G. and Lurz,R. (1981)
           Plasmid 5, 53.
8.   Oertel,W., Kollek,R., Beck,E. and Goebel,W. (1979)
           Mol. Gen. Genet. 171, 277.
9.   Orr,E. and Staudenbauer,W.L. (1981) Mol. Gen. Genet. 181, 52.
10. Hillenbrand,G. and Staudenbauer,W.L. (1982) Nucl. Acids Res.
           10, 833.

ANALYSIS OF Mu DNA REPLICATED ON CELLOPHANE DISCS

N. Patrick Higgins[1], Purita Manlapaz-Ramos and
Baldomero M. Olivera

Department of Biology
University of Utah
Salt Lake City, UT 84112, USA

[1]Department of Biochemistry
University of Alabama at Birmingham
Birmingham, Alabama, USA

INTRODUCTION AND BACKGROUND

The bacteriophage Mu is both a temperate bacterial virus and
a transposable element. During the Mu lytic cycle, uncontrolled replicative transposition takes place. As a result, a single copy of Mu
is amplified into ca. 100 copies each inserted at a different site
in the host chromosome. During lytic infection or induction, when Mu
replication is under viral control, replication is obligatorily
coupled to transposition (1, 2).

We have established an in vitro system in which bona fide Mu
replicative transposition events can be observed (3). The in vitro
system that we use (film lysates on cellophane discs) has features
that make it feasible to analyze Mu transposition. Foremost among
these is the fact that macromolecular components are present in highly
concentrated form, approaching intracellular concentrations. Since
intact host chromosomes are present in cellophane disc lysates, the
system is particularly favorable for analyzing events that occur during the induction of Mu.

Using this system, we have shown that: 1) the replication event
during transposition in the lytic cycle occurs within Mu sequence
boundaries. This means that Mu proteins control initiation and termination of DNA replication within these boundaries (4). 2) Replication
is semiconservative and semidiscontinuous. The discontinuous replica-

63

tion intermediates (Okazaki pieces) are similar in their size distribution to those found in an unperturbed host fork (3). 3) Replication in vitro is unidirectional and proceeds from the left end of Mu rightwards. As a consequence, Okazaki pieces hybridize primarly to the light strand of Mu, continuously synthesized DNA to the heavy strand (3, 5). 4) Initiation of replicative transposition does not normally take place on the disc; rather, Mu controlled replication forks initiated in vivo continue replicating Mu sequences and are terminated in vitro at the right end of Mu. Thus, the in vitro Mu replication system permits an analysis of the specificity of in vivo initiation events. However, one condition under which in vitro initiation apparently takes place has been defined (6).

SYNCHRONOUS Mu REPLICATIVE TRANSPOSITION IN VITRO FROM A DEFINED DONOR SITE

We are attempting to analyze replicative transposition events that originate from the same donor site in every cell. Synchronous transposition from a single site has been carried out in vivo by Pato and Reich (7). Their basic strategy was to induce a Mu lysogen in a thymine-requiring strain thereby allowing the synthesis of all Mu proteins necessary for replicative transposition. However, transposition was prevented by thymine deprivation. When thymine was added to the culture, synchronous replicative-transposition was observed. Our strategy in vitro is to harvest such induced but thymine-starved cells, and then carry out replicative transposition by providing deoxynucleoside triphosphates in vitro. All transposition events should be occurring from the original locus of insertion of the prophage.

When the first transposition event is complete, connections to a set of random acceptor sites from a defined donor (i.e., prophage insertion) site would be observed. Any replicated Mu sequences still attached to the defined donor site would yield specific restriction fragments at the left and right ends, since the donor sites would be the same in every cell. However, any Mu sequences attached to acceptor target sites would yield different restriction fragments in every cell; each host acceptor site would have a restriction site which is a different distance from the Mu/host junction.

The restriction enzyme EcoRI cuts Mu DNA at two sites; thus, Mu DNA which is inserted in the chromosome should show 3 Mu specific bands after EcoRI digestion. One restriction fragment will contain only Mu DNA ("middle fragment"); the other 2 restriction fragments however will contain both Mu and host DNA ("junction fragments"); the length of these fragments will depend on the position of Mu in the Escherichia coli chromosome. Thus- the restriction fragments corresponding to the left and right ends will be specific for each Mu lysogen.

A Southern transfer analysis of a particular Mu lysogen (the thymine-requiring strain MP505) is shown in Figure 1 (top). If the Mu lysogen is induced in the absence of thymine, a Southern transfer analysis yields the result shown in lane 1; intact Mu is found (possibly due to Mu-specific modification redering the DNA relatively resistant to restriction digestion). The same Mu specific restriction bands (labeled A, B, and C in lane 1) were also obtained in a Southern analysis of the uninduced lysogen (results not shown). Fragment A is an incomplete digestion product. B is the middle restriction fragment containing only Mu DNA. If Mu is induced in the presence of thymine, allowing extensive transposition to take place (lane 2), a considerable amplification is observed of restriction fragment B, the "middle EcoRI fragment". No amplification of restriction fragments A and C is detected. A control EcoRI digest of purified Mu viral DNA (lane 3) confirms that fragment B is the "middle restriction fragment".

Mu DNA synthesized on the cellophane disc from induced, thymine starved cells was analyzed by autoradiography (Figure 1, bottom). Lane 1 shows an autoradiogram of the labeled DNA from a thymine starved culture which was then allowed to synthesize DNA in vitro on the film lysate. It is seen that under these conditions fragments B and C that were seen on the Southern transfer analysis are labeled in the in vitro system. In addition, there are additional bands that are clearly detectable. However, in a control run with the parental strain, MP504, which does not contain a Mu prophage, the same set of bands (except B and C) were labeled (lane 2). This appears to be an artifact of the thymine starvation: for reasons that are not clear, when thymine starved cells are run in the in vitro system, a specific set of host restriction fragments become highly labeled. These restriction fragments are either unlabeled, or much more weakly labeled in vitro if cells growing in the presence of thymine are used. If band C is indeed a restriction fragment spanning both Mu and host DNA as postulated above, then in a different Mu lysogen, a different restriction fragment would be labeled. This experiment is shown in lane 4; a different band is visible in an EcoRI digest of DNA when a different lysogen (MP504-3) was used to synthesize DNA in the film lysate system. However, as expected, band B is still labeled in this strain.

These restriction analyses are consistent with the idea that a "one site" transposition event is taking place on the cellophane disc. The replication of a specific junction fragment in vitro is consistent with replicative transposition from the original prophage insertion site in all cells. It is worth commenting on our failure to detect two end fragments; only one is detectable. There are two technical reasons why left end fragments might not be detected in this type of an analysis. First, the amount of Mu DNA at the left end is relatively small, and therefore the label may be insufficient to detect under these experimental conditions. In addition, the residual pool of dTTP may be sufficient so that during the thymine

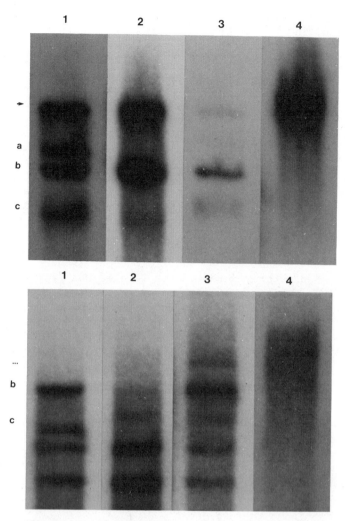

Figure 1:   Analysis of EcoRI digests. Top: Southern transfer analysis probed with nick translated (^{32}P)-Mu DNA. Lane 1; induced Mu Cts
lysogen (MP505) starved for thymine during induction. 2; the same
lysogen induced in the presence of thymine. 3; digested Mu viral DNA.
4; undigested MP505 DNA. Bottom: Autoradiogram of DNA synthesized in
vitro on cellophane disc lysates using (^{32}P)-dGTP as a label as described previously (3). DNA was restricted with EcoRI. Lane 1; DNA
synthesized in vitro by thymine starved MP505. 2; a similarly treated
lysate from the nonlysogenic bacterial strain MP504 (which contained
no Mu prophage). 3; an independently isolated lysogen of MP504,
MP504-3 and 4; MP505 DNA as in lane 1, except that it was not digested
with EcoRI. Induction and cellophane disc lysates (carried out in the
presence of NAD for 15 min) were as described previously (3,4). EcoRI
digestion, Southern transfer, and autoradiography were done by standard procedures (8).

starvation period, a minimal level of DNA synthesis takes place in vivo. If this were the case, then the left end fragments would be synthesized in vivo and there would be no significant ^{32}P label incorporated in vitro into the left end. These possibilities are currently being investigated.

SUMMARY AND CONCLUSIONS

We have presented restriction analyses which indicate that the first transposition event after induction can be carried out and analyzed in vitro using cellophane disc lysates. This would be a first step in describing in molecular terms a "wiring diagram" of replicative transposition.

ACKNOWLEDGEMENTS

This work was supported by Research Grants GM26107 and GM32462 from the NIH.

REFERENCES

1.   Bukhari,A. (1976) Ann. Rev. Genet. 10, 389.
2.   Toussaint,A. and Resibois,A. (1983) in: Mobile Genetic Elements,
          Shapiro,J. ed., Academic Press, New York, 105.
3.   Higgins,N.P., Moncecchi,D., Manlapaz-Ramos,P. and Olivera,B.M.
          (1983) J. Biol. Chem. 258, 4293.
4.   Higgins,N.P., Manlapaz-Ramos,P., Gandhi,R.J. and Olivera,B.M.
          (1983) Cell 33, 623.
5.   Higgins,N.P., Moncecchi,D., Howe,M., Manlapaz-Ramos,P. and
          Olivera,B.M. (1984) in: Mechanisms of DNA Replication
          and Recombination, Cozzarelli,N.R. ed., UCLA Symposia
          on Molecular and Cellular Biology, Alan R. Liss, Inc.,
          New York, Vol. 10, 187.
6.   Higgins,N.P. and Olivera,B.M. (1984) Mol. Gen. Genet., in press.
7.   Pato,M.L. and Reich,C. (1982) Cell 29, 219.
8.   Maniatis,T., Fritsch,E.F. and Sambrook,J. (1982) Molecular
          Cloning. A Laboratory Manual. Cold Spring Harbor
          Laboratory, 1982.

REPLICATION OF BACTERIOPHAGE Mu AND ITS MINI-Mu DERIVATIVES

Anne Resibois, Martin Pato[1], Patrick Higgins[2]
and Ariane Toussaint[3]

Laboratoire d'Histologie
Université Libre de Bruxelles
B-1000 Bruxelles, Belgium

[1]Department of Molecular and Cellular Biology
National Jewish Hospital and Research Center
Denver, Colorado 80206, USA

[2]Department of Biochemistry
University of Alabama
Birmingham, Alabama 35294, USA

[3]Laboratoire de Genetique
Université Libre de Bruxelles
B-1640 Rhode St. Genese, Belgium

INTRODUCTION

The temperate phage Mu behaves as a transposing replicon. During its whole replication phase, which starts 6-8 min after the onset of the lytic cycle, it remains attached to host DNA; the original Mu prophage remains in the host chromosome (1) and the bacterial nucleoid remains intact (2). Only Mu sequences are replicated in vivo (3) and in vitro (4, 5), implying that replication is initiated in or at the ends of the phage genome and terminates at the ends.

It has been known for a long time that during its lytic cycle Mu induces chromosomal rearrangements at high frequency and it is now clear that it is a direct consequence of its mode of replication. Several models have been formulated to account for the replicative transposition of prokaryotic transposable elements and replication of bacteriophage Mu (for a review see ref. 6). They share the pre-

diction that replication-transposition is initiated by a set of
nicks and ligations at the end(s) of Mu and at a target site on the
host genome which create replication fork(s) at the end(s) of the
phage DNA. How the replisome is then mobilized by the fork(s) remains
an open question.

Mu replication is largely dependent on host enzymes. It is
independent of DnaA and PolI and requires the products of the dnaC,
dnaB, dnaG, dnaE genes and most probably the gyrase (7, 8 and
McBeth and Taylor, unpublished results, this paper). Other host functions have not yet been tested. Mu replication also requires the Mu
encoded proteins A (the transposase) and B (of unknown function)(9)
and the ends of the phage genome. The level of replication is dependent on the Mu encoded function Arm, the role of which might be
to regulate expression of A and B (10).

Continuous synthesis of A is necessary during the lytic cycle
suggesting that the transposase is unstable and raises the possibility
that it is used stoichiometrically rather than catalytically (11).

The isolation of mini-Mu's, i.e. derivatives deleted for most
of the central portion of the phage genome which do not express any
maturation function but have intact ends and various portions of the
A, B and arm genes, has allowed further insights into Mu replication.
Mini-Mu with no A, B or arm genes are completely defective for replication. They are very poorly complemented by low amounts of A, B and
arm (provided by a non replicating A+, B+, arm+ c end fragment).
When large amounts of those proteins are provided in trans, for instance by a replicating Mu, the mini-Mu replicates but about 10 times
less efficiently than the helper Mu. Mini-Mu which carry a functional
A gene but no B or arm transpose about 100 times less than Mu but are
efficiently complemented for replication even by a non replicating
c end fragment (10). This suggests that the absence of complementation
of the mini-Mu A- is due to the limited amount of A provided by the
non replicating helper or to poor activity of the A protein in trans.
Mini-Mu with functional A, B and arm genes replicate almost as efficiently as a wild type Mu (10).

MINI-Mu AND Mu REPLICATION INTERMEDIATES

Thanks to the existence of mini-Mu's, replication intermediates
can be definitively identified. The most abundant are "key structures"
in which the mini-Mu genome is covalently linked to host sequences
and located either in the circle or in the tail with the replicating
section straddling the fork (see Figure 1) (12, 13). Such intermediates are also detected during in vivo replication of a normal Mu (14,
15). We have now examined the formation of replication intermediates
during in vitro replication of Mu on cellophane disks. In order to
increase the number of replicating prophages in the DNA preparation

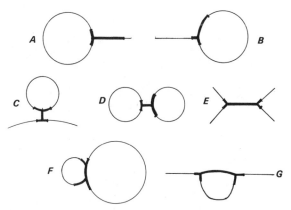

*Figure 1: Replication intermediates of Mud26. The thin lines repre-
sent host DNA, the thick lines, replicating Mud26 DNA. A and B: Keys
with the Mud26 replicating from one end and located in the tail and
the circle respectively; C: pending key; D: dumbbell; E: inverted
fork; F: partially fused circles; G: asymmetrical fork. In C, D, E,
F and G the mini-Mu replicates from both ends. According to Resibois
et al. (13, 16).*

to be analyzed by E.M., we synchronized the first round of Mu DNA
replication by inducing a Muct562 lysogen in the absence of thymine
(MP505, Pato and Reich, unpublished results). In order to distinguish
DNA synthesized in vivo and in vitro (^{3}H)thymidine was added in vivo
2 min prior to synchronization. In vitro synthesis was started 12 min
after synchronization and carried out in the presence of ^{32}P-dGTP and
dBUTP instead of dTTP; DNA isolated from samples harvested 5, 15 and
45 min after the beginning of in vitro synthesis were centrifuged on
CsCl density gradients; fractions were collected from the bottom of
the tube (for a detailed protocol see refs. 4 and 5). Fractions from
the intermediate density region between unreplicated (LL) DNA and
fully replicated (HL) DNA were pooled and mounted for E.M.. Keys
were (rarely) found in the intermediate density fractions at all
time points. No other abnormal configuration was seen.

After induction of a self replicating mini-Mu, which does not
express maturation functions as, Mud26, more complex structures are
found besides keys which again contain a mini-Mu covalently linked
to host DNA but contain two forks straddled by the mini-Mu which does
thus replicate from both ends (see Figure 2) (13, 16).

DIRECTION OF Mu AND Mud26 REPLICATION

The direction of replication of Mu DNA was inferred by
Wijffelman and van de Putte (17) and Goosen (18) to proceed from the

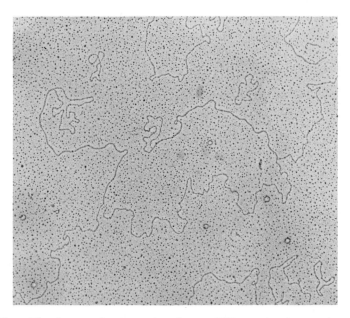

Figure 2:   Electron micrograph of a ssDNA segment renaturated on two sets of inverted repeats of Mud26.

c̲ towards the S̲ end. Their experiments showed that 80% of the Okazaki
pieces isolated after a short labeling period annealed to one of the
separated strands of Mu DNA (the L strand), while 20% bound to the
other (the H strand, which also binds Mu mRNA). Since the mRNA is
copied from c̲ to S̲ the continuous (5' to 3') strand of DNA would
also be copied from the H strand and Okazaki Pieces would anneal to
the L strand. They suggested that the 20% annealing of Okazaki pieces
to the H strand might involve inversion of the G region DNA and that
all the replication proceeds from the c̲ to the S̲ end. Similar results
were obtained in vitro (4). However, the S̲ end is required for replication although it is not clear whether it is necessary for initiation. In the replicating intermediates of Mud26 either the c̲ or the
S̲ or both ends are replicating (16). In agreement with that previous
observation we have found that Okazaki pieces isolated 50 min after
induction of a Mud26 lysogen do indeed hybridize essentially equally
well with both strands of Mu (see Table 1). It remains to be seen
whether replication from the S̲ end is an intrinsic property of the
mini-Mu. The most abundant intermediates of Mud26 replication are
keys in which only one end is replicating. Moreover in the intermediates which contain two forks the length of the replicated segments at
the c̲ and S̲ end are usually different. Thus initiation at the two
ends does not seem to be synchronized.

Table 1:   Annealing of Okazaki pieces to separated strands of
           Mu DNA*

Induced prophage(s)	Bacterial host	Fraction bound to	
		Mu light strand	Mu heavy strand
Mucts62	(rec$^+$)	.78	.22
	"	.80	.20
	(rec A)	.79	.21
Mucts62 Aam/Mud132		.74	.26
Mucts62 Bam/mud132		.82	.18
Mud26	(rec$^+$)	.61	.39
	"	.65	.35
	(rec A)	.53	.47
	"	.42	.58
	"	.51	.49

* To determine if Mud26 would replicate in a different manner than
Mu we isolated Okazaki pieces from induced Mud26 lysogens as well
as from Mucts62 lysogens. The mini-Mu lysogens were induced for 50
min before pulse labeling for 20 sec with (^{3}H)-thymidine. Mucts62
lysogens were induced for 30 min before labeling. Labeled DNA was
isolated and sedimented in 5% - 20% alkaline sucrose gradients in
a Spinco SW41 rotor at 40,000 rpm for 14-18 hrs. The gradients were
fractionated and samples corresponding to Okazaki pieces were pooled
and annealed to filters containing separated strands of Mu DNA.
About 80% of labeled Mu DNA from induced lysogens annealed to the
L strand as observed originally by Wijffelman and van de Putte (17).
This was also the case for the induced MuAam and MuBam lysogens.
However labeled DNA from the induced Mud26 lysogen annealed almost
equally well to either L or the H strand. The Mud132 prophage
(Mucts62 dJ-), the Aam1093 and the Bam1066 mutations were described
by Howe (19).

    There is some evidence that the recognition site(s) for the
transposase is not only at the ends of the Mu genome. The A proteins
of Mu and its close relative D108, which differ only by 72 aa at their
NH2-terminal ends are not functionally interchangeable. Mu and D108
are non homologous over a 1.2 kb segment located close to the c end
(at most 200 base pairs from the c end)(20) which extends 216 base
pairs into the beginning of the A gene, but the two phages have the

same c extremity (Bukhari pers. comm.) and their right ends are
functionally interchangeable, suggesting that the transposase might
recognize a site whithin the region of non homology rather than the
extremities (21). In view of this one could imagine that the transposase would contain at least two domains; one would lie in the NH2-
terminal portion of the proteins, be different in Mu and D108 transposases and bind preferentially to the DNA encoding it (in cis) at a
site located in the non homologous regions; from there it could move
along the DNA to find a Mu terminus. There, a second domain, similar
in the two transposases, would initiate transposition-replication.
With full length Mu DNA the A gene is located near the c end; hence
the c end is most frequently encountered first and replication will
initiate preferentially at the c end. By deleting most of the internal portion of the prophage DNA to form a mini-Mu the S end is brought
closer to the A gene and either terminus is likely to define the
origin of replication. To test this explanation in its simplest form
Mu Aam were complemented in trans using a c end non replicating fragment of Mu to supply the A product. Both termini of the Aam prophage
are presumably equally distant from the source of A and replication
from either end should be equally probable if the prophage ends are
really the A recognition sites. However the results in Table 1 show
that replication is still predominantly from the c to the S end,
and the same result was obtained with a Bam prophage. The existence
of a site near the c end for binding of the transposase preparatory
to selection for a terminus would again offer a simple explanation
for these results. The fact that the S end is not usually used as
an origin in full length Mu could as well be due to the presence on
the Mu genome of a site which blocks the movement of the A protein
and which is deleted in Mud26. If that is the case one should be
able to isolate mini-Mu's which still contain that site and have a
unidirectional replication. This is currently being tested.

LOCATION AND ARANGEMENT OF Mud26 REPLICAS ON THE HOST GENOME

We have analyzed by electron microscopy the total DNA extracted
from a Mud26 lysogen induced for 50 min at 42°C, denatured and self-
reannealed. In addition to long stretches of single-stranded (SS)
DNA, we found frequent stretches containing self reannealed double-
stranded (DS) segments of Mud26 length (10 kb), suggesting that several copies of the mini-Mu were integrated on the same SS segment
and arranged as inverted repeats. Figure 2 shows such a group of
sequential inverted repeats of Mud26. The average distance between
two inverted copies was 40 kb (36 measurements) and the most frequent
distance was around 20 kb (see Figure 3).

To see whether direct repeats of the mini-Mu were also present,
the DNA of the induced Mud26 lysogen was denatured and reannealed
in the presence of D108 DNA digested with BamHl which cuts it 16.9
kb from the c end. The D108 c end should hybridize with the c ends

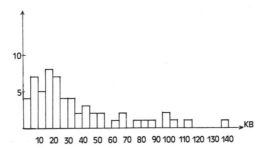

Figure 3: Histogram of the distances separating 2 adjacent inverted copies of Mud26. A total of 36 SS regions separating self reannealed mini-Mu's were measured. kb = kilobases.

of the mini-Mu which did not self reanneal, generating the characteristic 1.2 kb long bubble of non homology. Heteroduplexes between
D108 and Mud26 were very rare even in the SS segments which did not
contain inverted repeats of the mini-Mu. They were however frequent
when the bacterial DNA was sheared before denaturation so that the
number of mini-Mu on one DNA segment was reduced. This strongly
suggests that the large majority of the mini-Mu replicas are clustered on the host DNA and arranged as inverted repeats. This is consistent with our previous observation that replication intermediates
which lead to two inverted copies of the mini-Mu are far more frequent
than those leading to direct copies (16). This regular arrangement
of the mini-Mu replicas suggests that all the mini-Mu's are not
replicating simultaneously and would be easily explained if only the
outside copies replicate and do so only once. Analysis of Mu replication in vitro also suggests that only a few copies of Mu replicate
at a given time (5).

We also looked at self reannealed DNA extracted 50 min after
induction of dnaBts266, dnaCts325, dnaGts3, dnaEts486, dnaEts511 and
gurBts (LE316) strains lysogenic for Mud26. Inverted repeats of the
mini-Mu were essentially absent from DNA extracted from the dnaC,
dnaG and the gurB strains and only rarely seen in the DNA prepared
from the dnaB and dnaE strains suggesting that the mini-Mu replicates
very poorly in these host mutants. In agreement with these observations, pulse labeling with tritiated thymidine of the same strains
in the same conditions show reduced incorporation both in the host
and in the phage DNA. With dnaC and gurB mutants, Mu DNA replication
was inhibited more than host replication; e.g. 30 min after shifting
to 42°C total replication in the gurB was 40% the level in a gyrase
wild type, while Mu replication was down to 6% the level in the wild
type. With dnaB and dnaG mutants replication of both host and Mu DNA
was severely reduced.

It was found previously that Mu does not grow in dnaB, dnaC and dnaE hosts (17). Mucts62 also does not grow in dnaG and gurB hosts (data not shown). It thus seems that the inability of Mu to grow in those hosts does indeed result from the absence of replication, and that all these host functions are required for Mu replication-transposition.

ACKNOWLEDGEMENTS

This work was benefited from a NATO grant (No. 084.82).

REFERENCES

1. Ljungquist,E. and Bukhari,A.I. (1979) J. Mol. Biol. 133, 339.
2. Pato,M.L. and Waggoner,B.T. (1981) J. Virol. 38, 249.
3. Waggoner,B.T. and Pato,M.L. (1978) J. Virol. 27, 587.
4. Higgins,N.P., Moncecchi,D., Manlapaz-Ramos,P. and
        Olivera,B.M. (1983) J. Biol. Chem. 58, 4293.
5. Higgins,P.N., Manlapaz-Ramos,P., Gandhi,R.T. and Olivera,B.M.
        (1983) Cell 33, 623.
6. Bukhari,A.I. (1981) Trends Biochem. Sciences, Feb. 1981, 56.
7. Toussaint,A. and Faelen,M. (1974) Mol. Gen. Genet. 131, 209.
8. Teifel,J. and Schmieger,H. (1981) Mol. Gen. Genet. 184, 308.
9. Wijffelman,C. and Lotterman,B. (1977) Mol. Gen. Genet.
        151, 169.
10. Waggoner,B.T., Pato,M.L., Toussaint,A. and Faelen,M. (1981)
        Virology 113, 379.
11. Pato,M.L. and Reich,C. (1982) Cell 29, 219.
12. Harshey,R. and Bukhari,A.I. (1981) Proc. Natl. Acad. Sci. USA
        78, 1090.
13. Resibois,A., Colet,M. and Toussaint,A. (1982) EMBO J. 1, 965.
14. Waggoner,B.T., Gonzales,N.S. and Taylor,A.L. (1974)
        Proc. Natl. Acad. Sci. USA 71, 1255.
15. Schroeder,W., Bade,E.G. and Delius,H. (1974) Virology 60, 534.
16. Resibois,A., Toussaint,A. and Colet,M. (1982) Virology 117, 329.
17. Wijffelman,C. and van de Putte,P. (1977) in: DNA Insertion
        Elements, Plasmids and Episomes, Buhari,A.I., Shapiro,J.A.
        and Adhya,S.L. eds., Cold Spring Harbor Laboratory, Cold
        Spring Harbor, New York, 329.
18. Goosen,T. (1978) in: "DNA synthesis: Present and future",
        I. Molineux and M. Kohiyama, eds., Plenum Publishing
        Corporation, 121.
19. Howe,M.M. (1973) Virology 54, 93.
20. Gill,G.S., Hull,A.C. and Curtiss III,A. (1981) J. Virol.
        37, 420.
21. Toussaint,A., Faelen,M., Desmet,L. and Allet,B. (1983)
        Mol. Gen. Genet. 190, 70.

INITIATION OF DNA SYNTHESIS ON SINGLE-STRANDED DNA TEMPLATES IN

VITRO PROMOTED BY THE BACTERIOPHAGE λ O and P REPLICATION PROTEINS

Jonathan H. LeBowitz and Roger McMacken

Department of Biochemistry
School of Hygiene and Public Health
The Johns Hopkins University
Baltimore, Maryland USA 21205

ABSTRACT

The bacteriophage λ O and P protein replication initiators, in conjunction with six purified <u>Escherichia coli</u> replication proteins, replicate the single-stranded chromosomes of phages M13 and ϕX174 to a duplex form. Several discrete steps are involved in this DNA synthesis reaction. In an ATP-dependent step that preceeds priming, the λ O and P proteins interact with the <u>Escherichia coli</u> dnaJ and dnaK proteins to transfer the bacterial dnaB protein onto DNA coated with single-stranded DNA binding protein. This creates a stable prepriming intermediate, isolable by gel filtration, that is rapidly primed and replicated upon the addition of primase and DNA polymerase III holoenzyme. Each of the eight proteins required for this nonspecific single strand replication reaction also have physiological roles in the replication of the bacteriophage λ chromosome in vivo. We propose a scheme for the λ O and P protein-dependent initiation of DNA synthesis that may be relevant to strand initiation events occurring during λ DNA replication.

INTRODUCTION

Replication of the coliphage λ chromosome is initiated on superhelical DNA at a unique origin, form which replication forks are propagated bidirectionally (see references 1 and 2 for recent reviews of λ replication). Genetic and biochemical studies demonstrated that only two phage-encoded proteins, the λ O and P proteins, participate in this complex process. Instead, most of the required proteins are

77

manufactured by the bacterial host. These Escherichia coli proteins
include primase (dnaG protein), DNA polymerase III holoenzyme, single-
stranded DNA binding proteins (SSB), RNA polymerase, and DNA gyrase,
as well as the host dnaB, dnaJ and dnaK gene products.

Clues to the molecular roles of several of the required host
proteins first came from in vitro studies of the mechanisms by which
the chromosomes of the single-stranded (ss) phages M13, ϕX174, and
G4 were replicated. Several recent reviews detail the biochemical
properties of these proteins and discuss their probable functional
roles (3-5). The bacteriophage λ O and P initiators have also been
purified (6-9). Initial studies of these viral initiators suggest
that the λ O protein functions in the specific recognition of the
chromosomal replication origin (oriλ) and that the λ P protein acts
to direct the host replication machinery to the λ chromosome. The O
protein binds to a quartet of repeating sequences present at oriλ
(8, 10), while the P protein forms a stable complex with the Escherichia coli dnaB protein (9, 11, 12) and apparently interacts with
the bacterial dnaK protein as well (13).

The recent establishment of soluble in vitro systems that can
specifically replicate supercoiled plasmids containing oriλ (6, 14,
15) provides encouragement that the molecular events that occur in
the initiation of bidirectional λ DNA replication can ultimately be
described. This task will be difficult, however, since these replication systems are enormously complex. More than 20 gene products participate in the replication of λ DNA, acting in diverse types of biochemical reactions that take place concurrently during the replication process. For example, it is likely that events such as origin
recognition, transcriptional activation, protein transfer reactions,
DNA unwinding, RNA priming, DNA chain polymerization, DNA topoisomer-
ase-mediated topological alterations, RNA primer removal, filling of
ss gaps, replication termination, DNA ligation, segregation of daughter DNA molecules, and DNA supercoiling all occur in overlapping time
intervals.

The complexities of sorting out the multiple concurrent reactions involved in bidirectional λ DNA replication prompted us to
attempt to dissect this process into discrete leading strand and
lagging strand initiation events. Following the methodology pioneered
by D. Ray and coworkers for the isolation of rifampicin-resistant
single strand initiation determinants (16), we cloned candidate DNA
sequences from the oriλ region into phage M13 cloning vectors. We
tested such M13-λ hybrid chromosomes in the soluble λdv replication
system for their capacity to direct the initiation of complementary
strand DNA synthesis in the presence of rifampicin, a drug which specifically blocks chain initiation at the M13 complementary strand
origin. Through these studies we uncovered a novel DNA chain initiation mechanism that depends on the λ O and P replication initiators
and on key elements of the host replication machinery (17), but

surprisingly, does not require λ sequences in the template DNA. Our initial findings suggest that this single-strand replication reaction is a plausible model system for investigating many of the protein-protein interactions that occur during the initiation of λ DNA replication.

MATERIALS AND METHODS

The materials and experimental procedures have been or will be described in detail elsewhere (6, 8, 12, 13, 17 and LeBowitz et al., manuscript in preparation).

RESULTS AND DISCUSSION

We have established a soluble enzyme system that is capable of specifically replicating supercoiled plasmid DNA containing oriλ (6). Recently, we have found that this in vitro system can also convert the ss circular chromosomes of phages M13 and φX174 to a duplex form (12, 17). Although replication of these phage chromosomes in the presence of rifampicin was strictly dependent on both the λ O and P initiator proteins, there was no requirement for λ sequences in the ss template DNA. For ease of discussion, we have named this O and P protein-dependent replication of ss DNA the "λss replication reaction". The products of the replication of M13mp7 viral strand DNA consist of two species (Figure 1). A majority of the product (75%) is double-stranded supercoiled DNA (RF I), while the remainder is a circular duplex form (RF II) containing an interruption in the newly synthesized complementary strand. Further experiments indicated that the complementary DNA chains were initiated at multiple or random sites on the various ss templates tested (12, 17).

In addition to a requirement for the λ O and P initiator proteins, replication of the ss templates in the soluble in vitro system was found to depend on the presence of a crude Escherichia coli protein fraction, on an ATP-regenerating system, and on a hydrophilic polymer such as polyvinyl alcohol (17). Due to the crude nature of the bacterial protein fraction used in this system, it was necessary to use several different approaches to identify those Escherichia coli replication proteins that participate in the λ ss replication reaction. Although the replication of duplex λdv plasmid DNA depends on the action of both RNA polymerase and DNA gyrase (6), these enzymes do not function in the λ ss replication reaction, as judged by the resistance of the reaction to both rifampicin and coumermycin (17). Using antibodies directed against specific Escherichia coli replication proteins or soluble extracts prepared from thermosensitive Escherichia coli replication mutants, we identified several of the bacterial proteins that act in the λ ss replication reaction (12, 13, 17).

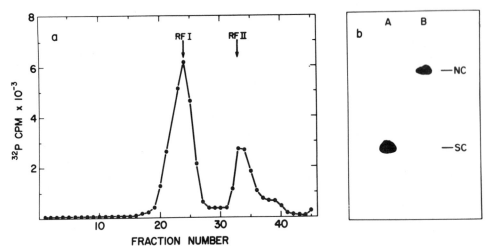

Figure 1: (a) Analysis of the products of the O and P protein-dependent replication of Ml3mp7 DNA in a CsCl/ethidium bromide equilibrium gradient. Ml3mp7 DNA was replicated in an in vitro system (17)
containing purified λ O and P proteins, an Escherichia coli ammonium
sulfate fraction, and (^{32}P)-labeled dNTPs. The product DNA was banded
in a CsCl/ethidium bromide gradient. The arrows denote the peak positions of (^{3}H)-labeled phage fd RF I (supercoils) and RF II (nicked
duplex circles) DNA markers. (b) Electrophoretic analysis of the
products of the λ ss replication reaction in neutral agarose gels.
Products of the λ ss replication reaction were banded in a CsCl/EtBr
gradient (see above), concentrated by isopropanol precipitation, and
electrophoresed in a neutral 1%-agarose gel. Lane A, product from
fractions 20-27 (Figure 1a); lane B, product from fractions 32-36.
The positions of unlabeled Ml3mp7 supercoiled (sc) and nicked-circu-
lar (nc) duplex DNAs, electrophoresed in a parallel lane, are denoted.

　　　　With these results as a guide, we discovered that we could replace the crude Escherichia coli protein fraction with a combination
of six extensively purified bacterial replication proteins. The required host proteins are dnaB protein, dnaJ protein, dnaK protein,
SSB, primase, and DNA polymerase III holoenzyme (Table 1). Thus, with
the inclusion of the λ O and P proteins, the λ ss replication reaction
can be reconstituted with eight purified proteins, making biochemical
analysis of the reaction feasible. Moereover, in the presence of sufficient concentrations of the purified proteins, both polyvinyl alcohol and an ATP regenerating system are no longer necessary for the
strand initiation reaction (Table 1). However, ATP itself clearly is
still required.

Table 1: Requirements for DNA synthesis in the λ ss replication
          system reconstituted with purified proteins.[a]

Component omitted	DNA synthesis (pmol)
None	117
λ O protein	4.9
λ P protein	1.8
dnaB protein	1.3
dnaJ protein	1.9
dnaK protein	1.7
primase	0.6
DNA polymerase III holoenzyme	0.4
O, P, dnaJ and dnaK	9.4
ATP	5.8
CTP, GTP and UTP	179
polyvinyl alcohol	143

[a] The complete reaction mixture contained 450 pmol of M13mp8 DNA
and saturating levels of eight extensively purified proteins: λ O
protein, λ P protein, SSB, dnaB protein, dnaJ protein, dnaK protein,
primase, and DNA polymerase III holoenzyme. Otherwise, assay conditions were similar to those previously described (6), except that
the reaction mixtures did not contain creatine phosphate and creatine kinase. SSB was present in all reaction mixtures to block the
nonspecific general priming reaction (3, 4) catalyzed by dnaB protein and primase.

As mentioned earlier, strand initiation in the λ ss replication
reaction occurs at various, perhaps random, sites on the ss template
DNA. One possible explanation for multiple sites of chain initiation
is the involvement in this reaction of a mobile and processive priming complex (a primosome (18, 19)). In fact, two of the key constituents of the φX174 primosome, the Escherichia coli primase and dnaB
proteins (18, 19), also participate in the λ ss replication reaction.
A diagnostic feature of primosome involvement in the replication of
ss φX174 DNA by bacterial proteins is the rate-limiting nature of
assembly of the preprimosome (4) on the viral DNA, after which the

template is rapidly primed and replicated (20). As shown in Figure
2A, a similar prepriming step is rate-limiting for the λ ss replication reaction in the purified protein system. Ml3 DNA is replicated
very rapidly if it is first preincubated with the required proteins
and ATP (Figure 2A). The kinetics of assembly of the prepriming complex, which resemble the kinetics of DNA synthesis in the normal one-
stage reaction, are depicted in Figure 2B. Our studies indicate that
formation of the prepriming complex depends on each of the proteins
needed for DNA synthesis, except for primase and DNA polymerase III
holoenzyme. Although the overall similarity of the λ ss replication

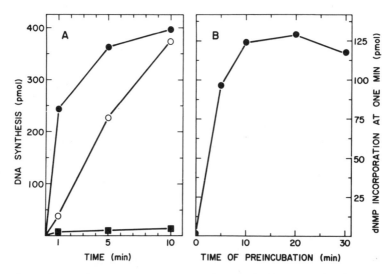

*Figure 2:   Formation of an activated replication intermediate in the
λ ss replication reaction. A: Time course of DNA synthesis in the
λ ss replication reaction in a system reconstituted with purified
proteins. Open circles, time course of DNA synthesis in a normal
one-stage reaction containing 450 pmol of Ml3mp8 DNA and saturating
levels of each of the eight required proteins (see text). Closed
circles, time course of DNA synthesis obtained after 450 pmol of
Ml3mp8 DNA was preincubated for 20 min at 30°C in the presence of
ATP and SSB, O, P, dnaB, dnaJ and dnaK proteins. Primase, DNA polymerase III holoenzyme, rNTPs and labeled dNTPs were added (zero time)
prior to a second-stage incubation. Closed squares, time course of
second-stage DNA synthesis when the O, P, dnaJ and dnaK proteins are
omitted from both the preincubation and replication stages. B: Time
course of the activation of Ml3mp8 DNA during preincubation of the
ss DNA at 30°C with purified, O, P, SSB, dnaB, dnaJ and dnaK proteins.
At the times indicated, aliquots containing 300 pmol of template DNA,
were supplemented with primase, DNA polymerase III holoenzyme, rNTPs
and labeled dNTPs. Rapid DNA synthesis obtained in a one minute
second-stage incubation at 30°C was determined.*

reaction to the mechanism used to replicate φX174 viral DNA is striking, the two reactions are clearly distinct. The sets of proteins required are only partially overlapping. In the λ ss reaction the λ O and P proteins and the Escherichia coli dnaJ and dnaK proteins replace the bacterial dnaC, i, n, n' and n" proteins that are necessary for replication of the φX174 chromosome.

Unlike assembly of the φX174 preprimosome (21), formation of the prepriming complex in the λ ss reaction apparently requires ATP hydrolysis (Table 2, experiment 2). Neither ATPγS nor App-(NH)-p, nonhydrolyzable analogoues of ATP, could support the assembly reaction. In control experiments neither analogue noticeably affected the complete reaction when ATP was also present (Table 2, experiment 1).

Table 2: Effect of ATP analogues on the formation of the activated prepriming complex during the λ ss replication reaction.[a]

Experiment 1, one-stage reaction		Experiment 2, two-stage reaction	
Nucleotide added	DNA synthesis (pmol)	Nucleotide present in stage one	Rapid DNA synthesis (pmol)
ATP	310	None	6
ATP + ATPγS	300	ATP	93
ATP + App-(NH)-p	320	ATPγS	1
		App-(NH)-p	2

[a] Experiment 1: The normal λ ss replication reaction was carried out in a system reconstituted with purified proteins (see the legend to Table 1). ATP and each of the nonhydrolyzable analogues of ATP were present at final concentrations of 0.5 mM.

Experiment 2: Standard two-stage reactions were carried out as described in the legend to Figure 2, except that, where indicated, nonhydrolyzable ATP analogues were present at a final concentration of 0.5 mM. In all cases ATP was present at 2.0 mM in the second-stage DNA synthesis reaction, which was terminated after 1 min at 30°C. ATPγS is adenosine-5'-O-(3'-thiotriphosphate); App-(NH)-p is 5'-adenylylimidodiphosphate.

An active prepriming nucleoprotein complex, formed in the λ
ss replication reaction, can readily be isolated free of unassociated
protein by agarose gel filtration (Figure 3). Several lines of evidence support our contention that the isolated material represents
a true prepriming complex. First, formation of the complex depends
on preincubation of the prepriming proteins with ss DNA in the presence of ATP. Also, in order to obtain DNA synthesis on the isolated
complex, just primase and DNA polymerase III holoenzyme needed to be
added, i.e., no further addition of any of the prepriming proteins
was necessary. Finally, DNA synthesis on the isolated complex was
very rapid, being essentially complete 5 minutes after the addition
of primase and DNA polymerase.

An examination of RNA primer synthesis on the isolated nucleoprotein replication intermediate revealed that each complex could
support the rapid synthesis of more than 120 ribonucleotide residues
per input template circle when the complex was incubated with primase
and rNTPs (Figure 4). In this experiment only about 20% of the starting ss DNA template was capable of being converted to a duplex form

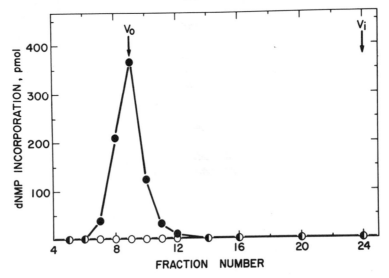

*Figure 3: Isolation by agarose gel filtration of the prepriming
nucleoprotein complex formed in the λ ss replication reaction. Closed
circles, 4500 pmol of Ml3mp8 DNA was preincubated for 20 min at 30°C
with the six required prepriming proteins and ATP to form the activated prepriming complex (Figure 2 legend). This reaction mixture was
chromatographed on a 1 ml column of Bio-Gel A-15m agarose. One drop
fractions were collected and supplemented with primase, DNA polymerase III holoenzyme, rNTPs and labeled dNTPs. DNA synthesis was measured after a 5 min incubation at 30°C. The open circles represent
the results of similar experiments in which either the first-stage
preincubation was omitted or primase was omitted from the second-
stage incubation.*

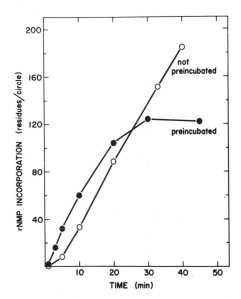

Figure 4: RNA primer synthesis in the λ ss replication reaction reconstituted with purified proteins. Open circles, RNA primer synthesis on M13mp8 DNA in a normal one-stage reaction at 30°C. The reaction mixture contained the six prepriming proteins (Figure 2 legend), primase and (^{32}*P)-labeled rNTPs. Closed circles, RNA primer synthesis on the isolated prepriming nucleoprotein complex. The prepriming protein complex, isolated as described in the legend to Figure 3, was supplemented with primase and labeled rNTPs and incubated at 30°C. Ribonucleotide incorporation was measured as previously described (18).*

following its isolation by gel filtration. Thus, the number of ribonucleotide residues incorporated per activated template may be much higher than 120. Electrophoresis of the RNA primers in polyacrylamide gels yielded a range of primer sizes (10 to 60 ribonucleotide residues), with an average length of about 30 nucleotides. We conclude, therefore, that in the absence of DNA replication at least 4, and perhaps as many as 20, RNA primers are synthesized on each template chromosome that is converted into an activated prepriming complex. It is not known, however, if the synthesis of multiple primers on activated template molecules reflects the assembly of multiple prepriming complexes on each circle or, as in the case of φX174 replication (18), multiple primers are produced by a mobile, processive, primosome-like complex.

Antibodies directed against specific prepriming proteins have been used to attempt to identify those proteins that are functional constituents of the nucleoprotein prepriming complex that is formed

Table 3: Effect of specific antibodies on the activity of the
          nucleoprotein prepriming complex formed in the λ ss
          replication reaction.[a]

Antibody	Activity (%)
None	100
Preimmune IgG	105
Anti-λ O protein IgG	78
Anti-λ P protein IgG	81
Anti-dnaB protein IgG	10
Anti-dnaJ protein IgG	109
Anti-dnaK protein IgG	145

[a] The prepriming nucleoprotein complex was formed and isolated as
described in the legend to Figure 3. Portions containing the isolated complex were incubated for 15 min at 0°C with individual antibody preparations. The activity of the treated complex was measured
in a second-stage DNA synthesis reaction as described in the legend
to Figure 3. The amount of each specific IgG added was sufficient
to cause > 90% inhibition of a standard one-stage λ ss replication
reaction.

in the λ ss replication reaction (Table 3). Of the antibodies tested,
only anti-dnaB protein immunoglobulin (IgG) blocked DNA synthesis
on prepriming complex isolated by gel filtration. This result suggests that the dnaB protein is an essential constituent of the prepriming complex, and we infer from its role in φX174 replication
(18, 19) that the dnaB protein participates in the RNA primer synthesis step of the overall replication reaction. In regard to the failure
of the other antibodies tested to block DNA synthesis on the isolated
replication intermediate, it is still possible that one or more of
the targeted antigens is required for function of the prepriming
complex. If so, then this polypeptide must exist in the prepriming
complex in a form that is inaccessible to the antibody directed
against it. More refined biochemical analyses will be necessary in
order to define the nature and stoichiometry of the polypeptides
present in the prepriming complex.

A plausible scheme for the λ O and P protein-dependent replication of ss DNA is presented in Figure 5. In the first step ss phage
DNA is coated with SSB in a stoichiometric fashion. Next, the λ O

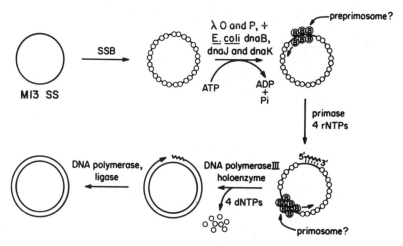

Figure 5: Hypothetical scheme for the λ O and P protein-dependent replication of ss circular DNA. For details see text.

and P initiators, in conjunction with the Escherichia coli dnaJ and
dnaK proteins, function to transfer the bacterial dnaB protein onto
the SSB-coated template. Formation of this nucleoprotein prepriming
complex requires hydrolysis of ATP, apparently occurs at multiple
sites, and is the rate-limiting step of the overall reaction. Once
formed, the prepriming complex, which may have the capacity to migrate
on the coated template strand, serves as a locus for the synthesis
of an RNA primer by primase. Extension of the primer into a nearly
full-length complementary strand is accomplished by the DNA polymerase III holoenzyme. In the crude enzyme system DNA polymerase I removes the RNA primer and fills the small gap present in the complementary strand of the duplex product. DNA ligase seals the remaining
interruption.

The assembly of the prepriming nucleoprotein replication intermediate is the most complex step of the overall reaction. In the case
of φX174 DNA replication, assembly of the preprimosome is potentiated
by the binding of Escherichia coli protein n' to a specific DNA sequence present at the φX174 complementary strand origin (22). We
suspect that the λ O protein initiator, which binds nonspecifically
to ss DNA, performs an analogous role in the λ ss replication reaction.
In this regard, we have found that the λ O protein binds to multiple
sites on SSB-coated M13 viral DNA and that a stable complex of O
protein, SSB and DNA can be isolated free of unassociated protein by
agarose gel filtration. Furthermore, the isolated ternary complex
supported DNA synthesis when it was supplemented with the other six

required proteins and essential substrates and cofactors (unpublished data). We presume, then, that the dnaB protein is transferred from a λ P protein-dnaB protein complex (9, 11, 12) onto the SSB-coated template strand through interactions with the dnaJ and dnaK proteins (13) and the carboxyl-terminal domain (2) of the DNA-bound O protein.

Although strand initiation in the λ ss replication reaction does not depend on the presence of λ DNA sequences in the ss template, it is highly probable that this reaction is physiologically relevant, since each of the 8 required proteins participates in replication of the λ chromosome in vivo. Further study of this in vitro system may clarify protein transfer events as well as leading and lagging strand priming reactions that occur during the initiation of bidirectional DNA replication at the λ chromosomal origin.

ACKNOWLEDGEMENTS

We gratefully acknowledge a collaboration with Drs. M. Zylicz and C. Georgopoulos of the University of Utah in portions of this work. They also supplied purified dnaJ and dnaK proteins and antibodies directed against each of these proteins. This work was supported by grants from the USPHS.

REFERENCES

1.   Skalka,A. (1977) Curr. Top. Microbiol. Immunol. 78, 201.
2.   Furth,M.E. and Wickner,S.H., Lambda DNA replication (1983) in:
          "Lambda II", Hendrix,R.W., Roberts,J.W., Stahl,F.W. and
          Weisberg,R.A., eds., Cold Spring Harbor Laboratory,
          Cold Spring Harbor, New York.
3.   Kornberg,A. (1980) "DNA Replication", Freeman,W.H., San Francisco.
4.   Kornberg,A. (1982) "1982 Supplement to DNA Replication",
          Freeman,W.H., San Francisco.
5.   Nossal,N.G. (1983) Ann. Rev. Biochem. 52, 581.
6.   Wold,M.S., Mallory,J.B., Roberts,J.D., LeBowitz,J.H. and
          McMacken,R. (1982) Proc. Natl. Acad. Sci. USA 79, 6176.
7.   Tsurimoto,T., Hase,T. Matsubara,H. and Matsubara,K. (1982)
          Molec. Gen. Genet. 187, 79.
8.   Roberts,J.D. and McMacken,R. (1983) Nucleic Acids Res.
          11, 7435.
9.   Mallory,J.B. (1983) Ph. D. dissertation, The Johns Hopkins
          University, Baltimore, Maryland.
10. Tsurimoto,T. and Matsubara,K. (1981) Nucleic Acids Res.
          9, 1789.
11. Klein,A., Lanka,E. and Schuster,H. (1980) Eur. J. Biochem.
          105, 1.

12. McMacken,R., Wold,M.S., LeBowitz,J.H., Roberts,J.D., Mallory,
        J.B., Wilkinson,A.K. and Loehrlein,C. (1983) in:
        "Mechanisms of DNA Replication and Recombination",
        Cozzarelli,N.R. ed., UCLA Symposia on Molecular and
        Cellular Biology, New Series, Vol. 10, Alan R. Liss,
        New York, 819.
13. Zylicz,M., LeBowitz,J.H., McMacken,R. and Georgopoulos,C.
          (1983) Proc. Natl. Acad. Sci. USA 80, 6431.
14. Anderl,A. and Klein,A. (1982) Nucleic Acids Res. 10, 1733.
15. Tsurimoto,T. and Matsubara,K. (1982) Proc. Natl. Acad. Sci.
          USA 79, 7639.
16. Ray,D.S., Hines,J.C., Kim,M.H., Imber,R. and Nomura,N.
          (1982) Gene 18, 231.
17. LeBowitz,J.H. and McMacken,R. (1984), submitted for publication.
18. McMacken,R., Ueda,K. and Kornberg,A. (1977) Proc. Natl. Acad.
          Sci. USA 74, 4190.
19. Arai,K. and Kornberg,A. (1981) Proc. Natl. Acad. Sci. USA
          78, 69.
20. Weiner,J.H., McMacken,R. and Kornberg,A. (1976) Proc. Natl.
          Acad. Sci. USA 73, 752.
21. Arai,K., Low,R., Kobori,J., Shlomai,J. and Kornberg,A. (1981)
          J. Biol. Chem. 256, 5273.
22. Shlomai,J. and Kornberg,A. (1980) Proc. Natl. Acad. Sci. USA
          77, 799.

II
In Vitro Eukaryotic DNA Replication Systems, Chromosomal and Viral Replication Studies

REPLICATION IN VITRO OF ADENOVIRUS DNA

Peter C. van der Vliet, Bram G.M. van Bergen,
Wim van Driel, Dick van Dam and Marijke M. Kwant

Laboratory for Physiological Chemistry
State University of Utrecht
Vondellaan 24A
3521 GG Utrecht, The Netherlands

INTRODUCTION

The replication of the human adenovirus (Ad) DNA in infected
cells is a highly efficient process. Within 40 hrs after infection
a total amount of viral DNA is synthesized which almost equals the
entire chromosomal DNA content of the cell. The abundance of replication proteins during infection has enabled the development of a
replication system that initiates Ad DNA synthesis in vitro (1). So
far, this is the only available system that replicates duplex DNA
in higher eukaryotes. Since this system utilizes both viral and
cellular proteins it may prove useful for the study of cellular DNA
replication as well.

The human adenovirus genome is a linear duplex DNA of about
35,000 base pairs (bp) which contains an inverted terminal repetition
of 103-164 bp, dependent upon the serotype (2). A 55,000 dalton terminal protein (TP) is covalently attached to the 5'-end of each
strand by a serine-dCMP phosphodiester linkage (3). A number of
studies in vivo and in vitro have revealed that Ad DNA replication
proceeds via a displacement mechanism (4, 5). Replication starts with
equal probability at either end of the molecule and proceeds by
displacement of a parental strand. The displaced single strand is
next converted into a duplex daughter molecule by complementary
strand synthesis. The latter process has not yet been mimicked in
vitro.

93

Initiation of DNA replication occurs by a protein-priming mechanism in which the 80,000 dalton precursor terminal protein (pTP) plays a crucial role. The pTP binds covalently a dCMP residue which functions as the 5'-terminal nucleotide in the new DNA (6, 7, 8). This reaction is dependent upon the presence of a suitable template and is catalyzed by an adenovirus-coded DNA polymerase which can also perform DNA chain elongation (9, 10). In addition, the reaction requires ATP and a nuclear 47,000 dalton host protein (11, 12).

Elongation of DNA chains from the pTP-dCMP primer requires DNA polymerase, the virus-coded 72,000 dalton DNA binding protein (DBP) and several host factors (13, 14). DBP also stimulates the pTP-dCMP formation but is not absolutely required for the initiation reaction.

Here we will describe the results obtained with nuclear extracts from mutant infected cells and with a partially reconstituted replication system. We will present evidence for a heat-stable host cell component which stimulates the initiation reaction. Special emphasis will be put on the DNA sequences in the origin region that are required for initiation of DNA replication.

RESULTS AND DISCUSSION

Ad5 DNA replication in vitro with nuclear extracts from wild type and mutant infected cells

Nuclear extracts from Ad5 infected HeLa cells provide a crude enzyme system that is suitable for initiation of DNA replication as well as elongation. We have used such extracts from cells infected at the permissive temperature with DNA-negative mutants from the two available complementation groups. The mutant H5ts125 is located in the C-terminal part of the DBP. The point mutation in H5ts125 leads to a pro → ser substitution at position 413 (15) which renders the DBP inactive at 38°C and causes increased proteolytic degradation in infected cells (16). The group N mutants H5ts36 and H5ts149 both are located in the DNA polymerase gene (10, 17). Cells infected with these mutants shut off viral DNA synthesis only slowly after a shift to the nonpermissive temperature suggesting that DNA chain elongation is not impaired (18).

Extracts prepared from H5ts125 or wild type infected HeLa cells, grown at 32°C, were pre-incubated at 37°C followed by incubation in the presence of ATP, dNTP's and (^{32}P)dCTP. As a template we added Ad5-DNA-TP, isolated from virions and pre-digested with the restriction enzyme XbaI. Digestion leads to 5 fragments, of which the C and E fragment are derived from the termini of the molecule. In wild-type extracts (13) or H5ts125 extracts at 32°C preferential labeling of these fragments is observed (Figure 1A). The newly synthesized C and E fragments had a reduced mobility in 1% agarose-0.1% SDS gels due to the presence of covalently bound pTP.

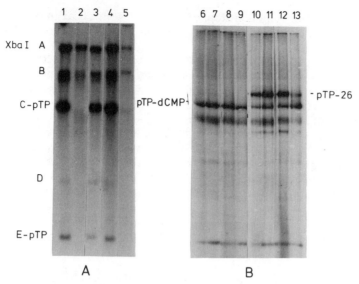

Figure 1: *DNA replication in vitro using extracts containing a temperature-sensitive DNA binding protein. Wild type or H5ts125 infected HeLa cells were grown at 32°C. Nuclear extracts were prepared (1, 13) and incubated at 32°C or 37°C with DNA-TP previously digested with XbaI (A) or intact DNA-TP (B). In A, incubation was in the presence of all four dNTP's. C and E are the terminal fragments. In B, lanes 6-9, only (^{32}P)dCTP was added while in lanes 10-13 (^{32}P)dCTP, dATP, dTTP and ddGTP was present to permit synthesis up to nucleotide 26. pTP-dCMP and pTP-26 designate the precursor terminal protein, linked to dCMP and to a 26-mer, respectively.*
lanes 1, 8, 12: H5ts125, 32°. lanes 2, 9, 13: H5ts125, 37°.
lanes 6 and 10: wild type, 32°. lanes 7 and 11: wild type, 37°.
lane 3: H5ts125 + 1 µg 72 kD DBP. lane 4: H5ts125 + 1 µg 47 kD DBP fragment. lane 5: H5ts125 + 1 µg 26 kD DBP fragment.

        Replication was strongly inhibited at 37°C in H5ts125 extracts
(Figure 1A). Addition of purified wt-DBP during incubation restored
the reaction. Complementation of the defective DBP could also be
demonstrated with the C-terminal proteolytic fragment of 47,000 and
even 34,000 dalton obtained by limited chymotrypsin digestion (13,
14). This indicates that all DNA replication functions are located
in the C-terminal, DNA binding domain of this multi-functional protein. Since the majority of the P-groups of the native DBP are located
in the N-terminal domain (15, 30), phosphorylation of DBP apparently
does not play an important role during DNA replication.

        In order to study the role of DBP in more detail we assayed
H5ts125 infected extracts for their capacity to form a pTP-dCMP

initiation complex. Incubation of wild-type nuclear extracts and
DNA-TP in the presence of (α-^{32}P)dCTP as the only nucleotide triphosphate leads to formation of a 80 kD labeled pTP-dCMP complex (Figure
1B). Addition of dATP, dTTP and 2'-3'-dideoxy-GTP (ddGTP) leads to
a partial elongation up to nucleotide 26, which is the first G-residue
in the newly synthesized DNA. Under these incubation conditions a
90 kD product is formed consisting of pTP covalently linked to a
26-mer (pTP-26). Thus, initiation and partial elongation can be
separately studied. Analysis of H5ts125 extracts shows that preincubation at the non-permissive temperature severely impaired the
capacity to form a pTP-26 product but did not inhibit the pTP-dCMP
reaction (Figure 1B). This leads to the conclusion that in this crude
system DBA is required already early in elongation but not in initiation. As will be shown later, a stimulation of pTP-dCMP formation
is obtained by DNP in more purified systems suggesting that in crude
extracts a host protein may substitute for DBP during pTP-dCMP synthesis.

Different results were obtained when H5ts149 or H5ts36 were
studied. Extracts from H5ts149 infected HeLa cells grown at the
permissive temperature were temperature-sensitive for the pTP-dCMP
(Figure 2). Extracts prepared from cells infected at 40°C were also
deficient in initiation. Only a weak band at the position of pTP-
dCMP is observed compared to wild type (Figure 2, lanes 5 and 6).
Addition of a purified complex of pTP and DNA polymerase (pTP-pol)
restored the capacity to synthesize a pTP-dCMP complex (Figure 2,
lane 7). This confirms the previous findings (9, 10) that the Ad
DNA polymerase catalyzes the deocycytidylation of the precursor
terminal protein during initiation.

Role of cellular proteins in a reconstituted system

From these genetic studies it appears that at least three viral
proteins are involved in the replication of Ad DNA. All three of
these proteins are coded by the early E2 region and are synthesized
from a single leftward promoter at 0.75 map units. At present no
other viral genes seems to exert a direct effect on DNA replication,
although indirect effects like the stimulation of E2 gene expression
by El products (2) can influence DNA replication in vivo.

To study the need for cellular proteins during replication we
have purified the 72 kD DBP and the pTP-pol complex according to
published procedures (13, 19) and compared their replication capacity
with that of crude nuclear extracts. As shown in Figure 3A, replication is completely dependent upon a nuclear extract from host cells.
Experiments by Nagata et al. (12) have indicated that these extracts
contain a 47 kD protein involved in initiation (factor 1). We have
partially purified factor 1 up to the phosphocellulose step and we
have observed that it can replace the need for a nuclear extract
completely (not shown).

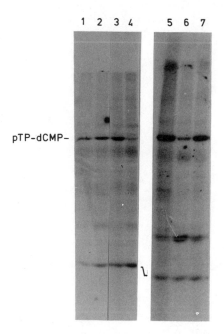

Figure 2:   Complementation of defective initiation in H5ts149 extracts
by Ad DNA polymerase. Wild type or H5ts149 infected HeLa cells were
grown at 32°C (1-4) or 40°C (5-7). Nuclear extracts were prepared as
described (1, 13) and incubated at various temperatures in the presence of DNA-TP, ATP and (^{32}P)dCTP (13). The reaction products were
analyzed by SDS-polyacrylamide gel electrophoresis and autoradiography. The position of the initiation complex is indicated.
1: wild type, 33°; 2: wild type, 39.5°; 3: H5ts149, 33°; 4: H5ts149,
39.5°; 5: wild type, 37°; 6: H5ts149, 37°; 7: H5ts149, 37° + purified
(19) pTP-pol.

Compared to crude extracts, this partially reconstituted system
is still deficient. Addition of a crude cytosol from uninfected cells
stimulates the reaction up to the level of crude nuclear extracts
from infected cells (Figure 3A). Initially we assumed that the cytosol
may contain similar factors as the nuclear extract which leaked out
of the nuclei during preparation. However, this explanation must be
rejected because the cytosol component is stable to inactivation for
15 min at 80°C in contrast to the nuclear component. We have not yet
characterized this cytosol factor in detail but preliminary experiments have shown that it is stable to phenol extraction and dialysis
and cannot easily be precipitated with ethanol. Sensitivity to both
ribonuclease and trypsin was observed. Thus, it is possible that we
are dealing with more than one factor which may influence DNA replication in different ways.

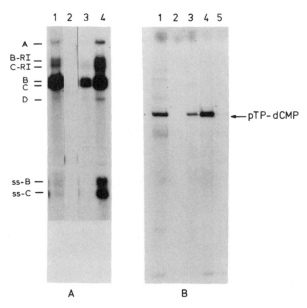

Figure 3A: Requirement for cellular proteins in a reconstituted
replication system. Ad5 DNA-TP was digested with XhoI and incubated
under various conditions. B and C are the terminal fragments. B-RI
and C-RI designate the replicative intermediates of these fragments
containing a single-stranded arm. ssB and ssC are the displaced
single-stranded fragments resulting from multiple rounds of replication. An autoradiogram of a 1% agarose gel containing 0.1% SDS is
shown.
1: crude nuclear extract from Ad5 infected HeLa cells; 2: only viral
proteins, pTP, pol and DBP; 3: pTP, pol, DBP and a nuclear extract
from uninfected cells; 4: same as 3, + cytosol from uninfected cells,
heated for 15 min at $80^{\circ}C$.

Figure 3B: Heat stable cytosol factor acts at the level of initiation. Synthesis of pTP-dCMP was studied in the presence of DNA-TP
as template (25). An autoradiogram of a 10% poly acrylamide-SDS gel
is shown.
1: crude nuclear extract from Ad5 infected HeLa cells; 2: only viral
proteins, pTP, pol, DBP; 3: pTP, pol, DBP and a nuclear extract from
uninfected HeLa cells; 4: same as 3, + cytosol factor, heated for
15 min at $80^{\circ}C$; 5: same as 4, but without nuclear extract.

        We have studied also the effect of the heat stable cytosol
factor on pTP-dCMP synthesis. The cytosol factor stimulates initiation but is not capable of pTP-dCMP synthesis without the nuclear
extract (Figure 3B). Both with and without DBP this stimulation was
observed (not shown). Based upon these results we assume that at

least two cellular factors are involved in the initiation of Ad DNA
replication.

Inhibition of DNA chain elongation by aphidicolin

Aphidicolin is generally thought to inhibit cellular DNA replication by its action on DNA polymerase α (20). Ad DNA replication
in vivo is resistant to aphidicolin at the concentrations used to
inhibit cellular DNA synthesis but at 300-400 fold higher concentrations (30 μM) aphidicolin also inhibits Ad DNA replication in vivo
(21, 22). Under these conditions the formation of the. pTP-dCMP complex
occurs normally, indicating that the drug acts on the level of chain
elongation (23). Direct evidence for such an effect has been obtained
in vitro as shown in Figure 4.

This result is quite surprising since the purified Ad DNA
polymerase is completely resistant to aphidicolin up to 600 μM (9,
Figure 4). One explanation of this discrepancy could be that aphidicolin has another target within the replication fork but such a
protein has not been found yet. Recently, Nagata et al. described
a nuclear topoisomerase I (factor 2) which is required for the synthesis of full length DNA, but this protein is also resistant to aphidicolin (24). Another possibility is that the adenovirus DNA polymerase
becomes sensitive to the drug by its interaction with other replication proteins.

Replication of origin containing DNA devoid of the terminal protein

Although the function of the precursor terminal protein as a
primer during initiation is well established, less is understood
about the role of the parental terminal protein. The TP has a strong
tendency to aggregate and this has led to the assumption that the
protein may form a focus-point for replication proteins, e.g. by
pTP-TP recognition. The need for an intact TP was strengthened by
the observation that pronase-treated DNA-TP is inactive as template
during replication (1). However, such a template still contains some
remaining amino acids which may act as an inhibitor of replication.

Recently, it was shown that the formation of a pTP-dCMP complex
and subsequent elongation of the template also occurs in the presence
of protein-free templates obtained from plasmids (25, 26). This indicates that nucleic acid-protein interactions occur during initiation
and that an interaction between the parental TP and initiation protein, if existing, is not sufficient. In order to provide a suitable
template, the origin region must be linearized. Circular DNA containing the first 1343 bp in an intact form is not active, nor is
linear DNA containing 29 bp ahead of the origin (26). The absence
of a recognition mechanism for internal origins suggests that denaturation of the origin area must occur in order to initiate. This
is an agreement with the finding that ss-DNA is able to sustain the
pTP-dCMP formation (12, 25).

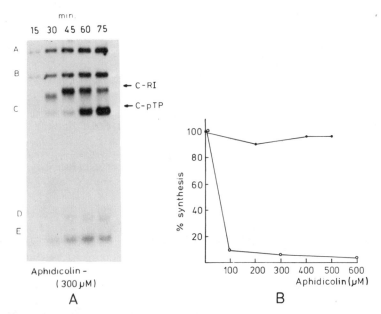

Figure 4A: Inhibition of fork movement by aphidicolin. Ad5 DNA-TP
was digested with XbaI and incubated for various periods in nuclear
extracts. Aphidicolin was present at 300 µM. The positions of the
XbaI fragments are indicated (A-E). C and E are the terminal fragments. C-RI is the replicative intermediate consitsting of newly
synthesized C-fragment containing in addition a full length single
strand. Molecules migrating between C-pTP and C-RI are in various
stages of replication. The rate of fork movement could be calculated
from the position of the RI in the gel and was linear for 45 minutes.

Figure 4B: Comparison of the effects of aphidicolin on DNA chain
elongation and on purified Ad DNA polymerase. The chain elongation
rate (o) was calculated from similar experiments as in A. 100% =
2100 nucleotides synthesized per minute per replication fork at 37°C.
The DNA polymerase activity (●) was assayed with activated calf
thymus DNA and a complex of pTP and pol, purified up to the ds DNA
cellulose step (19). 100% = 20 pmol dNTP per 20 min at 30°C.

        Although the activity of TP-free origin containing DNA can be
easily studied, it is a less efficient template. On a molar basis,
TP containing DNA is about 30-fold more active than protein-free
DNA (26). This may indicate that the terminal protein is required
for the stabilization of initiation complexes or may alter binding
constants between replication proteins and DNA. Alternatively, the
TP could serve to protect the 5' ends against exonuclease activity
or could suppress internal initiations which occur in protein-free
DNA.

An additional role of the terminal protein during elongation has been suggested based upon inhibition studies with anti TP-IgG. The immunoglobulin blocks elongation both in nuclear extracts and in isolated nuclei (27). This may indicate that the terminal protein co-migrates in the replication fork with the DNA polymerase. An attractive consequence of such a model is that at the end of the displacement synthesis the molecular ends of the displaced strand are still in close vicinity of each other thus facilitating the formation of a panhandle structure. This structure has been proposed as an intermediate in complementary strand synthesis (5).

Essential DNA sequences in the origin

The template activity of protein-free DNA opens the way to a functional analysis of the nucleotide sequences required for initiation and elongation. We have studied a number of Ad12 terminal fragments containing various deletions and mutations in the origin region (28). Comparison of the sequences of the inverted terminal repetitions from a number of adenovirus serotypes has revealed the presence of some conserved sequences. Within the main adenovirus groups A (e.g. Ad12), B, and C (e.g. Ad2) the base pairs 9-22, 34-41 and the distal hexanucleotide are identical and the sequence 9-18 is conserved among all human serotypes (2).

Our results (Figures 5) show that nuclear extracts from Ad5 infected cells were equally able to replicate both Ad2 and Ad12 origin sequences. A deletion extending into the conserved region destroyed the template activity. Some deletions in the first 8 nucleotides are permitted but others are not. The nucleotide sequence around the exact start position can vary considerably. Based upon the limited number or sequences studied so far, it appears that any G-residue in the 3' strand which is 4-8 bp away from the conserved sequence will provide a template, albeit with different efficiency. This result may be explained by a model (Figure 5), in which the pTP-DNA polymerase recognizes the internal conserved sequence followed by the use of an available G-residue in the template strand to form a pTP-dCMP complex. In this model, the distance between the G-residue and the conserved sequence rather than the sequence surrounding the G determines whether initiation can occur. The length of the spacer region may change the efficiency of the initiation reaction. More detailed modification of the origin region will be required to test this model further.

The interserotype replication in this system is not confined to protein-free templates. Various natural DNA-TP complexes isolated from different virions have been used succesfully as template for Ad2 nuclear extracts although their efficiency differed considerably (29).

Template	3' Sequence	Replication
Ad5 - TP	G T A G T A G T¦T A T T A T A T G G A A T¦A - -	+
XD7	G T A G T A G T¦T A T T A T A T G G A A T¦A - -	+
pAd12 RIC1	G T A C G A T A G A T¦T A T T A T A T G G A A T¦A - -	+
pAd12 RIC3	G T A C¦T A T T A T A T G G A A T¦A - -	+/-
pAd12 RIC7	G T A C T A G A T¦T A T T A T A T G G A A T¦A - -	-
pAsc2	G G G A T A G A T¦T A T T A T A T G G A A T¦A - -	+
pAd12 RIA	G T A C T G G A A T¦A - -	-
pLA 1	G G G T A G T A G T¦T A T T A T A T G G A A T¦A - -	+

```
            degenerated      9      conserved    22
       3'- - G - - - G - - -|T A T T A T A T G G A A T|- - - -
              |       |      |                       |
             -8      -4      0
            L_____J        L_____J
            accepted         possible pTP binding site
            distance
```

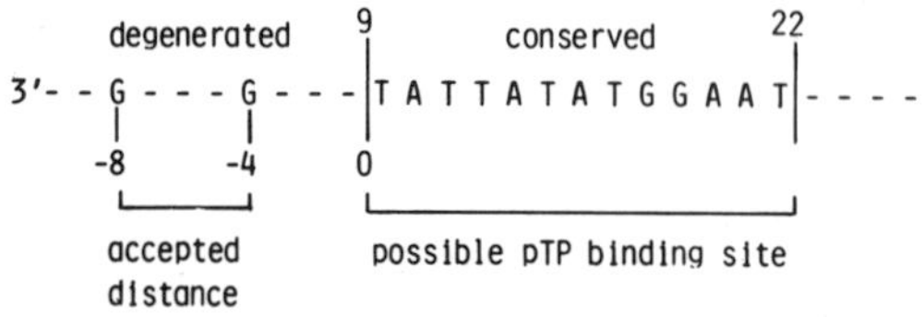

Figure 5: Nucleotide sequence in the origin required for DNA replication. The upper half of the figure shows the results of incubation
of various plasmid DNA molecules with Ad5 nuclear extracts. Details
of the incubation procedures have been described elsewhere (26). The
figure shows the 3'-sequence obtained after linearization of the
duplex DNA with Eco RI. The black triangles indicate the positions
where new strands can be initiated. The results of pLAl were from
ref. 25. The lower half of the figure indicates a possible explanation of the results. A distance of -4 to -8 nucleotides from the
conserved region seems permitted, but the efficiency with which a
G at -4 is used low (see pAd12 RI C3).

Concluding remarks

With the use of crude and partially reconstituted in vitro DNA
replication systems many details of the mechanism of adenovirus DNA
replication have been elucidated. A summary is shown in Figure 6. It
is clear that three viral proteins (precursor terminal protein, DNA
polymerase and DNA binding protein) play a crucial role. All three
participate in initiation and presumably also in elongation. Two
nuclear host proteins have been purified that are required for initiation and for synthesis of full length DNA, respectively, and a
heat stable component stimulates the initiation reaction. Information
about the essential nucleotide sequences is beginning to accumulate.
It seems clear that at least the conserved region 9-18 is of vital
importance for initiation, and a GC base pair within a short distance
of this functions as the actual starting place. Site directed mutagenesis and direct studies of binding sites will undoubtedly reveal more
important sequences within the inverted terminal repetition.

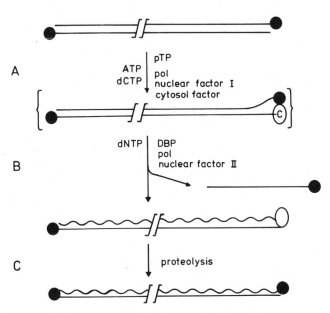

Figure 6: Model for the replication of adenovirus DNA. Initiation (A) requires the precursor terminal protein (pTP), DNA polymerase (pol) and a nuclear protein I while a heat stable cytosol factor stimulates the reaction about 7-fold. The viral DBP stimulates the reaction but is not essential. For elongation (B) DBP is necessary. In contrast to the initiation reaction elongation is aphidicolin and ddNTP sensitive. A nuclear factor II is required for complete elongation beyond 25% of the genome (24). The pTP is cleaved late in infection by viral protease.
● *terminal protein;* ○ *precursor terminal protein.*

It should be stressed that so far only the enzymatic mechanism of the displacement reaction has been studied. It is not excluded that nuclear extracts are also capable of complementary strand synthesis but this reaction has not been studied in detail due to the lack of a suitable template. We are currently trying to construct such a template and to study its properties in vitro.

ACKNOWLEDGEMENTS

Stimulating discussions with J.S. Sussenbach and H.S. Jansz are gratefully acknowledged. Aphidicolin was a kind gift of Dr. A.H. Todd (ICI, England). This work was supported in part by the Netherlands Foundation for Chemical Research (SON) with financial aid from the Netherlands Organization for the Advancement of Pure Research (ZWO).

REFERENCES

1. Challberg,M.D. and Kelly,T.J. (1979) Proc. Natl. Acad. Sci.
    USA 76, 655.
2. Tooze,J. (1981) Molecular Biology of Tumor Viruses 2, 2nd ed.,
    Cold Spring Harbor Laboratory, USA.
3. Rekosh,D.M.K., Russell,W.C., Bellett,A.J.D. and Robinson,A.J.
    (1977) Cell 11, 283.
4. Sussenbach,J.S., van der Vliet,P.C., Ellens,D.J. and Jansz,H.S.
    (1972) Nature New Biology 239, 47.
5. Lechner,R.L. and Kelly,T.J. (1977) Cell 12, 1007.
6. Lichy,J.H., Horwitz,M.S. and Hurwitz,J. (1981) Proc. Natl.
    Acad. Sci. USA 78, 2678.
7. Challberg,M.D., Desiderio,S.V. and Kelly,T.J. (1980)
    Proc. Natl. Acad. Sci. USA 77, 5105.
8. Stillman,B.W. (1981) J. Virol. 37, 139.
9. Lichy,J.H., Field,J., Horwitz,M.S. and Hurwitz,J. (1982)
    Proc. Natl. Acad. Sci. USA 79, 5225.
10. Stillman,B.W., Tamanoi,F. and Mathews,M.B. (1982) Cell 31, 613.
11. de Jong,P.J., Kwant,M.M., van Driel,W., Jansz,H.S. and
    van der Vliet,P.C. (1983) Virology 124, 45.
12. Nagata,K., Guggenheimer,R.A., Enomoto,T., Lichy,J.H. and
    Hurwitz,J. (1982) Proc. Natl. Acad. Sci. USA 79, 6438.
13. van Bergen,B.G.M. and van der Vliet,P.C. (1983) J. Virol.
    46, 642.
14. Friefeld,B.R., Krevolin,M.D. and Horwitz,M.S. (1983)
    Virology 124, 380.
15. Kruijer,W., van Schaik,F.M.A. and Sussenbach,J.S. (1981)
    Nucl. Acids Res. 9, 4439.
16. van der Vliet,P.C., Levine,A.J., Ensinger,M. and Ginsberg,H.S.
    (1975) J. Virol. 15, 348.
17. Friefeld,B.R., Lichy,J.H., Hurwitz,J. and Horwitz,M.S. (1983)
    Proc. Natl. Acad. Sci. USA 80, 1589.
18. van der Vliet,P.C. and Sussenbach,J.S. (1975) Virology 67, 415.
19. Enomoto,T., Lichy,J.H., Ikeda,J.E. and Hurwitz,J. (1981)
    Proc. Natl. Acad. Sci. USA 11, 6779.
20. Spadari,S., Sala,F. and Pedrali-Noy,G. (1982) Trends Biochem.
    Sci. 7, 29.
21. Longiaru,M., Ikeda,J., Jarkovsky,Z., Horwitz,S.B. and Horwitz,
    M.S. (1979) Nucleic Acids Res. 6, 3369.
22. Kwant,M.M., van der Vliet,P.C. (1980) Nucleic Acids Res.
    8, 3993.
23. Pincus,S., Robertson,W. and Rekosh,D. (1981) Nucleic Acids
    Res. 9, 4919.
24. Nagata,K., Guggenheimer,R.A. and Hurwitz,J. (1983) Proc. Natl.
    Acad. Sci. USA 80, 4266.
25. Tamanoi,F. and Stillman,B.W. (1982) Proc. Natl. Acad. Sci.
    USA 79, 2221.
26. van Bergen,B.G.M., van der Leij,P.A., van Driel,W.,
    van Mansfeld,A.D.M. and van der Vliet,P.C. (1983)
    Nucleic Acids Res. 11, 1975.

27. Rijnders,A.W.M., van Bergen,B.G.M., van der Vliet,P.C. and
    Sussenbach,J.S. (1983) Virology, in press.
28. Bos,J.C., Polder,L.J., Bernards,R., Schrier,P.I.,
    van den Elsen,P.J., van der Eb,A.J. and van Ormondt,H.
    (1981) Cell 27, 121.
29. Stillman,B.W., Topp,W.C. and Engler,J.A. (1982) J. Virol.
    44, 530.
30. Linne,T. and Philipson,L. (1980) Eur. J. Biochem. 103, 259.

CHARACTERIZATION OF IN VITRO DNA SYNTHESIS IN AN ISOLATED

CHLOROPLAST SYSTEM OF PETUNIA HYBRIDA

N. Overbeeke, J.H. de Waard and A.J. Kool

Department of Genetics, Biological Laboratory
Vrije Universiteit
Post Box 7161
NL - 1007 MC Amsterdam, The Netherlands

INTRODUCTION

The genetic organization and regulation of expression of chloroplast genes are extensively studied at this moment in order to answer questions about chloroplast biogenesis and the regulation thereof, a process in which both nuclear and chloroplast genes are involved (1, 2). However, until now only little attention has been paid to the process of the replication of the chloroplast genome and the regulation of this replication. To obtain more information about the proteins involved in these processes and their appearance during biogenesis, we started to analyze the DNA-synthesizing activity in isolated mature chloroplasts of Petunia hybrida. Not only because this plant is used by many research groups as a model plant for molecular and biological studies (3), but also because it has the advantage that several developmental processes in the biogenesis of chloroplasts can be induced in plant cell cultures (4, 5).

In this paper we describe the characterization of a DNA-synthesizing activity observed in isolated chloroplasts of Petunia hybrida. The enzyme activity, which could also be detected in the soluble fraction of lysed chloroplasts, resembles many but not all properties of the purified "γ-like" DNA polymerase from spinach chloroplasts (6) and the activity in crude extracts of maize chloroplasts (7). Furthermore, we show that after removal of the endogenous DNA the DNA-synthesizing activity could also use added purified chloroplast DNA as a template.

MATERIALS AND METHODS

Isolation and incubation of chloroplasts and chloroplast fractions

Chloroplasts were isolated from 30 gr of leaves of 7 weeks old Petunia hybrida plants as described previously (4), with the exception that the homogenization (H)-buffer consisted of 330 mM Sorbitol, 50 mM Tris-HCl pH 8.0, 5 mM NaCl, 10 mM $MgCl_2$ and 2 mM DTT. Chloroplasts were washed, resuspended in 3 ml H-buffer and used for the deoxynucleotide triphosphate (dNTP) incorporation assay. For isolation of the soluble fraction, chloroplasts from 30 gr of leaves were resuspended in 1 ml H-buffer and lysed by the addition of 1 volume of lysis (L)-buffer (50 mM Tris-HCl pH 8.0, 0.7 M NaCl, 2 mM DTT, 20% glycerol and 50 µg/ml phenylmethanesulfonyl fluoride and 50 µg/ml p-toluene-sulfonyl fluoride both dissolved at 5 mg/ml in iso-propanol). After incubation for 30 min on ice with occasionally mixing, the suspension was centrifuged for 20 min at 20,000 x g. The supernatant was desalted by passing 2 times over 7,5 ml Sephadex G25 resin equilibrated with L-buffer without NaCl and packed in a 10 ml syringe (elution was performed by centrifugation at 2000 x g for 5 min).

The standard assay of DNA-synthesizing activity was performed at 37°C as follows: To 100 µl of chloroplast suspension or soluble fraction was added an equal volume of incubation buffer consisting of 50 mM Tris-HCl pH 8.0, 10 mM $MgCl_2$, 2 mM DTT, 4 mM ATP and 100 µM of each dATP, dCTP, dGTP and dTTP, with the understanding that one of the dNTPs was replaced by a radioactive one: 10 µCi/100 µl (^{3}H) TTP (44 Ci/mmol) or 3 µCi/100 µl (α^{32}P) dCTP (3000 Ci/mmol). The incubation was stopped by addition of 1/10 volume of 50 mM Tris-HCl pH 8.0, 200 mM EDTA and 10% Triton X-100 followed by placing on ice. Appropriate samples were spotted on filters (Schleicher and Schüll no. 2316), precipitated with TCA and the radioactivity was determined as described (4). When DNA had to be isolated after incubation, the chloroplasts were spun down in a microfuge and resuspended in 50 mM Tris-HCl pH 8.0, 0.35 M sucrose, 7 mM EDTA, 5 mM 2-mercaptoethanol. DNA was isolated according to the procedure described for rapid screening of cp-DNA (8).

Other procedures

Procedures for restriction enzyme digestion, agarose gel electrophoresis and autoradiography are described (9). Radioactive chemicals were from the Radiochemical Centre, Amersham; DTT, NTP's and dNTP's from Boehringer, Mannheim; N-ethylmaleimide and ethidium bromide from Sigma Chemical Company, St. Louis and aphidicolin from Wako Pure Chemical Industries (Japan).

RESULTS AND DISCUSSION

Kinetics and characteristics of the DNA synthesizing activity in isolated chloroplasts

The kinetics of the incorporation of (^{3}H) TMP into TCA-insoluble products by isolated chloroplasts is linear for at least one hour with a chloroplast protein concentration of up to 1000 µg/ml. For incorporation the presence of all four dNTP's and ATP or GTP is essential (Table 1). The presence of Mg^{2+}-ions is also required (Table 1) and an optimum concentration was found at 10 mM. Mg^{2+}-ions cannot be replaced by Mn^{2+}-ions in any of the concentrations tested (0-50 mM). The enzyme activity increases with increasing temperature up to 37°C, but drops sharply at incubation temperatures above 45°C. A broad pH-optimum was found around pH 7.5, resulting in an enzyme activity of 80% or more between pH values from 7.0 to 8.0. Incorporation of dNTP's into TCA-insoluble products was sensitive to DNAse treatment, but not to RNAse treatment. Analysis of the DNA isolated from incubated chloroplasts on agarose gels showed (Figure 1) that the radioactive label was incorporated into high molecular weight material which co-migrates with purified chloroplast DNA and that the DNA-fragment pattern obtained after digestion of the DNA with endonuclease Bam HI was identical to the pattern obtained from cp-DNA purified from leaves. It can therefore be concluded that the radioactive dNTP's are incorporated into chloroplast DNA.

Table 1: Requirements for the incorporation of (^{3}H) TMP into DNA by isolated chloroplasts

Components	% incorporation$^\alpha$
complete system$^\alpha$	100
+ rNTP's (2 mM)	102
− MgCl$_2$	5
− dATP/dCTP/dGTP	5
− ATP	10
− ATP, + GTP	100

$^\alpha$Chloroplasts were isolated and incubated as described in Materials and Methods. The 100% value in this complete system was 0.03 pmol (^{3}H) TMP/hr in an incubation system with a final chloroplast protein concentration of 300 µg/ml. Samples of 20 µl were taken after 30 min of incubation.

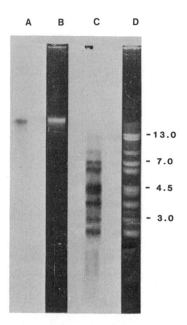

*Figure 1: Analysis of the ${}^{32}P$-labelled product synthesized in iso-
lated chloroplasts after incubation with ($\alpha^{32}P$) dCTP. DNA was isolated from these chloroplasts and analyzed by electrophoresis on a
1% agarose gel and subsequent autoradiography (lanes B and D) and
compared with the ethidium bromide stained pattern of purified cp-DNA
(lanes A and C) either directly (lanes A and B) or after digestion
with restriction endonuclease Bam HI (lanes C and D). Molecular
weights are indicated at the right in kilobasepairs.*

       The DNA synthesizing activity was further characterized with
several inhibitors (Table 2), specific for various DNA polymerases
from animal or plant origin (10). The DNA synthesizing activity in
Petunia hybrida chloroplasts is insensitive to aphidicolin (Table 2),
a well-known inhibitor of DNA polymerase α (11). The enzyme activity
is also insensitive to dideoxy-NTP's, while ethidium bromide, N-ethylmaleimide and the presence of high concentrations of salt (200 mM
KCl) are inhibitory (Table 2). The properties of the Petunia hybrida
chloroplast enzyme activity resemble those of the animal DNA polymerase γ found in mitochondria, with respect to its sensitivity to
ethidium bromide and N-ethylmaleimide and its insensitivity to
aphidicolin. However, the sensitivity for high concentrations of salt
and the insensitivity for ddNTP's are in disagreement with the properties of the mitochondrial DNA polymerase γ. On the other hand, a
"γ-like" DNA polymerase purified from spinach chloroplasts (6) is also
insensitive for ddNTP's, as is the enzyme activity in crude extracts
from maize chloroplasts (7). Also the broad pH-optimum and the requirement for Mg^{2+} and not Mn^{2+} agrees with the results obtained with

Table 2: Characterization of the DNA-synthesizing activity of
isolated chloroplasts by various inhibitors

Inhibitor	concentration/condition	% incorporation
none		100^{α}
N-ethylmaleimide	2 mM	50
aphidicolin	20 µg/ml	95
ethidium bromide	100 µM	15
dideoxy NTP	ratio ddNTP/dNTP 10/1	80
KCl	100 mM	50
heat inactivation	preincubation 10' 45°C	100

αThe incorporation of (α^{32}P) dCTP in the control incubation without
inhibitors was set at 100% (0.02 pmol/hr; protein concentration
400 µg/ml. Samples of 20 µl were taken after 10 min of incubation.

maize chloroplasts extracts, but is in contrast with the properties
of the purified "γ-like" polymerase from spinach. In spite of these
two differences it is very likely that the DNA-synthesizing activity
in isolated Petunia hybrida chloroplasts is the result of a "γ-like"
DNA polymerase.

DNA synthesizing activity of a soluble fraction from Petunia hybrida
chloroplasts

Isolated chloroplasts were lysed in 0.7 M NaCl and both, the
soluble and insoluble fraction were assayed for DNA-synthesizing
activity. The results show that almost all of the DNA-synthesizing
activity is present in the soluble fraction.

Furthermore, the soluble fraction still contains chloroplast
DNA because no template DNA has to be added to the incubation medium
to get incorporation of (α^{32}P) dCTP while treatment with nuclease
destroys this incorporation. The properties of the soluble fraction
with respect to its DNA synthesis were identical to the enzyme activity described for intact chloroplasts. Interestingly, inactivation
of the nuclease after the treatment followed by addition of purified
cp-DNA at a concentration of 2µg/ml restores the incorporation of
(α^{32}P) dCTP. This enables us to use this organelle free chloroplast
system for the analysis of the template specificity of the enzyme
activity. Not only synthetic templates but also specific chloroplast
DNA-fragments carrying a so-called Autonomously Replicating Sequence

(ARS), which are recently cloned in our laboratory (12), will be studied to determine whether the DNA-synthesizing activity in chloroplasts has a preference for specific chloroplast -ARS's. Such a preference has been reported for a DNA-synthesizing system of yeast, which initiates DNA replication of yeast-ARS plasmids at a specific site (13).

REFERENCES

1.   Witfield,P.R. and Bottomley,W. (1983) Ann. Rev. Plant Physiol.
         34, 279.
2.   Bohnert,H.J., Crouse,E.J. and Schmitt,J.M. (1983) in:
         Encycl. Plant Physiol. vol. 14B, B.Parthier and D.Boulter
         eds., Springer Verlag, Berlin.
3.   Hanson,M.R. (1980) Plant Mol. Biol. Newslett. 1, 37.
4.   Colijn,C.M., Kool,A.J. and Nijkamp,J.J. (1982) Planta 155, 37.
5.   Colijn,C.M., Sijmons,P., Mol,J.N.M., Kool,A.J. and Nijkamp,
         H.J.J. (1982) Curr. Genet. 6, 129.
6.   Sala,F., Amileni,A.R., Parisi,B. and Spadari,S. (1980)
         Eur. J. Biochem. 112, 211.
7.   Zimmermann,W. and Weissbach,A. (1982) Biochemistry 21, 3334.
8.   Kumar,A., Cocking,E.C., Bovenberg,W.A. and Kool,A.J. (1982)
         Theor, Appl. Genet. 62, 377.
9.   Bovenberg-W.A., Kool,A.J. and Nijkamp,H.J.J. (1980)
         Nucl. Acids Res. 9, 503.
10.   Kornberg,A. (1980) DNA replication, Freeman and Co., San
         Francisco.
11.   Spadari,S., Sala,F. and Pedrali-Noy,G. (1982) Trends in
         Biochem. Sci. 7, 29.
12.   Overbeeke,N., Haring,M.A. and Kool,A.J., submitted for
         publication.
13.   Celniker,S.E. and Campbell,J.L. (1982) Cell 31, 201.

IN VITRO SYNTHESIS OF CAULIFLOWER MOSAIC VIRUS DNA IN VIROPLASMS

J.M. Bonneville, M. Volovitch[1], N. Modjtahedi[1],
D. Demery and P. Yot

Laboratoire de Biologie Moléculaire CNRS-INRA
Groupement scientifique "Microbiologie" de Toulouse
BP 12
31320 Castanet-Tolosan, France

[1]Section de Biologie, Institut Curie
26, rue d'Ulm
75231 Paris Cedex 05, France

INTRODUCTION

Cauliflower Mosaic Virus (CaMV) is a DNA plant virus, and as such an attractive model for studying the molecular biology of plants (1, 2). Its genome is a circular double-stranded molecule, 8 kb in length. The encapsidated form possesses three single-stranded discontinuities (3) where the 5' and 3' ends of the interrupted strand overlap one another (4, 5). One interruption is on the transcribed strand α and the other two on the opposite strand β. Viral DNA is also found in the nucleus as covalently closed supercoiled molecules associated with histones (6). The resulting minichromosomes are actively transcribed (7) into at least two polyadenylated transcripts: the 19S RNA, messenger for the viroplasm matrix protein, and the 35S RNA spanning the whole genome with a terminal redundancy of 180 nucleotides. A model for the replication of CaMV DNA which involves reverse transcription of the 35S RNA is emerging (refer to Pfeiffer et al, in this volume).

Viroplasms are the main cytological modification observed upon infection of Brassica leaves by CaMV: they are cytoplasmic inclusions formed of a protein matrix inside which the virions accumulate. They are thought to play a role in the assembly of viral particles (8). In order to test whether viroplasms are a storage site of replicated

molecules or an active site for DNA synthesis, we have developed an
in vitro system able to polymerize nucleotides into viral DNA sequences.

A cell-free system for DNA synthesis

The system is made of a leaf crude extract from infected plants
containing the main candidates for the replication of CaMV DNA, i.e.
nuclei and viroplasms, and the other heavy organelles from infected
leaves, chloroplasts and amyloplasts, as checked by electron microscopy (not illustrated). Optimal conditions for DNA synthesis, without the addition of any template-primer, have been reached by resus-
pending the pellet of organelles in a mixture of 37 mM Tris-HCl,
pH 8.0, 12 mM $MgCl_2$, 30 mM KCl, 5 mM DTT, 7 mM EGTA, 2 mM ATP, 25
µg/ml tRNA, and 40 µM of each of the unlabeled dNTPs. Sucrose (0.3M)
was added and incorporations were performed at $30^{\circ}C$ to maintain
organelle integrity. Polymerization of deoxyribonucleotides depends
on the amount of added extract: kinetics are linear at least for
30 min and incorporation goes on up to 3 hours; products of the
reaction are sensitive to DNase but not to RNase treatments (not
shown); the reaction is abolished by boiling the extract and requires
the presence of Mg^{2+} and of the four deoxyribonucleotides (Table 1).
These results demonstrate that this endogenous reaction is indeed
a true DNA synthesis, not a mere addition of single nucleotides at
scattered nicks.

Table 1: Polymerization of dTMP by a crude extract of organelles
          from infected leaves

Conditions	Activity[a]	%
Complete mixture [b]	16	100
5 min heating at $100^{\circ}C$	0.9	5
Minus 3 dNTPs	0.6	4
Minus tRNA	4.5	33
Minus $MgCl_2$	1.6	10

(a) Activity is measured from data of the kinetics (0, 4, 8 and 12
    min) of incorporation of acid-insoluble (^{3}H) dTMP, and expressed
    in pmols of dTMP incorporated per hour for 20 µl of extract.
(b) 25 µM (^{3}H)dTTP (2000 cpm/pmol); for other components, see text.

DNA synthesis occurs partly on viral templates

    Labeled DNA recovered after incubation of the crude extract
with $(\alpha-^{32}P)dCTP$ has been shown to hybridize against a Southern blot
of virion DNA. In a control experiment, no hybridization was observed
using ^{32}-P-DNA synthesized by a crude extract from healthy plants.
The relative proportion of viral sequences synthesized in this endogenous reaction has been deduced from quantitative hybridization experiments: only 3% of the bulk of DNA synthesis correspond to viral
sequences (Modjtahedi et al., submitted).

Viroplasms are sites of accumulation of newly-synthesized molecules

    The distribution of labeled molecules between the different
organelles was studied by autoradiomicrography. Experiments were
performed with the rinsed pellets recovered after incubation; the
maximum incubation time (90 min) was chosen to allow sufficient
retention of synthetic activity inside the organelles. After incubation for various times in the presence of $(^{3}H)dTTP$, fractions were
treated for electron microscopy. A significant labeling was observed
at the level of viroplasms. For a quantitative analysis, two different
experiments were carried out, the second one on a larger scale.
Results are presented in Table 2. Two points should be noted: (i)
the proportion of labeled nuclei is higher than the proportion of
labeled chloroplasts; (ii) viroplasms are more frequently labeled
than nuclei, and they are also, on average, more heavily labeled
than the latter.

Newly-synthesized viral sequences are preferentially found in
viroplasms

    In order to test whether in vitro synthesized CaMV DNA was
preferentially associated with one type of organelle, we proceeded
as follows: at the end of a preparative incubation carried out in
the presence of $(\alpha-^{32}P)dTTP$ and $(\alpha-^{32}P)dCTP$, a 12,000 g centrifugation provided a supernatant S2 and a pellet P2, from which semi-purified viroplasms (Vp) and a solubilized chloroplast-enriched fraction
(Ct) were prepared by centrifugation over a saturated sucrose cushion.
DNA was extracted from these three fractions (S2, Ct and Vp), and
hybridized to nitrocellulose strips upon which were fixed either total
cellular DNA from healthy plants digested with EcoRI, or cloned CaMV
DNA digested with SalI. The results, illustrated in Figure 1, indicate
that radioactive viral DNA is more abundant in the Vp fraction, enriched in viroplasms.

Highly purified viroplasms contain and synthesize only viral sequences

    We achieved the purification of viroplasms which retain DNA
synthesis ability. Such preparations of purified viroplasms are devoid
of chloroplasts, amyloplasts and intact nuclei. Contrary to semi-

Table 2: In vitro labeling of the organelles from infected leaves observed by autoradiomicrography

	Chloroplasts			Nuclei			Viroplasms		
	T(a)	1	%	T	1	%	T	1	%
1st exp 0 min	251	1	0.4	5	0	0	6	0	0
30 min	1256	86	7	50	0	0	24	1	4
60 min	1836	48	3	17	2	12	25	8	32
90 min	968	172	18	20	10	50	25	14	56
2nd exp 90 min	3962	196	5	558	134	24	1020	511	50
Silver grains/organelle (b)	from 1 to 2			from 1 to 8			from 1 to 17		

(a) For each organelle, (T): total number of observed organelles; (1) number of labeled organelles;
(%): percentage of labeled organelles.
(b) The number of silver grains per organelle has been counted for 30-100 organelles.

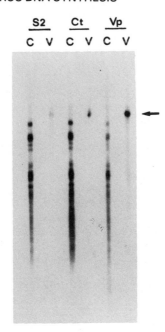

Figure 1: Analysis of labeled DNA from different subcellar compart-ments. After 3 hours the incubation mixture was divided into: supernatant-released subfraction S2, chloroplast subfraction Ct and viroplasm subfraction Vp (see text). DNA extracted from these sub-fractions was hybridized against Southern blots of either cellular (tracks C) or viral (tracks V) DNA and visualized by autoradiography. The same radioactive input has been used for all three samples. The arrow indicates the position of electrophoresed linear CaMV DNA.

purified viroplasms prepared as before, they contain only viral DNA (Figure 2a). In consequence, direct analysis of this DNA is made possible by gel electrophoresis, particularly after in vitro synthesis (Figure 2b).

Products of discrete size are formed in these purified viroplasms. Effects of inhibitors

Purified viroplasms were allowed to synthesize ^{32}-P-DNA in the presence of several inhibitors. Nucleic acids were then extracted and analysed by gel electrophoresis (Figure 3). In addition to circu-lar and linear material of 8 kb, an important band of fast-migrating material appears. Its migration is slightly increased by RNase treat-ment (lane 1) and its synthesis is completely inhibited by Actinomycin D (lane 4), suggesting a RNA primer and a DNA template. Since the whole pattern of radioactive material is identical for control sample (lane 0) and for samples corresponding to the treatments with either aphidicolin (lane 2) or N-ethylmaleimide (lane 3), a role for α-like polymerases of the plant cells can be excluded.

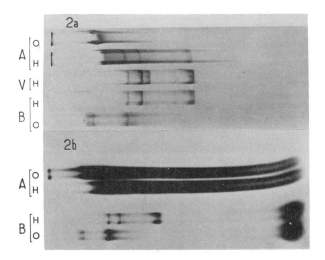

Figure 2: Analysis of DNA after viroplasm purification. Nucleic acids from semi-purified and purified viroplasms incubated in the presence of (α-32P)dCTP were extracted and electrophoresed in a 1% agarose gel after : (O) No additional treatment, (H) HpaI digestion. (a) Ethidium bromide staining; (b) Autoradiograph. (A) Semi-purified viroplasm DNA; (V) Virion DNA; (B) Purified viroplasm DNA.

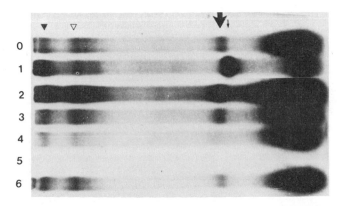

Figure 3: Analysis by electrophoresis (1.4 % agarose) of the DNAs synthesized by purified viroplasms. (O) Control incubation. Incubations in the presence of : (1) RNase, 100 µg/ml; (2) Aphidicolin, 120 µg/ml; (3) N-Ethylmaleimide, 25 µM; (4) Actinomycin D, 100 µg/ml; (5) Ethidium bromide, 50 µM; (6) Dideoxy TTP, 25 µM, and dTTP, 1 µM. ▼ and ▽ : positions of circular and linear material of 8 kb, respectively.
➤ : material of about 600 bp; → : same material after RNase treatment.

CONCLUSION

On the basis of cytochemical and autoradiographic observations from in vivo experiments, Favali et al. (9) suggested a two step process for CaMV DNA replication, the first step taking place in the nucleus and the second one in the viroplasms. The former would correspond to the results of Ansa et al. (10), obtained with purified nuclei. This step would be related to the formation of minichromosomes in the model for CaMV DNA replication proposed by Pfeiffer and Hohn (11). The latter step would correspond to our observations. Work is in progress to characterize the replicative complexes present in our purified system, at the level of both their templates and proteins. Preliminary results suggest that both RNA- and DNA- directed DNA syntheses occur in these viroplasm preparations.

REFERENCES

1. Shepherd,R.J. (1979) Ann. Rev. Plant Physiol. 30, 405.
2. Howell,S.H. (1982) Ann. Rev. Plant Physiol. 33, 609.
3. Volovitch,M., Dumas,J.P., Drugeon,G. and Yot,P. (1977)
        in: "Nucleic Acids and Proteins Synthesis in Plants",
        L. Bogorad and J.H. Weil, eds., Colloq. No 261,
        Centre National de la Recherche Scientifique, Paris.
4. Franck,A., Guilley,H., Jonard,G., Richards,K. and Hirth,L.
        (1980) Cell 21, 285.
5. Richards,K.E., Guilley,H. and Jonard,C. (1981)
        FEBS Lett. 134, 67.
6. Menissier,J., De Murcia,G., Lebeurier,G. and Hirth,L.
        (1983) EMBO J. 7, 1067.
7. Olszewski,N., Hagen,G. and Guilfoyle,T.J. (1982) Cell 29, 395.
8. Conti,G.C., Vegetti,G., Bassi,M. and Favali,M. (1972)
        Virology 47, 694.
9. Favali,M.A., Bassi,M. and Conti,G.C. (1973) Virology 53, 115.
10. Ansa,O.A., Bowyer,J.W. and Shepherd,R.J. (1982) Virology
        121, 147.
11. Pfeiffer,P. and Hohn,T. (1983) Cell 33, 781.

SEARCH FOR THE ENZYME RESPONSIBLE FOR THE REVERSE TRANSCRIPTION

STEP IN CAULIFLOWER MOSAIC VIRUS REPLICATION

P. Pfeiffer, P. Laquel and T. Hohn[1]

Institut de Biologie Moléculaire et Cellulaire du CNRS
15 rue Descartes
67084 Strasbourg Cédex, France

[1]Friedrich Miescher Institut
Postfach 273
CH-4002 Basel, Switzerland

Unlike the vast majority of plant viruses, Cauliflower Mosaic Virus (CaMV) possesses DNA rather than RNA as genetic material. The study of the life cycle of this virus has long been hampered by low virus yields, slow multiplication and lack of adequate systems capable of supporting synchronous multiplication and permitting efficient radiolabelling.

However, the advent of molecular cloning and hybridization techniques together with the surge of interest in plant genetic engineering - for which CaMV could be seen as a candidate vector (1) - boosted studies on CaMV in the recent years. A number of unusual features emerged then: the genomic DNA is packaged in virions as a relaxed circle containing three single stranded interruptions (2) with overhanging sequences (3). One of these (Δ1) is in the coding strand and has therefore to be crossed by the transcribing RNA polymerase II (4) to produce the large 35 S transcript which encompasses the whole viral genome plus a 180 nucleotide terminal repeat.

A covalently closed, supercoiled form of CaMV has been described by Ménissier et al. (5) and found to form, probably in association with plant histones, a transcriptionally active minichromosome structure (6, 7): this explained how the 35 S RNA could be generated.

By analogy with small animal DNA viruses for which the minichromosome serves as a template for both transcription and replication, the CaMV minichromosome could also provide a suitable template

121

for CaMV replication. To test this hypothesis we decided to study
in vitro CaMV replication using nuclear extracts that contain actively
transcribing minichromosomes and possibly, replicating minichromosomes.

RESULTS

Cauliflower Mosaic Virus does not replicate via a minichromosome
structure

     Minichromosome-containing nuclear extracts prepared from CaMV-
infected turnip leaves were fractionated by sucrose gradient centrifugation, and transcriptional activity assayed along the gradient.
A peak of CaMV-specific activity was detected, migrating at about
100 S. This corresponds to CaMV minichromosomes (6, 7) being transcribed by RNA polymerase II (4, 6). Transcription is entirely as-
symetrical, with only the α strand (or minus DNA strand) being
transcribed (8). RNA elongated in vitro hybridizes to the whole set
of Eco RI restriction fragments of CaMV DNA (9), suggesting that
minichromosomes support the synthesis of the 35 S transcript.

     However, when replicative activity was tested on these gradient
fractions, no incorporation of deoxyribonucleotide triphosphate was
detected at the level of the leading edge of the minichromosome peak,
i.e. where replicating minichromosomes (10) were expected to migrate.

     Instead, synthesis of viral-specific DNA sequences was entirely
confined to the 20 S region of the gradient (9), suggesting that
CaMV DNA does not replicate as a nucleoprotein complex.

Characterization of the CaMV replicative complexes

     The nucleic acids present in the fractions directing the synthe-
sis of virus-specific DNA sequences were extracted, submitted to gel
electrophoresis, and analyzed by molecular hybridization. No supercoiled DNA was detected there, but only free genomic DNA possessing
the single strand interruptions together with heterodisperse viral
RNA. When replicative complexes were treated with RNAase before incubation with radiolabelled precursors, incorporation dropped by up
to 50%. Similarly, about half of the DNA synthesis was resistant to
Actinomycin D, a drug that inhibits selectively DNA synthesis from
DNA template. This set of evidence suggested that part of the reaction used RNA rather than DNA as a template.

A model for replication of CaMV via reverse transcription

     In parallel to mounting experimental evidence for an unorthodox
replication pathway, a perfect site of homology spanning 14 nucleotides of the 3' end of plant met tRNA and localized at the level of

Δl in CaMV DNA was detected independently in several laboratories
(9, 11, 12): this RNA could base-pair with one of the strands of
CaMV DNA, but also with the 35 S RNA and therefore serve as a primer
to initiate DNA synthesis. This discovery provided the decisive clue
to elaborate the model for CaMV replication depicted in Figure 1.

Enzymes involved in CaMV replication

        Nuclear repair enzymes, e.g. DNA polymerase β, are certainly
involved in the early steps leading to the conversion of the relaxed
infecting genomic DNA into covalently closed circular molecules:
overhanging sequences are probably digested off and the resulting
nicks or gaps resealed. Once converted into a minichromosome, CaMV
DNA is transcribed by RNA polymerase II to yield -among others- the
large 35 S transcript: this is so far the only step in the replication cycle of CaMV that has been ascribed unambiguously to a defined
enzyme.

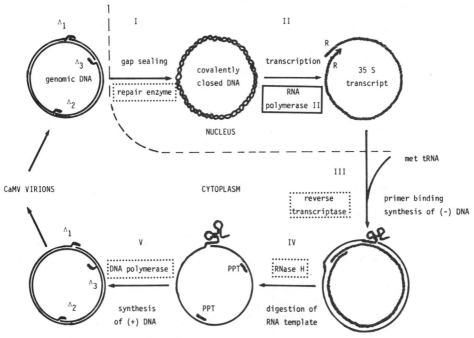

*Figure 1: Current model of the CaMV life cycle. The main steps have
been indicated, together with the relevant enzymes. Enzymatic activities unidentified so far are in dotted boxes. Those which have been
demonstrated to direct specific steps are in solid boxes. Steps III
to V may be directed by a single multifunctional enzyme. The primers
used for DNA synthesis are met tRNA for the (-) strand, and polypurine
tracts (PPT) that resist RNAase H digestion for the (+) strand. R =
terminal 180 nucleotide repeat of the 35 S RNA.*

The key enzyme responsible for the subsequent steps is reverse transcriptase. By analogy with better studied systems such as retroviruses, a single multi-functional enzyme may achieve positioning of the primer tRNA on the template (13), elongation of the nascent DNA chain (14)- subsequent digestion of the RNA template by means of its RNAase H activity (15) and finally, synthesis of the second strand of DNA. Alternatively, some of these steps can be performed by independent enzymes.

The identity of the reverse transcriptase involved in CaMV replication is currently subject to much speculation: it could be a host enzyme that is either constitutive, activated or modified by addition of a virus-encoded subunit. Alternatively, this enzyme could be encoded by the viral genome. For the time being, the only experimental material available to test these hypotheses are replicative complexes (9): these are probably a population of structures elongating minus DNA on an RNA template and plus DNA on the previously synthesized minus DNA. Liquid phase molecular hybridization of the DNA synthesized in vitro to cloned single stranded CaMV DNA permits to follow variations of DNA synthesis, both at the quantitative and qualitative levels (i.e. minus vs. plus strand synthesis).

Using specific inhibitors, we were able to rule out a reaction directed by the α-like DNA polymerases B and C (16), since the synthesis of neither strand was affected by aphidicolin. On the other hand, the response of the DNA synthesis reaction to dideoxyTTP, glycerol, Mn^{++} and KCl suggests the involvement of an activity very similar to the γ-like DNA polymerase A (17), an host enzyme to which no biological role had been assigned so far in healthy plants.

A large, unassigned open reading frame, "gene V" (18), exists however in CaMV DNA, and the 79 kilodalton protein it can encode is a possible candidate for a reverse transcriptase function, provided that the virus does not use a host enzyme.

As a matter of fact, Toh et al. (19) presented very recently evidence for a significant degree of homology between the amino acid sequences of the reverse transcriptase of several retroviruses (the pol gene product) and the putative polymerase gene products of Hepatitis B virus and Cauliflower Mosaic Virus who both include a reverse transcription step in their life cycle.

It seems therefore that Cauliflower Mosaic Virus shares an increasing number of characteristics with the retroid virus family it has just joined. If these viruses derive from a common ancestor as argued by Toh et al. (19) then the apparent inability for CaMV to integrate into its host genome may have to be reexamined or explained.

Gene V is currently being cloned, and it is hoped that its expression product can eventually be compared with the enzymatic activity directing CaMV DNA synthesis in in vitro systems.

REFERENCES

1.   Hohn,T., Richards,K. and Lebeurier,G. (1982) Curr. Topics
        Microbiol. and Immunol. 96, 194.
2.   Volovitch,M., Drugeon,G. and Yot,P. (1978) Nucleic Acids Res.
        5, 2913.
3.   Richards,K.E., Guilley,H. and Jonard,G. (1981) FEBS Lett.
        134, 67.
4.   Guilfoyle,T.J. (1980) Virology 107, 71.
5.   Ménissier,J., Lebeurier,G. and Hirth,L. (1982) Virology
        117, 322.
6.   Olszewski,N., Hagen,G. and Guilfoyle,T.J. (1982) Cell 29, 395.
7.   Ménissier,J., De Murcia,G., Lebeurier,G. and Hirth,L. (1983)
        EMBO J. 2, 1067.
8.   Howell,S.H. and Hull,R. (1978) Virology 86, 468.
9.   Pfeiffer,P. and Hohn,T. (1983) Cell 33, 781.
10.   Edenberg,H.J. (1980) Nucleic Acids Res. 8, 573.
11.   Guilley,H., Richards,K.E. and Jonard,G. (1983) EMBO J. 2, 277.
12.   Hull,R. and Covey,S.N. (1983) Trends in Bioch. Sciences 8, 119.
13.   Litvak,S. and Araya,A. (1982) Trends in Bioch. Sciences 7, 361.
14.   Verma,I. (1977) Biochim. Biophys. Acta 473, 1.
15.   Grandgenett,D.P., Gerard,G.F. and Green,M. (1973)
        Proc. Natl. Acad. Sci. USA 70, 230.
16.   Hevia-Campos,E. (1982) Thèse de 3e Cycle, University of
        Bordeaux (France).
17.   Castroviejo,M., Tarrago-Litvak,L. and Litvak,S. (1975)
        Nucleic Acids Res. 2, 2077.
18.   Franck,A., Guilley,H., Jonard,G., Richards,K.E. and Hirth,L.
        (1980) Cell 21, 285.
19.   Toh,H., Hayashida,H. and Miyata,T. (1982) Nature (London)
        in press.

CHROMATIN STRUCTURE AND DNA REPLICATION

Wolfgang Pülm and Rolf Knippers

Fakultät für Biologie
Universität Konstanz
D-7750 Konstanz

SUMMARY

The structure of (^{3}H) thymidine pulse labeled, replicating chromatin differs from that of non replicating chromatin by several operational criteria which are related to the higher nuclease sensitivity of replicating chromatin. We summarize the structural changes that we observe using replicating chromatin as substrate for micrococcal nuclease. The data suggest a more extended configuration of replicating compared to non replicating chromatin. We use these data to discuss a model of the chromatin structure in the vicinity of replication forks.

Finally, we present data to show that the reversion of structural changes in replicative chromatin depends on continued DNA replication.

INTRODUCTION

The in vivo substrate for the eukaryotic replication apparatus is chromatin rather than DNA.

A number of observations suggest that the structure of chromatin has some influence on the replication of a eukaryotic genome. For example, heterochromatic centromeric regions replicate late in the DNA replication phase (S-phase) of the cell cycle as does the heterochromatic in comparison to the euchromatic X-chromosome of female mammalian cells (1, 2). Chromatin structure may affect the initiation of replication as well as replicative chain elongation.

127

For our present discussion we consider two levels of chromatin structure (3, 4).

- The 10 nm-filament, a repeating array of spacer DNA sections alternating with "chromatosomes", particles, containing 166 bp (base pairs) DNA wrapped in two superhelical turns on the outside of the histone octamer. Histone Hl, if present, is thought to clamp the DNA termini emerging from the chromatin subunits. The flat faces of the chromatosome discs are probably more or less in parallel to the filament axis.

- The 30 nm-solenoid-structure, a configuration representing the majority of eukaryotic chromatin in vivo. According to current models of chromatin about six nucleosomes are radially organized on one solenoid turn with a tilt angle of 20-30° between the solenoid axis and the flat faces of the chromatosome discs. Histone Hl is assumed to be essential for the maintenance of this type of higher order chromatin structure.

In vitro experiments have shown a reversible and cation control-led interconversion between the 10 nm filament and the 30 nm solenoid (5). We shall argue below that a similar inter-conversion occurs in the vicinity of replication forks.

CHROMATIN STRUCTURE AT REPLICATION START POINTS

Tightly structured heterochromatin could block the entry site for replication factors and the assembly of the replication apparatus.

However, very little is known about the relation of replication start points to specific chromatin structures.

One exception may be the paradigmatic eukaryotic replicon, the Simian Virus 40 (SV40) minichromosome. Work by Yaniv and his colleagues (6) as well as by others (7, 8) has shown that nucleosomes are not evenly distributed in SV40 chromatin. Instead, a histone free region of about 400 bp of DNA exists which includes the viral origin of replication as well as part of the in vitro binding sites for the viral initiator protein (the gene A protein or T antigen, T-ag). We recently started to investigate a type of T-ag-DNA inter-action which is characterized by the unusual stability of this complex at salt concentration < 1M (ref. nr. 9). As described in detail elsewhere (9, 10), we think that this type of T-ag binding to DNA may be a necessary step in initiation of DNA replication. Electro-microscopical evidence suggests that a salt stable contact of T-ag with DNA occurs before or at the time of replication initiation in a region between nucleotides 150 and 300, i.e. well within the nucleo-some free region of the SV40 chromosome (6).

It is not known, however, whether the binding of T antigen or other regulatory proteins is the cause rather than a consequence of this nucleosome free region, or whether the overlap of both sites is quite fortuitous. These questions remain important points in studies aimed at an understanding of the relationship between initiation of replication and chromatin-structure.

CHANGES OF CHROMATIN STRUCTURE AT THE REPLICATION FORK

The structure of replicating chromatin

More is known about the structural changes that chromatin undergoes during progression of replication forks. Experiments to investigate this question have been performed in several laboratories. The results are more or less compatible. In the following section we summarize some data from our laboratory.

The experimental approach includes: (i) (^{3}H)thymidine-pulse labeling of SV40 infected or uninfected cells, (ii) preparation of nuclei or chromatin and (iii) the comparison of pulse labeled and continuously labeled bulk chromatin as substrates for endonucleases, usually micrococcal nuclease. In the experiments reported below, soluble chromatin fragments are investigated either directly by sucrose gradient centrifugation or after extraction of DNA by gel electrophoresis.

This approach allows, as we shall see, conclusions concerning the structure of chromatin, containing pulse labeled, i.e. newly replicated DNA.

Regardless of whether we use SV40-chromatin or chromatin from various uninfected mammalian cells, we obtain the following results (11-17):
-DNA in newly replicated chromatin is more sensitive against nuclease attack than bulk chromatin;
-under conditions of mild nuclease treatment, bulk chromatin is degraded to give comparatively large supranucleosomal structures, sedimenting with 30 S and more and containing groups of 6 and more nucleosomes, possibly representing substructures of the 30 nm-solenoid. Under the same conditions, pulse labeled chromatin is digested to give monomeric nucleosomes (Figure 1).

These results may be explained by the assumption that the spacer DNA between adjacent nucleosomes in newly replicated chromatin is more exposed and, therefore, more accessible to nuclease attack than spacer DNA in non replicating chromatin.

The monomeric nucleosomes, released from replicating chromatin (Figure 1), could in principle arise from two fractions:
(i) from the fraction of newly assembled chromatin; an
(ii) from a second fraction of replicated chromatin which had received the parental set of nucleosomes. To investigate the spacing of chro-

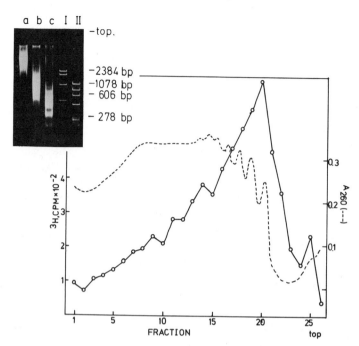

Figure 1: Replicative chromatin. Bovine lymphocytes were activated
with Concanavalin A (14) and pulse labeled for 30 sec with (^{3}H)
thymidine as described (17). Nuclei (6x10^7) were prepared from these
cells, washed and resuspended in nuclease buffer (12) before incubation for 40 sec at 37°C with 10 units micrococcal nuclease (Boehringer,
Mannheim). The nuclei were then lysed with EDTA. Soluble material was
centrifuged through a linear 30% to 10% sucrose gradient. After centrifugation, the gradient was pumped through a quartz cuvette to
monitor the absorbance at 258 nm (broken line). Aliquots from each
fraction were precipitated in trichloroacetic acid to determine the
distribution in this gradient of (^{3}H) radioactivity (o).
Insert: DNA was phenol-chloroform extracted from aliquots of the
combined fractions 5-10 (a), 11-15 (b) and 15-20 (c), respectively.
The extracted DNA was analysed on a 1% agarose gel and identified
by ethidium bromide staining. Hae II-restricted Col EI-DNA (I) and
Hae III-restricted φX 174-RF-DNA (II) served as size markers.

matin subunits in this second fraction, we pretreated cells with
cycloheximide until more than 95% of protein synthesis, including
histone synthesis, was inhibited. Since DNA replication continues
in the presence of cycloheximide (although at a reduced rate) most
newly formed chromatin must contain parental nucleosomes. We find
that chromatin, prepared from cycloheximide treated, (^{3}H)thymidine
pulse labeled cells, has properties which are very similar to that

prepared from untreated cells. In particular, monomeric nucleosomes are released by mild treatment with micrococcal nuclease from (^{3}H) thymidine pulse labeled chromatin at very similar rates regardless of whether the cells had been pretreated with cycloheximide or not (not shown; see ref. nr. 17).

Since histone Hl is assumed to play an essential role in maintaining the higher order structure of chromatin it was obvious to ask whether Hl may be absent in newly replicated chromatin. The most direct evidence supporting this assumption was provided by Schlaeger (16) who was able to show that most monomeric chromatin subunits, produced by nuclease treatment of pulse labeled chromatin, travel during electrophoresis like Hl- (and HMG-) depleted histone octamers.

We describe now another property which characterizes newly replicated chromatin. This property becomes detectable after a more extensive treatment of chromatin with micrococcal nuclease. Under these conditions, a considerable fraction of bulk chromatin is degraded to give nucleosomal core particles, i.e. chromatin fragments consisting of 145±5 bp long DNA in association with histone octamers, but lacking histone Hl. Under the same conditions much of the pulse labeled chromatin fragments are cut at internal sites to give sub-nucleosomal DNA fragments of 60-70 bp length (Figure 2). The 60-70 bp-subnucleosomal DNA is loosely bound to histones and dissociates from the histone octamer in 0.5 M NaCl when typical nucleosomal core particles remain intact (Figure 3).

A detailed interpretation of this finding which was obtained with primary lymphocytes cultures (Figure 3) as well as with established cell lines (Figure 2b) and with replicating SV40 chromatin (Figure 2a) is not yet possible. But it could well indicate a less stringent histone-DNA interaction in a fraction of replicating chromatin.

The structural changes of replicating chromatin are reversible

When the (^{3}H) thymidine pulse is followed by a chase (in the presence of an excess of non radioactive thymidine) labeled chromatin can be observed to quickly resume the structure of bulk chromatin (15-17).

After a few minutes of chase the internal nicking of nucleosomal DNA and the production of subnucleosomal DNA fragments is no longer detectable. At about the same time, (^{3}H) labeled monomeric nucleosomes begin to electrophorese like bulk nucleosomes, i.e. they contain bound histone Hl.

Somewhat later, monomeric nucleosomes are no longer preferentially released from (^{3}H) labeled chromatin (compare: Figure 1) and, finally, the overall nuclease sensitivity of labeled chromatin disappears.

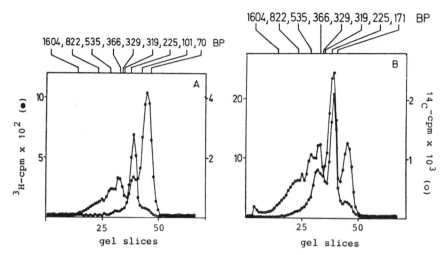

Figure 2: Subnucleosomal DNA from replicative chromatin. Analysis
by gel electrophoresis.
(A) SV40-infected CV1-cells were continuously labeled for 24-36 h
after infection with (14C)thymidine and then pulse labeled with
(3H)thymidine (15). Nuclei were prepared from these cells and incubated to elute most labeled viral chromatin which was further purified
by sucrose gradient centrifugation (15). Fractions, containing SV40
chromatin, were combined and digested for 2.5 min with 15 units
micrococcal nuclease. DNA was extracted with phenol-chloroform and
ethanol precipitated before electrophoresis in a 1.7% agarose gel.
The gel was cut into 2-mm-slices to determine the distribution of
14C- and 3H-cpm. Hae II/Hae III-restricted SV40-DNA served as size
marker.
(B) Uninfected, semiconfluent CV1-cells were labeled for 12 h with
(14C)thymidine and pulse-labeled with (3H)thymidine. Nuclei (2x10⁶)
from these cells were treated for 2 min at 37°C with 75 units micrococcal nuclease. The DNA was extracted and investigated as under (A).
(o) (14C) continuously labeled DNA-fragments; (●) (3H) pulse labeled
DNA-fragments. (Data from: K.H. Klempnauer, 1980, Dissertation,
Universität Konstanz).

A summary of the structural changes in pulse labeled chromatin

        We propose that newly replicated chromatin exists in an extended
state very much like the structure of chromatin at low ionic strengths
as observed by Thoma et al. (5). The structural changes in replicative
chromatin may be due to a dissociation of histone H1. They are reversed by reassociation of H1 at some, as yet unspecified, distance
from the replication fork (Figure 4).

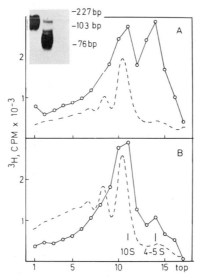

*Figure 3: Subnucleosomal DNA. Analysis by sucrose gradient centri-
fugation. Isolated nuclei from Concanavalin A acitivated lymphocytes
were incubated for 45 sec under conditions optimal for DNA replication (13). Nuclei were pelleted and washed to remove unincorporated
nucleotides and incubated with micrococcal nuclease until 20% of
the DNA was acid soluble. The soluble digestion products were divided
into two equal parts. One part was sedimented through a sucrose gradient containing 0.5 M NaCl (A). The second part was sedimented
through a sucrose gradient without salt (B). After centrifugation,
the position of bulk chromatin fragments was determined by their
absorbance of 258 nm-ultraviolet light (broken line) and the distribution of (^{3}H) labeled fragments by precipitation in trichloroacetic
acid (o).*
*Insert: DNA was extracted from fractions 9-11 and 14-16, respectively
of the gradient shown in (A) and investigated for length determination
by polyacrylamide gel (6%) electrophoresis and fluorography.*

How does parental chromatin react when it is located immediately
in front of an approaching replication fork?

First, we assume that the section of chromatin in close proxim-
ity ahead of the replication fork also exists in an extended configuration (Figure 5). This assumption is an extrapolation of the findings
reported above showing that newly replicated chromatin lacks histone
Hl which, of course, means that histone Hl dissociates from chromatin
somewhere in the vicinity of the replication fork, most probably from
the section of unreplicated chromatin. If this assumption is correct
the chromatin structure ahead of the replication fork may look more
or less like a mirror image of the structure on the replicated branch

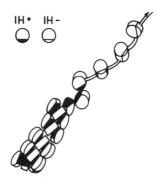

Figure 4: The distribution of parental chromatin subunits on repli-
cated DNA. An interpretation (see text). (The arrow head indicates
the growing DNA strand.)

behind the fork (Figure 5). Presently, a mechanism which could induce Hl dissociation is unknown. It should be recalled though that
Hl becomes highly phosphorylated during the DNA replication phase
of the cell cycle (18) and that phosphorylated Hl has DNA binding
properties which are different from those of unmodified Hl (19).

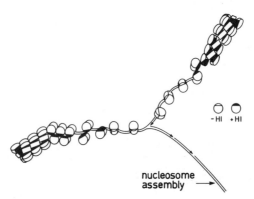

Figure 5: The structure of chromatin in the vicinity of a replica-
tion fork. A model. Details are described in the text. In this figure,
we assume that parental histone octamers are transferred asymmetrical-
ly to one branch of the replication fork (20). However, this may not
be generally valid. In some cells types, parental octamers may be
transferred randomly or in clustered groups to both emerging DNA
strands (21-23).
It is also not established whether new chromatin subunits are always
assembled on replicated DNA or on non replicated DNA (24, 25, 26)
or both (28).

Second, we like to argue that the replication enzymes are able to pass histone octamers. In fact, DNA polymerase α can be detected in substantial quantities on chromatin subunits from replicating cells (Figure 6). Moreover, it has been reported that histone octamers are transferred as intact and conserved entities from the unreplicated stem to the replicated branches of chromatin (27). Furthermore, McGhee and Felsenfeld (28) have shown that the bases in nucleosomal DNA are nearly as accessible to methylation by dimethylsulfate as the base in free DNA showing that DNA in chromatin may be suspended on the surface of chromatin subunits in a way which interfers as little as possible with the biochemical activity of DNA. It is also known (29) that single stranded DNA binds effectively to histone

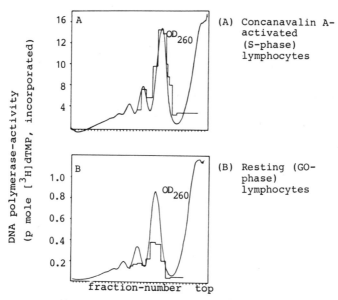

(A) Concanavalin A-activated (S-phase) lymphocytes

(B) Resting (GO-phase) lymphocytes

Figure 6: DNA polymerase on nucleosomes. Nuclei (10^8) were prepared from resting (bottom) and from Concanavalin A-activated lymphocytes (top). The nuclei were washed several times to remove all detectable free DNA polymerase. The nuclei were then resuspended in nuclease buffer (12) and digested with micrococcal nuclease. The nuclei were then lysed by EDTA. The soluble chromatin fragments were sedimented through sucrose gradients as detailed under Figure 3 (except that the sucrose gradient buffer contained 0.1 M NaCl rather than 0.5 M NaCl). After centrifugation, the distribution of chromatin fragments was monitored by uv-absorbance (258 nm; smooth line). DNA polymerase activity (horizontal bars) was determined as described (35). More than 90% of the polymerase activity were sensitive against N-ethyl-maleimide or aphidicolin. Free DNA polymerase α when centrifuged under identical conditions in a parallel gradient showed an activity maximum at fraction 35.

cores suggesting that chromatin subunits may well be transiently
associated with single stranded DNA when the replication apparatus
travels along the chromatin fiber.

We do not know whether the less stringent histone DNA interaction that we observe in chromatin particles with pulse labeled DNA
(Figure 2 and 3) is a requirement for or a consequence of the presence of replication enzymes. We have pointed out above that the
chromatin structure, as observed in newly replicated chromatin,
could well be very similar to the extended conformation that chromatin
acquires under conditions of very low ionic strengths (5). This analogy might even hold for the single chromatin subunit. Martinson et
al. (30) described a loosening of histone-histone interaction at low
ionic strengths. A similar change may also occur in subunits of
replicating chromatin and could be the cause for the internal nuclease
sites in replicating nucleosomes.

CHROMATIN REPLICATION IN VITRO

We now present some results from our in vitro studies on
chromatin replication to demonstrate a functional relation between
chromatin and DNA replication.

DNA replication in isolated nuclei

It is well established that limited semiconservative DNA replication occurs in nuclei, isolated from cells in the S-phase of the
cell cycle. Some years ago several groups were able to show that DNA,
replicating in isolated nuclei, is associated with chromatin particles
like in vivo replicated DNA (13, 31). In fact, all structural alterations of replicating chromatin are also observed in isolated nuclei.
Moreover, replicative chromatin in isolated nuclei matures in the
same sequential order as described above although at a reduced rate.

This in vitro system was used to investigate the relationship
between replication fork movement and chromatin maturation.

For this purpose, nuclei were prepared from lymphocytes in
S-phase and incubated for 45 sec in vitro with desoxynucleosidetri-
phosphates and ATP under conditions known to be optimal for DNA
replication in this system. The nuclei were then pelleted to remove
unincorporated nucleotides. After resuspension, the nuclei were
distributed into several fractions. One fraction received an excess
of non radioactive nucleotides and was incubated for 20 min under
replication conditions. Another fraction was treated under identical
conditions except that nucleotides were omitted from the reaction
mixture. After incubation, the chromatin structure was investigated
using micrococcal nuclease. We observe (Figure 7) a maturation of
chromatin only when the incubation was performed in the presence of
DNA precursors and ATP.

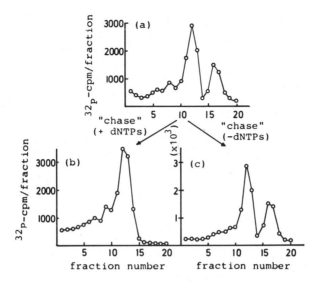

Figure 7:   Replication and maturation of chromatin.
(a) The lymphocyte nuclei were prepared and incubated for in vitro
DNA replication in the presence of (^{3}H)dTTP as described (13, 14,
17). After 45 sec of in vitro DNA replication, the nuclei were pel-
leted, washed to remove radioactive dNTPs and distributed into three
equal parts. One part was used directly for micrococcal nuclease
treatment and sucrose gradient centrifugation (in 0.5 M NaCl) of
soluble chromatin fragments.
(b) Another part of the pulse labeled nuclei was resuspended and
incubated for 20 min in the presence of an excess of non radioactive
dNTPs. The nuclei were then treated with micrococcal nuclease. Sucrose
gradient centrifugation was performed as in (a), i.e. in the presence
of 0.5 M NaCl.
(c) The third sample was processed exactly as in (b) except that
dNTPs were omitted from the reaction mixture.
We show results of sucrose gradient centrifugations. The sucrose
gradients, made up in 0.5 M NaCl; 5 mM Tris-HCl, pH 8; 1 mM EDTA
were linear from 30% to 10%. Centrifugation was performed in the
Beckman SW40-rotor at 38 000 rpm and 7°C for 3,5 h.

We conclude that chromatin maturation depends on continued DNA
replication and, furthermore, that chromatin maturation occurs in a
spatial, not temporal distance from a replication event (17).

Replication of SV40 chromatin in vitro

We have also used the SV40 minichromosome as a template for
chromatin replication in vitro. Replicating SV40 chromatin is separated from non replicating SV40 chromatin by sucrose gradient centri-

fugation. Using these separated chromatin fractions it could be
shown that DNA polymerase α and probably other replication proteins
are specifically associated with replicating chromatin (32). However,
when replicating SV40 chromatin was incubated with desoxyribonucleo-
tides and ATP we detect under optimal conditions only very limited
DNA synthesis (Figure 8) leading to small DNA fragments. A soluble
cell extract must be added to achieve a completion of a replication
cycle. We have prepared protein extracts from infected and uninfected
CVl cells, from mouse cells and from Xenopus laevis eggs (33) and
found comparable results suggesting that no species specific factor
is present in the complementing protein mixture. This does not
necessarily mean that there are no specific factors involved in the
reaction. In fact, the viral T antigen is present on most replicative
intermediate forms of SV40 chromatin (10) and may play a role during
replicative chain elongation since deproteinized replicative SV40-
DNA is a poor template in the in vitro reaction.

In the presence of a complementing protein extract we find a
faithful completion of replication rounds as well as maturation of
SV40 chromatin. Thus, the protein extracts contain some factors
essential for DNA replication and for chromatin maturation.

We have started to analyze the complementing protein extract
by isolation of a DNA polymerase-primase (34, 35). It would be interesting to identify additional replication enzymes as well as factors,
which may be necessary for chromatin maturation.

CONCLUSION

How does chromatin structure affect the replication of the
eukaryotic genome?

It appears likely that nucleosome depleted regions of chromatin
are required for the assembly of the replication machinery and as
initiation sites for replication. However, no data are available
to prove or disprove this assumption except, may be, for SV40 chromatin. However, even in this case it is unknown whether the nucleosome
free section of chromatin defines the replication start site or
whether specific initiation proteins, like the SV40 T antigen, first
bind to DNA and thereby exclude an association of histones.

Somewhat more is known about the chromatin structure around
the replication fork. Probably most replicative chromatin exists
in a more extended form like the chromatin investigated in vitro at
low ionic strengths. Note that during transition from the 30 nm-
solenoid to the 10 nm-filament radially clustered groups of nucleosomes are being oriented approximately parallel to the filament axis.
This will certainly facilitate the progression of the replication
fork. It is quite possible that a reversible dissociation of histone

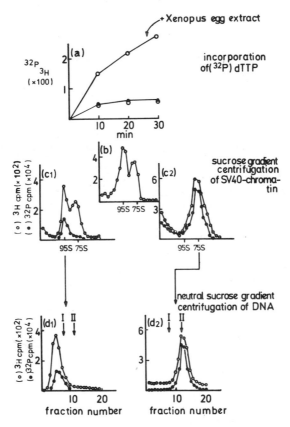

Figure 8:   In vitro replication of SV40 chromatin. SV40-infected CVl cells were pulse labeled with (^{3}H) thymidine at about 40 h after in- fection. SV40-chromatin was prepared from isolated nuclei according to published protocols (31) and purified by sucrose gradient centri- fugation (15).

(a) Purified SV40-chromatin was incubated with ATP and optimal con- centrations of α(^{32}P)-labeled dNTPs in the presence (o) and absence (•) of an extract from Xenopus laevis eggs (33). The data are ex- pressed as ratios of incorporated (^{32}P) counts over (^{3}H) template counts.

(b) Sedimentation profile of (^{3}H) labeled SV40 chromatin, incubated in the absence of ATP and dNTPs. The faster sedimenting 95S-peak contains replicating intermediates, the slower sedimenting 75S peak mature viral chromatin (15, 31).

(c) Sucrose gradient purified SV40 chromatin, (^{3}H) pulse labeled in vivo, was incubated -for 20 min- with ATP and α(^{32}P) labeled dNTPs in the absence (C1) or in the presence (C2) of an Xenopus egg extract as shown under (a). After incubation, the chromatin was directly centrifuged through a sucrose gradient under conditions used for the control experiment in (b).

(d) An experiment was performed with in vivo pulse labeled viral

*chromatin exactly as described under (c), except that the viral DNA
was extracted by phenol and chloroform after 20 min incubation in
the absence (d1) or presence (d2) of an extract prepared from Xenopus
laevis eggs. We show the sedimentation properties of purified DNA.
Marker form I and form II SV40-DNA was sedimented in a parallel tube
of the same centrifugation run (data from A. Richter, 1981, Dissertation, Universität Konstanz).*

H1 is a key element in the regulation of these transient structural
changes. We think that replication enzymes like DNA polymerase α
travel around intact octamers whose protein-protein and protein-DNA
contacts may be more relaxed than the corresponding interactions in
non replicating chromatin. One may ask whether these transient structural alterations of chromatin are a requirement that affects the
progression of the replication fork.

ACKNOWLEDGEMENT

        The experimental work performed in the authors' laboratory
was supported by DFG/SFB 138 and by "Fonds der Chemischen Industrie".

REFERENCES

1.   Belasz,I., Brown,E.H. and Schildkraut,C.L. (1974) Cold Spring
         Harbor Symp. Quant. Biol. 38, 239.
2.   Willard,H.F. (1977) Chromosoma 61, 61.
3.   Igor-Kemenes,T., Hörz,W. and Zachau,H.G. (1982)
         Ann. Rev. Biochem. 51, 89.
4.   McGhee,J.D., Nickel,J.M., Felsenfeld,G. and Rau,D.C. (1983)
         Cell 33, 831.
5.   Thoma,F., Koller,T. and Klug,A. (1979) J. Cell Biol. 84, 403.
6.   Saragasti,S., Cereghini,S. and Yaniv,M. (1982) J. Mol. Biol.
         160, 133.
7.   Varsharsky,A.J., Sundin,O.H. and Bohn,M. (1979) Cell 16, 453.
8.   Jacoborits,E.B., Bratosin,S. and Aloni,Y. (1980) Nature
         285, 263.
9.   Stahl,H., Bauer,M. and Knippers,R. (1983) Eur. J. Biochem.
         134, 55.
10.   Stahl,H. and Knippers,R. (1983) J. Virol. 47, 65.
11.   Fanning,E., Klempnauer,K.H., Otto,B., Schlaeger,E.H. and
         Knippers,R. (1978) in: "DNA Synthesis, Present and Future",
         I. Molineux and M. Kohiyama, eds, Plenum Publ. Comp. N.Y.
12.   Schlaeger,E.J. and Klempnauer,K.H. (1978) Eur. J. Biochem.
         89, 567.
13.   Schlaeger,E.J. (1978) Biochem. Biophys. Res. Comm. 81, 8.
14.   Schlaeger,E.J. and Knippers,R. (1979) Nucl. Acids Res. 6, 645.
15.   Klempnauer,K.H., Fanning,E., Otto,B. and Knippers,R. (1980)
         J. Mol. Biol. 136, 359.

16. Schlaeger,E.J. (1982) Biochemistry 21, 3167.
17. Schlaeger,E.J., Pülm,W. and Knippers,R. (1983)
        FEBS Lett. 156, 281.
18. Hohmann,P., Tobey,R.A. and Gurley,L.R. (1976)
        J. Biol. Chem. 25, 3685.
19. Knippers,R., Otto,B. and Böhme,R. (1978) Nucl. Acids Res.
        5, 2113.
20. Seidman,M.M., Levine,A.J. and Weintraub,H. (1979) Cell 18, 439.
21. Seale,R. ((1976) Cell 9, 423.
22. Russev,G. and Hancock,R. (1982) Proc. Natl. Acad. Sci. USA
        79, 3143.
23. Jackson,V., Granner,D. and Chalkley,R. (1976)
        Proc. Natl. Acad. Sci. USA 73, 2266.
24. Seale,R. (1976) Proc. Natl. Acad. Sci. USA 73, 2270.
25. Murphy,R.F., Wallace,R.B. and Bonner,J. (1980)
        Proc. Natl. Acad. Sci. USA 77, 3336.
26. Hancock,R. (1979) Proc. Natl. Acad. Sci. USA 75, 2130.
27. Laffak,I.M., Grainger,R. and Weintraub,H. (1977) Cell 12, 837.
28. McGhee,J.D. and Felsenfeld,G. (1979) Proc. Natl. Acad. Sci.
        USA 76, 2133.
29. Palter,K.B., Foe,V.E. and Alberts,B.M. (1979) Cell 18, 451.
30. Martinson,H.G., True,R.J. and Burch,J.B.E. (1979)
        Biochemistry 18, 1082.
31. Kaufmann,G., Anderson,S. and DePamphilis,M.L. (1977)
        J. Mol. Biol. 116, 549.
32. Otto,B. and Fanning,E. (1978) Nucl. Acids Res. 5, 1715.
33. Richter,A., Otto,B. and Knippers,R. (1981) Nucl. Acids Res.
        15, 3793.
34. Riedel,H.D., König,H., Stahl,H. and Knippers,R. (1982)
        Nucl. Acids Res. 10, 5621.
35. König,H., Riedel,H.D. and Knippers,R. (1983)
        Eur. J. Biochem. 135, 435.

INHIBITION AND RECOVERY OF THE REPLICATION OF DEPURINATED

PARVOVIRUS DNA IN MOUSE FIBROBLASTS

J-M. Vos[a], B. Avalosse, Z.Z. Su and J. Rommelaere[b]

Laboratory of Biophysics and Radiobiology
Université Libre de Bruxelles
B-1640 Rhode St. Genèse, Belgium

ABSTRACT

Apurinic sites were introduced in the single-stranded DNA of parvovirus minute-virus-of-mice (MVM) and their effect on viral DNA synthesis was measured in mouse fibroblasts. Approximately one apurinic site per viral genome, is sufficient to block its replication in untreated cells. The exposure of host cells to a sublethal dose of UV-light 15 hours prior to virus infection, enhances their ability to support the replication of depurinated MVM. Cell preirradiation induces the apparent overcome of 10-15% of viral DNA replication blocks. These results indicate that apurinic sites prevent mammalian cells from replicating single-stranded DNA unless a recovery process is activated by cell UV-irradiation.

INTRODUCTION

Effects of DNA depurination

Spontaneous depurination of DNA occurs at a significant rate under physiological conditions. It was estimated that the number of purines lost from the DNA of mammalian cells is of the order of 10^4 per generation (1). The rate at which apurinic (AP) sites are produced can be increased by a variety of DNA damaging agents. In particular, DNA alkylation at the N-3 and N-7 positions of purines

----

[a]Aspirant and [b]Chercheur Qualifié du Fonds National de la Recherche Scientifique de Belgique.

143

greatly destabilizes the N-glycosylic bond connecting the base to
the sugar-phosphate backbone (2). One would expect AP sites to be
noninstructive, i.e. to be devoid of information for base pairing.
In vivo, the bacterium Escherichia coli has a low spontaneous capacity
to bypass AP sites during DNA replication, with a resultant lethal-
but little mutagenic effect of these lesions (3). In vitro, eukaryotic
DNA polymerases were shown to achieve higher levels of total incorporation and of misinsertion of nucleotides when copying templates
containing AP sites, compared with prokaryotic enzymes (4). The
greather tolerance of eukaryotic DNA polymerases to AP sites led
Kunkel et al. (4) to predict a highly mutagenic response to these
lesions in mammalian cells. The treatment of bacteria with various
genotoxicants triggers the expression of a complex set of functions
which is termed the SOS system and includes mutator and recovery
process(es) (5). In particular, this system appears to induce bacteria
to accommodate a small fraction of AP sites (3, 4). Similarly,
genotoxicants such as UV light, were shown to confer mutator and
repair phenotypes to mammalian cells (5). However, the molecular
mechanism responsible for these acquired phenotypes is ill-defined
(6) and its role in the tolerance of AP sites by mammalian cells
remains to be determined.

The effects of AP sites on in vivo DNA replication in mammalian
cells has not been assessed so far. This question was investigated
in the present study by comparing untreated and sublethally UV-
irradiated rodent cells for their susceptibility to the inhibition
of replication by AP sites.

Autonomous parvoviruses as probes of DNA replication in eukaryotic cells

The autonomous parvovirus minute-virus-of-mice (MVM) (7) was
used as a probe to measure the sensitivity of host cells to the inhibitory effect of AP sites.

The genome of parvoviruses is a predominantly single-stranded,
linear DNA molecule containing about 5×10^3 nucleotides. DNA replication of autonomous parvoviruses takes place in the cell nucleus and
involves three steps: (i) the conversion of single-stranded DNA to
double-stranded replicative forms (Figure 1A), (ii) the replication
of double-stranded DNA (Figure 1B) and (iii) the asymmetric production of progeny single-stranded DNA from replicative forms (Figure
1C).

DNA replication of autonomous parvoviruses appears to be
closely dependent on the replication machinery of host cells. In
particular, the conversion of the input single-stranded genome to
a duplex form is likely to rely entirely on cellular functions, some
of which are expressed specifically during the S phase (7). Moreover
a series of evidence supports the involvement of DNA polymerase α

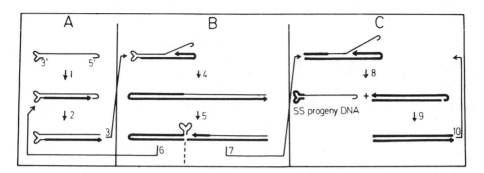

Figure 1:   Model for MVM DNA replication (after Astell et al (10)).
A: Palindromic sequences exist in the form of hairpin duplexes at
   both termini of viral single-stranded DNA. Input DNA is first
   converted into a monomer-length double-stranded replicative form
   (RF) by unidirectional and continuous elongation of the 3' hairpin terminus that serves as a natural primer and links covalently
   the viral and complementary strands (step 1). This process eventually extends and duplicates the 5' terminal palindrome (step 2).
B: RF DNA is then replicated by strand displacement synthesis which
   is primed by the copied 5' terminal palindrome (step 3) and produces concatemeric DNA species (step 4). Site-specific cleavage
   of concatemers and strand displacement elongation of the parental
   viral strand generates two types of monomeric RF molecules, both
   with the 3' hairpin region in its original orientation (step 5).
   One type of RF DNA contains only one copy of the 3' palindrome
   and may be recycled through the RF DNA pool (step 6). The other
   type of RF DNA contains a replicated 3' hairpin and can be used
   for progeny strand synthesis (step 7).
C: Self-primed elongation of the copied 5' terminal palindrome of
   full-length RF DNA (step 7) causes the displacement of a viral
   single-strand and the concomitant formation of an RF molecule
   with viral and complementary strands attached covalently through
   the 5' palindrome (step 8). A hairpin transfer process allows the
   replication of the 5' palindrome (step 9) and regenerates an
   extended RF molecule which can be used for the formation of additional single-strands (step 10). The 5' terminal hairpin region
   in the pool of progeny DNA will contain 2 possible sequence
   orientations because this palindrome is not perfect and its unique
   sequences are inverted by the hairpin transfer process.

in DNA replication of autonomous parvoviruses (8). In those respects,
parvoviral DNA can be used as a model system for determining the
effects of DNA lesions on continous DNA synthesis on a leading strand
(6).

In this work, MVM was exposed to a DNA depurination treatment and the residual level of viral DNA replication was measured after infection of normal or UV-irradiated cells. The rationale for using parvoviruses as probe of cellular responses to AP sites, relies on (i) the great stability of parvoviruses, in particular to heating (9), allowing one to expose whole virions to the DNA depurination treatment and (ii) the lack of an excision of AP sites from unreplicated viral DNA, as a result of its single-strandedness (11). Moreover, UV light was shown to enhance the ability of rodent cells to reactivate UV-damaged MVM (6, 12). Therefore, this virus appears especially suitable to probe the effects of AP sites on DNA synthesis catalyzed by the enzymatic machinery of normal and UV-treated cells.

RESULTS AND DISCUSSION

Inhibition of MVM replication by depurination of viral DNA

Heating MVM virions under acid conditions produced alkali-labile sites in the viral genome (Figure 2A). Alkali-labile lesions were identified as AP sites on the basis of (i) the alkali conditions required to convert them to DNA strand breaks (13) and (ii) their sensitivity to an endonuclease cleaving DNA specifically at AP sites (data not shown). The treatment of MVM at pH 5 and 60°C introdouced one AP site per viral genome every 10-12 min (Figure 2B). The formation of AP sites correlated with an inhibition of overall viral DNA synthesis as measured by in situ hybridization (Figure 3A). The level of residual viral DNA synthesis decreased exponentially with depurination time (Figure 3B). Under the experimental conditions used, input viral DNA was barely detectable and parental replicative forms contributed minimally to the total amount of viral DNA measured in infected cells. Thus, residual viral DNA synthesis gave an estimate of the fraction of input single-stranded genomes whose conversion to double-stranded forms went to completion (Figure 1A) and which eventually underwent subsequent DNA amplification steps (Figure 1B and C). This fraction was used to calculate the average number of replication blocks per parental viral genome. As shown in Figure 3C, the number of replication blocks was very similar to that of apurinic sites. Exposure of DNA to heat and acid produces other kinds of damage in addition to AP sites. However, these non-depurination types of lesions are minor products which will have no detectable effect on the survival of DNA molecules (3). Therefore our results suggest that approximately one AP site in viral sigle-stranded DNA is sufficient to block its replication on mouse cells. Consistently, AP sites constituted lethal hits for virus infectivity (data not shown), similarly to the results obtained with single-stranded DNA phages propagated in bacteria (3).

Altogether, our data indicate that AP sites prevent mammalian cells from replicating single-stranded DNA. In contrast, purified

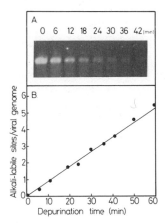

Figure 2:   Quantitation of apurinic sites in DNA of parvovirus MVM. Depurination was achieved by heating MVM suspensions (pH 5.0) for various times at 60°C, as described (1).

A: Electrophoretic profiles of depurination MVM DNA. Virions were exposed to the depurination conditions for the times indicated on top of lanes. DNA was both released from virions and broken at apurinic sites by treatment with alkali (0.1 M NaOH, 37°C, 1h). The average number of breaks per DNA molecule was estimated by electrophoresis of samples (1.5 μg DNA) through an agarose gel, as described previously (6). The photograph of the ethidium bromide-stained gel is shown. The decrease in the intensity of the band of full length DNA can be ascribed to the formation and subsequent hydrolysis of alkali-labile sites since the same amount of DNA was applied to each lane and no breakage of depurinated viral DNA was detected in the absence of alkali treatment (data not shown).

B: Formation of alkali-labile sites as a function of time of depuri- nation. The average number of alkali-labile sites was calculated from the fraction of total DNA devoid of these lesions and migra- ting as full molecules under alkaline conditions (see panel A), assuming a Poisson distribution. A good quantitative correlation was found between the numbers of sites sensitive to alkali and to an apurinic specific endonuclease (data not shown).

eukaryotic DNA polymerases are able to bypass AP sites with a signif- icant frequency when copying primed single-stranded templates in vitro (4). This apparent discrepancy raises the possibility that additional cellular factors contribute to the lack of in vivo toler- ance of mammalian cells to non-coding lesions such as AP sites. These factors might be part of the cellular replicating complex and confer high fidelity requirements to the DNA polymerase. Alternatively, these factors might interact with AP sites and make them unavailable for replication. Assuming similar requirements for parvoviral and cellu-

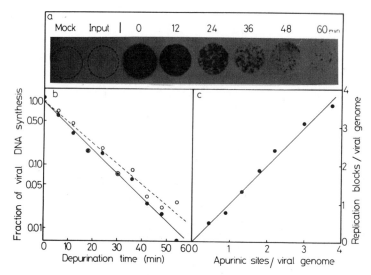

Figure 3:   Inhibition of MVM DNA replication by depurination.
A: Measurement of viral DNA synthesis by in situ hybridization. Cultures of mouse A9 cells were infected with 0 (Mock) or 0.3 (others)
   MVM plaque-forming-units/cell, as described previously (12). Prior
   inoculation, virus was exposed to depurination conditions for increasing times(min), as indicated. Cells were analyzed at 2
   (Input) and 24 (others) hours post-infection to measure parental
   and amplified viral DNA, respectively. Infected cells were trapped
   on nitrocellulose filters (1.5-10^5 cells/2.5 cm filter), denatura-
   ted and hybridized against ^{32}P-labeled MVM DNA (2x10^8 cpm/μg;
   1.5x10^6 cpm/ml), essentially as described by Winocour and Keshet
   (14). The autoradiogram of the filters is shown. Since all filters
   were incubated in the same hybridization mixture, the comparison
   of their radioactivities gives the relative amounts of intracellular viral DNA.
B: Residual MVM DNA replication with increasing time of depurination.
   The amount of viral DNA synthesized was quantitated by liquid
   scintillation counting of the filters (see A). A9 host cells were
   exposed (-o-) or not (-●-) to UV-light (5 Jm^{-2}) 15 h prior to infection.
C: Number of DNA replication blocks as a function of the number of
   apurinic sites in the viral genome. The average number of replication blocks was calculated from the residual fraction of viral
   DNA synthesis in unirradiated cells, assuming a Poisson distribution. The straight line is the theoretical function expected if
   every apurinic site constitutes a block to DNA replication.

lar DNA replication, our results also suggest that in vitro incorporation data might not be extrapolated directly to predict lethal and
mutagenic effects of DNA lesions in vivo.

UV-enhanced reactivation of depurinated MVM

We showed recently that UV-light enhances the ability of mouse
cells to reactivate the infectivity (12) and DNA replication (6) of
UV-damaged MVM. UV-enhanced reactivation of MVM is optimal for
sublethal UV-doses given to cells 15 h prior to virus infection.
These irradiation conditions were used in the present study to investigate whether mammalian cells could be induced to tolerate a
fraction of AP sites in MVM DNA. As shown in Figure 3B, the replication of depurinated MVM DNA was inhibited to a lesser extent in
cells pre-exposed to UV-light than in control cultures. It was calculated that 10-15% of viral DNA replication blocks were overcome
as a result of the cell inducing treatment (Table 1). Consistently,
depurinated MVM recovered from a similar proportion of lethal hits
when virus infectivity was measured in UV-irradiated cells (data
not shown).

Therefore UV-irradiation appears to generate intracellular
conditions which enhance the cell competence in accommodating AP
sites. These conditions might prevent the conversion of a fraction
AP site into less tolerated secondary products. Alternatively, UV-
light might modulate DNA replication or repair activities. An intri-
guing possibility would be that UV-irradiation induces the replicating
complex to behave like purified DNA polymerases, i.e. to copy past
some AP sites. A parallel can be drawn between our data and features
of depurination-dependent mutagenesis in prokaryotes. AP sites in
single-stranded phage DNA are not detectably mutagenic unless the
host bacterias are induced to express the SOS regulatory system, in
particular by UV-irradiation (3). It was inferred that UV-light
activates a function allowing bacteria to replicate a small fraction

Table 1: UV-enhanced reactivation of lethal hits and DNA replication
          blocks caused by depurination

DNA replication blocks / virus / min[a]

Normal cells	Induced cells[b]	% reactivation[c]
0.088 ± 0.004	0.077 ± 0.006	13 ± 6

---

[a] Calculated from the slope of the inactivation curve of viral DNA
    synthesis (Figure 3). Average values and standard deviations from
    3 experiments.
[b] Cells were UV-irradiated (5 Jm^{-2}) 15 h prior to virus infection.
[c] % blocks overcome as a result of cell pre-irradiation.

of AP sites, such as a bypass activity (15). However, in contrast
with our results, the fraction of AP sites tolerated as a result of
UV-irradiation of bacteria, is too small to cause a significant increase in the survival of depurinated phages (3). It remains to be
determined whether this discrepancy is merely quantitative or reflects
more basic differences between pro- and eukaryotic responses to AP
sites.

REFERENCES

1. Lindahl,T. and Nyberg,B. (1972) Biochemistry 11, 3610.
2. Singer,B. (1979) J. Natl. Cancer Inst. 62, 1329.
3. Schaaper,R.M. and Loeb,L.A. (1981) Proc. Natl. Acad. Sci. USA
        78, 1773.
4. Kunkel,T.A., Schaaper,R.M. and Loeb,L.A. (1983) Biochemistry
        22, 2378.
5. Defais,M.J., Hanawalt,P.C. and Sarasin,A.R. (1983)
        Adv. Radiat. Biol. 10, 1.
6. Rommelaere,J. and Ward,D.C. (1982) Nucleic Acids Res.
        10, 2577.
7. Ward,D.C. and Tattersall,P.J. (1982) in: "The Mouse in
        Biomedical Research", Academic Press, New York.
8. Hardt,N., Dinsart,C., Spadari,S., Pedrali-Noy,G. and
        Rommelaere,J. (1983) J. gen. Virol. 64, in press.
9. Rose,J.A. (1975) Comprehensive Virology 3, 1.
10. Astell,C.R., Thomas,M., Chow,M.B. and Ward,D.C. (1982)
        Cold Spring Harbor Symp. Quant. Biol. 47, in press.
11. Teebor,G.W. and Frenkel,K. (1983) Adv. Cancer Res. 38, 23.
12. Rommelaere,J., Vos,J-M., Cornelis,J.J. and Ward,D.C. (1981)
        Photochem., Photobiol. 33, 845.
13. Teebor,G.W. and Brent,T.P. (1981) in: "DNA repair", E.C.
        Friedberg and P.C. Hanawalt, eds., M. Dekker Inc.,
        New York and Basel.
14. Winocour,E. and Keshet,I. (1980) Proc. Natl. Acad. Sci. USA
        77, 4861.
15. Schaaper,R.M., Kunkel,T.A. and Loeb,L.A. (1983) Proc. Natl.
        Acad. Sci. USA 80, 487.

THE GENETICS OF ADENO-ASSOCIATED VIRUS

Nicholas Muzyczka, Richard J. Samulski,
Paul Hermonat, Arun Srivastava and Kenneth I. Berns

Department of Immunology and Medical Microbiology
University of Florida College of Medicine
Gainesville, Florida 32610, USA

INTRODUCTION

Adeno-associated virus (AAV) is a defective parvovirus which
is absolutely dependent on coinfection with a helper virus for a
productive lytic cycle (for a review, see ref. 1). Either adenoviruses
or herpes viruses can act as helpers (2-5). In the absence of a helper
virus, AAV efficiently integrates into host cell chromosomes via its
inverted terminal repeats (6, 7). Integrated AAV genomes are essentially genetically stable and do not express their genes (8). An interesting feature of AAV biology is the fact that when a cell line
that is latently infected with AAV is superinfected with a helper
virus, the integrated AAV genome is rescued and proceeds through a
normal lytic cycle (6, 8). It is likely that the inverted terminal
repeats of AAV are involved in the rescue of the genome as well as
its integration, but as yet relatively little is known about these
processes. There is substantial evidence, however, to support the
idea that the AAV terminal repeat is the origin for DNA replication
(1, 9-11) and a reasonable model (1, 9) which describes the mode of
AAV DNA replication has been proposed (Figure 1). In the model the
terminal repeat, which contains palindromic sequences, acts as a
hairpin primer to initiate DNA replication. The hairpin is resolved
by cleavage of the parental strand opposite the primer position. This
is followed by repair synthesis to produce a complete duplex DNA molecule. In the next round of synthesis, the terminal repeat sequence is
denatured so that the hairpin is reformed and DNA synthesis can proceed. This leads to the displacement and packaging of a single-stranded progeny molecule (12, 13). The enzymes that are presumably involved in this model, namely a site-specific nuclease, a DNA poly-

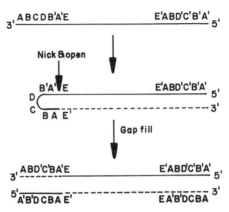

Figure 1:   Model for AAV DNA replication. C and D represent two different palindromic sequences which are themselves within a larger palindrome, AB...B'A'. DNA synthesis proceeds from the 3'-terminal hairpin to the end of the molecule, essentially forming a "shortened" DNA molecule. A putative site-specific enzyme makes a single nick on the parental strand opposite the original 3' end of the parental strand. This results in inversion of the terminal repeat and transfer of the parental palindromic sequence to the daughter strand. The parental DNA is then extended to give the full-length duplex genome.

merase and an enzyme which destabilizes the terminal sequences, have not been identified. Indeed, it is not known whether these activities are coded within the host, AAV, or helper genomes. With respect to adenovirus helper activity, it appears that the adenovirus E1B and E4 regions are likely to be directly involved in AAV duplex DNA replication (14-19). Adenovirus early regions which code for the adenovirus DNA polymerase or terminal protein do not appear to be essential.

Because AAV does not plaque by itself, it has been difficult to study the genetics of AAV. In an attempt to circumvent this problem, we have cloned AAV in the procaryotic vector pBR322. When the recombinant AAV/pBR322 plasmid, pSM620, is transfected into human cells with an adenovirus coinfection, the AAV sequences are rescued and replicate normally (20, 21). The frequency of this event is comparable to that seen in cell lines which are latently infected with AAV. Thus, the recombinant plasmid itself is a model for the AAV latent infection. We have taken advantage of this fact by constructing mutations within the AAV terminal and non-repeated internal sequences and examining their phenotypes.

RESULTS AND DISCUSSION

Mutations within the terminal repeat sequences of AAV

        Two types of terminal deletion mutants have been isolated (21):
mutants with deletions in either one or the other terminal repeat
and mutants with deletions in both terminal repeats. To examine the
phenotype of these mutants, we transfected recombinant plasmids into
human cells (HeLa or Ad transformed 293 cells) which were then coin-
fected with adenovirus (Ad2) virions. When we transfected the mutants
which had a deletion in only one of the terminal repeats (Table 1
and Figure 2, pSM621 and pSM802), the AAV sequences were rescued,
replicated, and infectious virions were produced. Examination of the
AAV virion DNA produced in these transfections by restriction enzyme
analysis indicated that the rescued projeny DNA was identical to wild
type DNA (not shown). Mutants which had extensive deletions in both
termini, on the other hand, were defective for rescue and/or replication (Table 1 and Figure 2, pSM703). We concluded that the terminal
sequences were essential for rescue or replication of the AAV genome,
but that only one intact terminal repeat was needed to rescue. In
addition, we concluded that the terminal repeats of AAV must be capable of interacting with each other by some mechanism which allowed
the correction of the missing sequences in the deletion mutants.

Table 1: Properties of recombinant plasmids

| Plasmid | Deletion (bp) | | Virus[b] |
	Left	Right	
pSM620	0[a]	0[a]	+
pSM621	80	0	+
pSM802	0	100	+
pSM703	100	95	−
pSM704	116	90	−
pSM803	111	83	−
pSM609	113[a]	9[a]	+
pSM1205	113[a]	20[a]	−

[a]In these mutants the deletion has been determined by DNA sequencing.
In the remaining mutants the size of the deletion was inferred from
restriction enzyme analysis.

[b]Production of virus was assayed as described in reference 21.

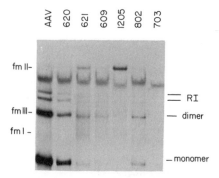

Figure 2: Rescue of AAV DNA from terminal deletion mutants. One microgram of covalently closed circular recombinant plasmid DNA, or 0.5 μg of wild type duplex linear AAV DNA was transfected into HeLa cells in the presence of Ad2 helper virus. After 48 hr, low molecular weight DNAs were isolated by Hirt extraction and were fractionated on 1.4% agarose gels. The DNA was transferred to nitrocellulose by Southern blotting and hybridized to nick-translated ^{32}P-labeled AAV DNA. Each lane represents the DNA recovered from one 10 cm dish. FmI is the position of form I covalently closed circular DNA (input DNA); FmII and FmIII are the positions of nicked circular and linear plasmid DNA respectively. Monomer and dimer refer to the forms of linear duplex AAV DNA.

The most instructive mutant which was examined was the double mutant pSM609 (Figure 3). The mutant contained a deletion of the first 113 base pairs (bp) of the left terminal repeat and the first 9 bp at the right end. In spite of this the mutant could be rescued (Figure 2) and the DNA sequence was corrected to that of wild type AAV as judged by both restriction enzyme analysis (Figure 4) and DNA sequencing (Figure 5). This presumably occurred in spite of the fact that the two deletions overlapped by 9 bp (Figure 3) because the 9 bp of overlap is a portion of the terminal palindrome. Thus, there are a total of 4 copies of this 9 bp sequence in wild type AAV DNA of which two copies are retained in pSM609.

In order to be certain that we were observing a gene conversion event we engineered a mutant, pSM1205, in which it should have been impossible to generate the wild type AAV DNA sequence. This was done by removing an additional 11 bp from the right terminus of the parental plasmid pSM609 (Figure 3). The 11 bp palindromic sequence was then completely misssing from both termini of pSM1205 and as expected pSM1205 could not generate wild type virions (Table 1) and was severely deficient in rescue and replication (Figure 2).

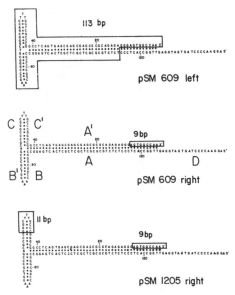

Figure 3: Nucleotide sequences of the inverted terminal repeats of pSM609 and pSM1205. The terminal AAV sequences are represented as a single-stranded sequence in a form which illustrates the position and size of the three palindromes (A, B, C) within the terminal repeat. The remaining D sequences are not palindromic but are part of the total inverted terminal repetition. The boxes indicate the positions of the deletions in the left and right termini of pSM609 and the right terminus of pSM1205. The DNA sequence of pSM1205 is identical to that of pSM609, its parent, except for an additional 11 bp deletion in the C palindrome at the right end of pSM1205. By convention (27) the flip orientations contain the terminal palindromic sequences in the order 3'A'B'BC'CAD'5'. The right ends of both pSM609 and 1205 are in the flop orientation (i.e., 3'A'C'CB'BAD'). Because of the magnitude of the deletion in the left end of pSM609 and 1205, no orientation can be assigned. For comparison, the sequence of the pSM609 left end is written as if it were in the flip orientation.

Our studies of pSM609 suggest several things about the possible mechanism by which AAV DNA is rescued and corrected. First, because both of the AAV terminal sequences (at the junctions with pBR322) have been deleted in pSM609, it is unlikely that the terminal AAV sequences themselves direct the initial steps in the rescue of the genome. Thus, either a recognition sequence elsewhere in the terminal repeat or some aspect of the secondary structure of the AAV terminal sequence directs the rescue events. Second, because the efficiency with which the deletion mutants are rescued (Figure 2) is quite high (1/2 to 1/20 that of wild type) it is unlikely that recombination

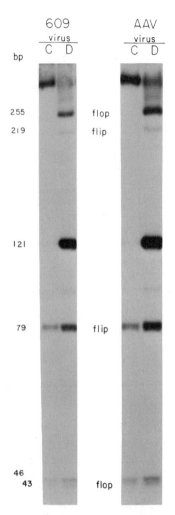

Figure 4: Characterization of the AAV PstI terminal fragments from
rescued pSM609 virion DNA. The AAV PstI C (left end) and PstI D
(right end) terminal fragments were isolated from rescued pSM609
viral DNA (left panel) or AAV duplex DNA (right panel). The fragments were then digested with BglI, labeled at the 3' ends with ^{32}P,
and fractionated on an 8% polyacrylamide gel. Flip and flop indicate
the fragments which are generated from these two orientations.

between two or more AAV/pBR322 plasmids is responsible for correcting
the terminal deletions. It is likely, in fact, that the correction
phenomenon is the result of an intermediate which occurs frequently
during the normal course of DNA replication. Finally, because the

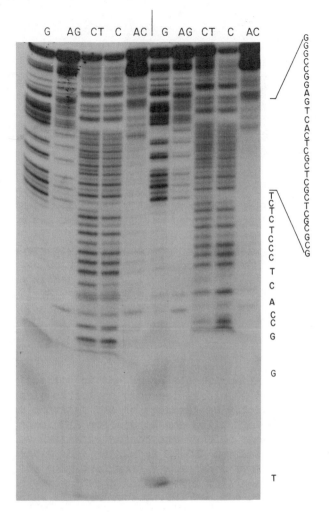

Figure 5:   Terminal sequence of the left and right ends of AAV virion DNA rescued from pSM609. Both the flip and flop orientations of the terminal BglI fragments were isolated independently from each end of rescued AAV virion DNA following transfection with pSM 609. The 5' end-labeled fragments were strand separated and sequenced. The sequence of the left and right flip fragments and the left and right flop fragments was identical to those of wild type AAV DNA (27). Only the flip orientations are shown in the figure.

terminal sequences of pSM609 have retained only 3 bp of tandem repetition, it would be difficult to explain AAV rescue by invoking homologous recombination between the tandemly terminal sequences. In-

deed, the products of such a recombination event, namely AAV and
pBR322 DNA molecules, have not been observed. With these considerations in mind, we have illustrated one of several possible models
for rescue and correction of AAV DNA (Figure 6). A key feature model
is the formation of a single-stranded pan-handle intermediate (bottom
of Figure 6) by hydrogen bonding between the terminal sequences. This
intermediate has been suggested to occur during normal AAV DNA replication (22) and could account for the correction of AAV terminal se-

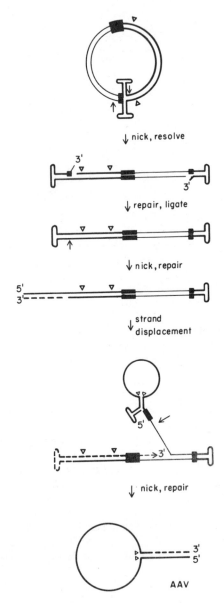

quences. It is worth noting that similar intermediates have been
suggested to occur during the replication of two other virus families,
herpes (23) and adenoviruses (24), and both of these viruses are capable of efficiently correcting mutations within their terminal repeats (25, 26). Thus, the formation of this intermediate may be a
general mechanism for the maintenance of the terminal sequences of
mammalian DNA viruses.

Mutations within the coding regions of AAV

Because AAV does not plaque by itself it has been very difficult to isolate AAV mutants. This has left some basic questions about
AAV biology unanswered, namely, how many genes does AAV have and are
any of them involved in DNA replication. Two recent developments have
essentially eliminated this impasse. The first is the determination
of the complete nucleotide sequence of AAV by Srivastava et al. (27)
and the second is the discovery that AAV recombinant plasmids are
almost as infectious as AAV duplex DNA isolated from virions (20).
We have been taking advantage of these two developments by construc-

*Figure 6:   A model for the rescue and correction of AAV sequences
from the recombinant plasmid pSM609. In the diagram, the thin lines
represent pBR322 sequences and the thick lines represent AAV sequences. The long and short black boxes indicate the relative positions
of the large left hand (113 bp) and small right (9 bp) deletions of
pSM609. They represent sequences, therefore, which are not present
in pSM609 and must be corrected. The triangles represent the end of
the total inverted repetition (145 bp) of AAV and the arrows represent the position of nicks. In the first step of the model the circular plasmid is converted to a linear molecule by virtue of nonspecific nicks near the right hand AAV/pBR322 junction. The nature
of the enzyme(s) which makes these nicks is unknown. However, an interesting possibility relies on the fact that even though there is
a small deletion (9 bp) at the left hand pBR322/AAV junction, most
of the AAV palindromic sequences remain intact. Thus, if the cruciform structure shown in the figure can form, it might be recognized
as a Holliday structure leading to nicks near the base of the cruciform. In the next step, the small right hand deletion is repaired
by using its complementary sequence as a template. A second nick
then occurs on the parental strand which leads to the synthesis of
a normal duplex end and the inversion of the terminal palindromic
sequences. The second nick as well as the inversion of the terminal
palindrome have been suggested to occur during normal AAV DNA synthesis in lytic infections (9). In the next round of synthesis, a single
strand of the AAV/pBR322 plasmid is displaced allowing the formation
of a panhandle intermediate. The 5' strand in the panhandle intermediate serves as a template for correction of the deleted 3' strand.*

ting mutations within the known AAV potential coding regions and by examining the phenotypes of these mutants following DNA transfection into human cells. The preliminary results can be summarized as follows. There are two major open reading frames in the AAV genome and several minor ones (27). The two largest reading frames occupy approximately half of the genome each but both of these reading frames are represented in more than one messenger RNA species. Therefore, both potential coding regions are likely to be represented in more than one protein. The major open reading frame on the right side of the genome almost certainly codes for the three AAV capsid proteins, although it is not yet clear how the three proteins are synthesized from the same coding region (27). Mutants within this coding region are generally defective in the production of single-stranded progeny DNA and infectious virions. They do, however, rescue from the recombinant plasmid and are capable of DNA replication. Thus, capsid mutants of AAV behave much the same as capsid mutants in the single stranded bacteriophage ϕX174 (28). Mutants within the left-hand open reading frame were found to be generally defective for rescue and replication of AAV. Although it is not known yet how many proteins the replication defective mutants define, it is now clear that there is at least one AAV coded protein involved in DNA replication.

In summary, the cloning of AAV has facilitated the genetic analysis of AAV and a preliminary genetic map of AAV will be available soon. In light of the heavy dependence of AAV on host and helper virus proteins for its life cycle, it would seem that AAV and DNA replication is now ripe for the kind in vitro enzymology that has been so successful in the study of single-stranded prokaryotic virus DNA replication.

REFERENCES

1.   Berns,K.I. and Hauswirth,W.W. (1979) Adv. Virus Res. 25, 407.
2.   Atchison,R.W., Casto,B.C. and Hammon,W.McD. (1965)
          Science 194, 754.
3.   Hoggan,M.D., Blacklow,N.R. and Rowe,W.P. (1966)
          Proc. Natl. Acad. Sci. USA 55, 1457.
4.   Parks,W.P., Melnick,J.L., Rongey,R. and Mayor,H.D. (1967)
          J. Virol. 1, 171.
5.   Buller,R.M.L., Janik,J.E., Sebring,E.D. and Rose,J.A. (1981)
          J. Virol. 40, 241.
6.   Hoggan,M.D., Thomas,G.F. and Johnson,F.B. (1972)
          Proc. 4th Lepetit Colloq., North-Holland, Amsterdam, 243.
7.   Cheung,A.K-M., Hoggan,M.D., Hauswirth,W.W. and Berns,K.I.
          (1980) J. Virol. 33, 738.
8.   Berns,K.I., Cheung,A.K-M., Ostrove,J.M. and Lewis,M. (1982)
          in: Virus Persistence, Hurison,A.C. and Barby,G.K.,
          eds., Massachusetts, Cambridge University Press, p.p.249.
9.   Straus,S.E., Sebring,E. and Rose,J.A. (1976) Proc. Natl. Acad.
          Sci. USA 73, 742.

10.   Lusby,E., Fife,K.H. and Berns,K.I. (1980) J. Virol. 34, 402.
11.   Hauswirth,W.W. and Berns,K.I. (1977) Virology 79, 488.
12.   Rose,J.A., Berns,K.I., Hoggan,M.D. and Koczot,F.J. (1969)
              Proc. Natl. Acad. Sci. USA 64, 863.
13.   Mayor,H.D., Torikai,K., Melnick,J. and Mandel,M. (1969)
              Science 166, 1280.
14.   Berns,K.I., Muzyczka,N. and Hauswirth,W.W. (1984) in:
              Human Viral Diseases, B. Fields et al., eds., Raven
              Press, New York, in press.
15.   Laughlin,C.A., Jones,N. and Carter,B.J. (1982) J. Virol
              41, 868.
16.   Janik,J.E., Huston,M.M. and Rose,J.A. (1981) Proc. Natl.
              Acad. Sci. USA 78, 1925.
17.   Richardson,W.D. and Westphal,H. (1981) Cell 27, 133.
18.   Straus,S., Ginsburg,H. and Rose,J.A. (1976) J. Virol. 17, 140.
19.   Carter,B., Marcus,C., Laughlin,C. and Ketner,G. (1983)
              Virology 125, 505.
20.   Samulski,R.J., Berns,K.I., Tan,M., Muzyczka, N. (1982)
              Proc. Natl. Acad. Sci. USA 79, 2077.
21.   Samulski,R.J., Srisvastava,A., Berns,K.I. and Muzyczka,N.
              (1983) Cell 33, 135.
22.   Berns,K.I. and Kelly,Jr.,T.J. (1974) J. Mol. Biol. 82, 267.
23.   Roizman,B. (1979) Cell 16, 481.
24.   Lechner,R.L. and Kelly,Jr.,T.J. (1977) Cell 12, 1007.
25.   Stow,N.D. (1982) Nucl. Acids Res. 10, 5105.
26.   Knipe,D.M., Ruyechan,W.T., Honess,R.W. and Roizman,B. (1979)
              Proc. Natl. Acad. Sci. USA 76, 4534.
27.   Srivastava,A., Lusby,E.W. and Berns,K.I. (1983) J. Virol.
              45, 555.
28.   Hayashi,M. (1978) in: "The Single-Stranded DNA Phages",
              Denhardt,D.T., Dressler,D. and Ray,D.S. eds., Cold
              Spring Harbor Laboratory, pp. 531.

RELATIONSHIP BETWEEN THE ORGANIZATION OF DNA LOOP DOMAINS AND

OF REPLICONS IN THE EUKARYOTIC GENOME

Monique Marilley and Mario Buongiorno-Nardelli[1]*

LA 179 - Faculté des Sciences de Luminy
13288 Marseille cedex 9, France

[1]Department of Genetic, Facoltà di scienze M.F.N.
Università di Roma
Rome, Italy

The DNA in the eukaryotic nucleus has been found to be arranged in a series of supercoiled loops. This supercoiled loop-organization results from DNA attachment to a structure which is mainly proteic, the so-called nuclear matrix (1-9). This DNA organization remains stable throughout the cellular cycle as far as the size of DNA loops and the anchorage sites are concerned (5, 10, 11). It is worthwhile noting that this type of organization has been found in all cells whatever their state of differentiation (early embryonic stage up to fully differentiated cells (21)).

Strong evidence now suggests a functional involvement of super-coiled loop domains in basic chromosome functions, for instance: replication, expression, segregation of genetic material.

Concerning replication, several experiments (4, 6-9, 13, 15, 28) have suggested that DNA replication occurred on the nuclear matrix. These results led PARDOLL et al, 1980 (6) to formulate a dynamic model of replication, in which DNA replication occurred by DNA reeling through a couple of replicative complexes which were themselves fixed on the nuclear matrix. With this model, the multireplicon organization and the bidirectionality of replication may be easily explained.

On the other hand, there is more and more evidence suggesting a relationship between the organization of DNA in supercoiled loop domains and the organization for gene expression.

---

* This work is dedicated to the memory of Professor Mario Buongiorno-Nardelli.

Several sets of observations may be discerned:
- There is a relationship between potentially transcribed sequences and distance to the nuclear matrix (5, 16, 15).
- Nascent RNA is localized at the nuclear matrix (18).
- Topology of DNA and transcription efficiency are related (19-21).

In fact, we do not exclude the possibility that organization for DNA replication and organization for gene expression are related. We agree with LASKEY and HARLAND (22) that "the use of the same structural units for replication and transcription would offer a clear benefit for patterns of chromatin segregation". Indeed, it is suggestive to observe that, during Xenopus laevis development, we can follow parallel changes in replicon size and loop size at the precise time when big changes in expression patterns are occuring (12).

On these grounds was formulated a general model of replication (12), in which the loop organization and replicon organization may be superposed. This dynamic model (Figure 1) was contructed in such a way as to fulfil the following conditions:
- Replication is bidirectional.
- growing forks are localized on the nuclear matrix (4, 6-9, 13, 15, 28).
- There is a correspondence of size between loops and replicons. The mean size of a replicon is the size of two mean DNA loops (12).
- Origin regions are permanently bound to the nuclear matrix (23).

It may be seen (Figure 1) that, in order to suggest a way in which not only DNA sequences but also the DNA loop organization is replicated, it is sufficient to postulate a mechanism for duplication of anchorage sites of the matrix. In this way the conservation of loop organization in metaphasic chromosomes and the transmission of that organization from one cell to daughter cells may be easily explained.

First evidence of a relationship between DNA loop organization and replicon organization at the level of a specific gene (rDNA in Xenopus laevis) has been obtained. (This work will be published extensively elsewhere.)

Previous work (24) realised on rDNA from Xenopus laevis incicated that all (or at least most) of the repeating units of rDNA cluster contained an origin of replication. We therefore planned a dissection of the DNA loops in order to test, at the level of a specific gene, the relation between replicon organization and loop organization.

One of the strategies used to make this molecular dissection of the loop, consisted in introducing a single cut into each supercoiled DNA loop using S1 endonuclease from Aspergillus oryzae. This

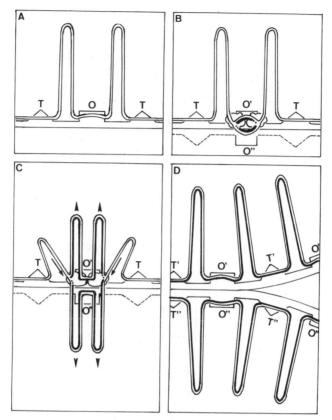

Figure 1:   General model for replication.
A – Before activation: One replicon is viewed as a double looped
    structure (anchorage site for origin ⌐‾‾⌐, anchorage sites
    for termination ⌐‾⌐).
B – Early DNA synthesis: A couple of replicative complexes (⬤ ⬤)
    are associated to the nuclear matrix. Growing forks (◁═⬤═▷)
    are localized on the nuclear matrix. Replication occurs bidirec-
    tionally. The new replicated origin binds to a newly formed an-
    chorage site.
C – Mid DNA synthesis: DNA synthesis occurs by reeling through repli-
    cative complexes. Origins remain attached to the matrix.
D – DNA synthesis has been completed. Termination anchorage sites
    have been duplicated. The two-looped replicon organization has
    been restituted. The newly formed matrix may segregate from the
    old one. DNA sequences are replicated, the organization has been
    duplicated concomitantly.

enzyme is known to introduce a site-specific cut in covalently-closed
supercoiled circular DNA (25). This action is explained by the presence of unstably paired regions resulting from tension accumulated

in the domain. Since one may consider that, from a topological point of view, a closed DNA loop resembles a circle, we performed S1 digestion on erythrocyte nucleoids from Xenopus laevis. That is, our experiment was performed on negatively supercoiled DNA anchored on a residual proteinaceous structure that was verified to maintain the 90 Kb loop organization characteristic of erythrocytes from Xenopus laevis. S1 digestion was followed by isolation of DNA from the nuclear matrix. The linear fragments of DNA obtained in this way separated according to their size by agarose gel electrophoresis. DNA was transferred to nitrocellulose filters (26) and then hybridized with a radioactive rDNA homologous probe (Figure 2). In parallel, total nuclear DNA from the same individual was treated by Hind III restriction nuclease and run on agarose gel. Both experiments gave a band hybridizing with the rDNA probe, in the range of 15 Kb (the exact size depends on the individual used for experimentation).

Figure 2:   Lane A: Supercoiled DNA (nucleoids from Xenopus laevis erythrocytes) was cut with S_1 nuclease. After cleavage, DNA was purified, run on agarose gel 0.3%, transferred to nitrocellulose filter and then probed with rDNA. The resulting autoradiogram is shown.
Lane B: idem A, without S_1. Non digested rDNA is localized on high molecular weight fragments (>200 Kb). After S_1 digestion, rDNA hybridizes to a band whose size is the size of one unit of repetition. The thickness of this band is due to heterogeneity in length of the repeating units (heterogeneous molecular weigth of the non-transcribed spacer). There is no band of the size of 2 or 3 repeats. Some material maintains a high molecular weight.

Since it is known that there is one Hind III site per unit of repetition in rDNA from <u>Xenopus laevis</u>, we may conclude that all (or at least most) of the rDNA tested had a monogenic loop organization whose size is the size of a replicon.

This first result gives new support to the idea that loop organization and replicon organization are strongly related. However, the size relationship expected from the general model is not the one found in rDNA material, since we found that rDNA replicon had the same size as rDNA loop instead of the size of two DNA loops.

To explain this result we may envisage two possibilities:
1: rDNA in non-dividing cells(erythrocytes) has a different organization from rDNA of cells maintaining the ability to divide.
2: rDNA behaves differently from "bulk" DNA.

For lack of space we will limit the discussion here to the second point.
- The size of an rDNA loop is very different from that of "bulk" DNA: about 15 Kb instead of 90 Kb.

This result is indirectly confirmed by an observation made by PARDOLL and VOGELSTEIN (27). These authors working on "sequence analysis of nuclear matrix associated DNA from rat liver" found that, in comparison with bulk DNA, genes for rRNA are significantly enriched in the residual matrix DNA. They tentatively explained their result by the fact that genes for rRNA may have more attachment per unit length of DNA than the bulk of DNA. Their experimental result gave a 6-fold enrichment. This is the exact amount we can predict from our experiments (90 Kb/15 Kb = 6).

- rDNA replicon is organized in a one-loop structure; "bulk" DNA seems to be organized in a double-loop structure.

This result is consistent with E.M.measurement of replicon sizes. The values published by BOZZONI et al. (24) concerning <u>Xenopus laevis</u> (differentiated cells) distinguished a value of about 60 μm for "bulk" DNA and a value which is the same as that of a unit of repetition for rDNA (which may be about 5 μm for a 15 Kb unit). In this case the ratio between the two sizes is 12, and to obtain the same ratio between loop-sizes we must speculate a double loop organization for "bulk" DNA (90 Kb x 2 = 180 Kb, 180 Kb/15 Kb = 12).

In conclusion, these results indicate that differently specialized parts of the genome (rDNA opposed to bulk DNA) may have a different loop organization. Moreover, our results lead us to envisage the possibility that rDNA and "bulk" DNA do not follow the same pattern for replication.

REFERENCES

1. Laemmli,U.K., Cheng,S.M., Adolph,K.W., Paulson,J.R., Brown,J.A.
      and Baumbach,W.R. (1977) Cold Spring Harbor Symp. Quant.
      Biol. 42, 351.
2. Comings,D.E. (1978) in: the Cell Nucleus (H. Bush ed.)
      Acad. Press, New York, 4, 345.
3. Georgiev,G.P., Nedospasov,S.A. and Bakayev,V.U. (1978) in:
      the Cell Nucleus (H. Bush ed.) Acad. Press, New York,
      6, 3.
4. Shaper,J.H., Pardoll,D.M., Kaufmann,S.H., Barrack,E.R.,
      Vogelstein,B. and Coffey,D.S. (1979) in: Advances in
      Enzyme Regulation (G. Weber ed.), Pergamon Press, Oxford,
      17, 213.
5. Cook,P.R. and Brazell,I.A. (1980) Nucl. Acids Res. 8, 2895.
6. Pardoll,D.M., Vogelstein,B. and Coffey,D.S. (1980) Cell, 19,527.
7. Berezney,R. and Buchholtz,L.A. (1981) Exp. Cell Res. 132, 1.
8. McCready,S.J., Godwin,J., Mason,D.W., Brazell,I.A. and Cook,P.R.
      (1980) J. Cell Sci. 46, 365.
9. Vogelstein,B., Pardoll,D.M. and Coffey,D.S. (1980) Cell, 22, 79.
10. Razin,S.V., Mantieva,V.L. and Georgiev,G.P. (1979)
      Nucl. Acids Res. 7, 1713.
11. Wray,V.P., Elgin,S.C.R. and Wray,W. (1980) Nucl. Acids Res.
      8, 4155.
12. Buongiorno-Nardelli,M., Micheli,G., Carri,M.T. and Marilley,M.
      (1982) Nature, 298, 100.
13. Dijkwel,P.A., Mullenders,L.H.F. and Wanka,F. (1979)
      Nucl. Acids Res. 6, 219.
14. Nelkin,B.D., Pardoll,D.M. and Vogelstein,B. (1980)
      Nucl. Acids Res. 8, 5623.
15. Buckler-White,A.J., Humphrey,G.W. and Pigiet,V. (1980)
      Cell, 22, 37.
16. Robinson,S.I., Nelkin,B.D. and Vogelstein,B. (1982)
      Cell, 28, 99.
17. Cook,P.R., Lang,J., Hayday,A., Lania,L., Fried,M., Chiswell,D.J.
      and Wyke,J.A. (1982) EMBO J. I, 447.
18. Jackson,D.A., McCready,S.J. and Cook,P.R. (1981) Nature,
      292, 552.
19. Smith,G.R. (1981) Cell, 24, 599.
20. Carnevali,F., Caserta,M. and Di Mauro,E. (1983) J. Mol. Biol.
      165, 59.
21. Pedone,F., Felicetti,P. and Ballario,P. (1982) Nucl. Acids
      Res. 10, 5197.
22. Laskey,R.A. and Harland,R.M. (1981) Cell, 24, 283.
23. Carri,M.T. and Buongiorno-Nardelli,M. Exp. Cell. Res., in press.
24. Bozzoni,I., Baldari,C.T., Amaldi,F. and Buongiorno-Nardelli,M.
      (1981) Eur. J. Biochem. 118, 585.
25. Lilley,D.M.J. (1980) Proc. Nat. Acad. Sci. USA, 77, 6468.
26. Southern,E.M. (1975) J. Mol. Biol. 98, 503.
27. Pardoll,D.M. and Vogelstein,B. (1980) Exp. Cell Res. 128, 466.
28. Hunt,B.F. and Vogelstein,B. (1981) Nucl. Acids Res. 9, 349.

APHIDICOLIN AND EUKARYOTIC DNA SYNTHESIS

Silvio Spadari, Francesco Sala and Guido Pedrali-Noy

Istituto di Genetica Biochimica ed Evoluzionistica
del C.N.R.
Via S. Epifanio 14
27100 Pavia, Italy

INTRODUCTION

Animal cells are endowed with two sites containing DNA: nucleus and mitochondrion. Plant cells are more complex in this respect: they contain also plastids. The DNAs of nucleus, mitochondrion and plastid differ in structural organization as well as in the mode of replication. Furthermore both animal and plant cells are able to repair damaged nuclear DNA. So it is not surprising that different DNA polymerases are involved in the different types of DNA synthesis occurring in the three distinct cellular site (1).

Great efford has been devoted in the past 10-15 years to the characterization of the DNA polymerases present in animal cells. They contain at least three DNA polymerases named α, β and γ (1-2). A fourth DNA polymerase found in calf thymus and erythroid hyperplastic rabbit bone marrow, named δ, resembles DNA polymerase α but contains an intrinsic 3'→5' exonuclease with "editing properties" similar to those of prokaryotic DNA polymerases (2). DNA polymerases α and β are found in the nucleus. DNA polymerase γ is the mitochondrial enzyme.

Plant cells have been studied more recently. They contain a DNA polymerase α (3), a chloroplast γ-like DNA polymerase (4), a mitochondrial DNA polymerase (5, 6) and, possibly, a β-like enzyme, associated with chromatin (7).

This paper (i) reviews the effect of aphidicolin (8), a specific inhibitor of DNA synthesis in eukaryotic cells (9, 10) on DNA poly-

merases purified from animal and plant cells and on their activity
in intact cells; (ii) describes studies on the use of aphidicolin to
synchronize the cell cycle of animal and plant cells grown in culture
or in vivo.

Aphidicolin inhibits growth and DNA replication of eukaryotic cells without interfering with RNA and protein synthesis

Aphidicolin inhibits the growth of both animal and plant cells
(10). The addition of the drug to growing cells results in an immediate inhibition of DNA synthesis (Figure 1). In fact, in the presence
of 5 μM aphidicolin, the synthesis of DNA, as assessed by measuring
^{3}H-thymidine incorporation into acid-insoluble material, is depressed
to approximately 2-3% of the control in animal cells, or to approximately 15-20% in plant cells. As shown in the next paragraph the residual aphidicolin-resistant DNA synthesis is ascribable to mitochondrial (in animal cells) or to both mitochondrial and plastid (in plant
cells) DNA synthesis.

It is of interest to notice that, in the case of animal cells,
higher backgrounds of thymidine incorporation in the presence of inhibiting concentrations of the drug were observed only when cells were
contaminated by mycoplasma: in this case, although nuclear DNA synthesis is inhibited in the presence of aphidicolin, perinuclear incorporation of ^{3}H-thymidine continues, due to mycoplasma activity (11).

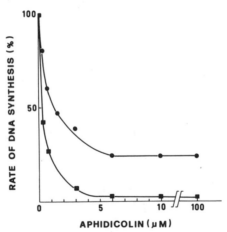

Figure 1: Effect of increasing concentrations of aphidicolin on the
rate of DNA synthesis in HeLa (■——■) and Oryza sativa (●——●) cells
grown in suspension culture. The aphidicolin resistant incorporation
is due to mitochondrial DNA synthesis in HeLa cells (11) whereas in
the case of O. sativa this is due in part to organellar DNA synthesis
and in part to thymidine metabolization and subsequent utilization
for different cellular synthesis (12).

In the case of plant cells the aphidicolin insensitive ^{3}H-thy-
midine incorporation may also be higher, but this represents, in part,
the utilization of the precursor for synthesis other than DNA (12).

Unlike most other inhibitors of DNA synthesis the drug does not
interfere with other important metabolic activities of the cell, such
as RNA and protein synthesis (Figure 2) or with the synthesis of
nucleic acids precursors or with the accumulation of DNA polymerase
α itself (13).

Aphidicolin specifically inhibits the replicative DNA polymerase α of eukaryotic cells

Preliminary experiments ruled out direct binding of aphidicolin
to DNA, at variance from most other DNA synthesis inhibitors. Thus the
attention was pointed to the study of its effect on the DNA polymerases present in eukaryotic cells and indeed, the in vitro assays with

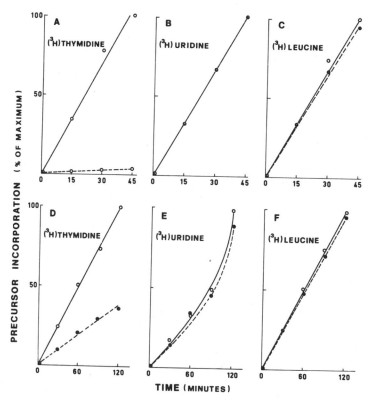

*Figure 2:   Effect of aphidicolin on DNA (A, D), RNA (B, E) and pro-
tein (C, F) synthesis in suspension cultured cells of HeLa (A, B, C)
and O. sativa (D, E, F). (●——●) 15 μM aphidicolin; (○——○) no aphidicolin. The aphidicolin resistant thymidine incorporation is accoun-
ted for as in Figure 1.*

purified enzymes have shown that aphidicolin specifically inhibits
DNA polymerase α in both animal and plant cells (Figure 3A and C)
but has no effect on the other nuclear DNA polymerase (the β-enzyme),
on the animal mitochondrial γ-polymerase (panel A) or on the cloro-
plast γ-like DNA polymerase (panel C). The plant cell mitochondrial
DNA polymerase is also resistant to the drug (6).

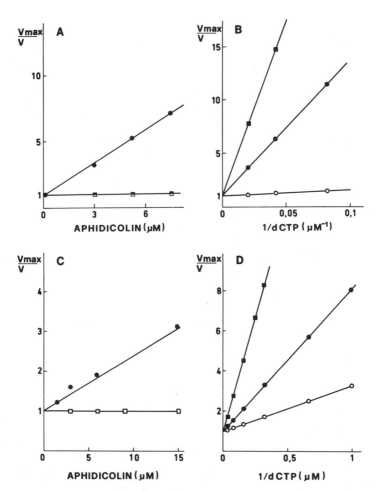

Figure 3:   Effect of aphidicolin on eukaryotic DNA polymerases (A
and C).   (A) HeLa cells DNA polymerases α (●——●), β (■——■) and γ
(□——□). (C) O.sativa cells DNA polymerase α (●——●) and spinach
chloroplast DNA polymerase γ-like (□——□). Double reciprocal plot of
the effect of aphidicolin on the polymerization rate of the HeLa cells
α-polymerase (B) and O.sativa cells α-polymerase (D) in the presence
of varying concentrations of (³H)dCTP. The other three non-radioactive
dNTPs were 100 μM.

Aphidicolin compets with dCTP for the binding to DNA polymerase α and inhibits both DNA-primed and RNA primed-DNA synthesis

Studies on the mechanism of action of aphidicolin have shown that the inhibition of DNA polymerase α is due to binding of the drug to the enzyme. The inhibition is non competitive with the DNA precursors dATP, dGTP, dTTP and with Mg^{++}, uncompetitive or non-competitive with DNA but competitive with dCTP (14 - 16). In fact, with increasing concentrations of dCTP the drug affects the apparent Michaelis constant and not the V_{max} of both animal and plant α-polymerase (Figure 3B and D). To distinguish between a pure and partially competitive type of inhibition the V_{max}/V values were plotted versus the aphidicolin concentration. This gave a linear plot thus indicating a pure competitive mode of inhibition (17).

These data, and others comparing the effect of aphidicolin and of arabinosyl CTP (an inhibitor of DNA polymerases which truly competes with dCTP) on several DNA or RNA-directed or template-independent bacterial, bacteriophage, viral, animal and plant DNA polymerases (15), involved in DNA replication or repair have allowed us to conclude that most likely aphidicolin recognizes a binding site which is specific for the nuclear replicative DNA polymerase of eukaryotes: this binding site is probably so near to or even overlapping with the binding site for dCTP that the drug mimicks a competitive effect with dCTP although it shares no similarities with dCTP itself.

The results shown in Table 1 indicate that aphidicolin in vitro inhibits both DNA-primed and RNA-primed DNA synthesis. Therefore in vivo it could inhibit the initiation of new replicons as well as their elongation.

The lack of inhibition of DNA polymerase when utilizing synthetic templates (including poly dI:oligo dC where only dCMP is incorporated) suggests the hypothesis that the extent and mode of inhibition of DNA polymerase α by aphidicolin may depend on a particular conformation that the enzyme assumes upon interaction with the template. This is also supported by the kinetics of inhibition of DNA polymerase α at increasing DNA concentrations: the inhibition is non-competitive at high and uncompetitive at low concentrations (15). It is therefore possible that at high DNA concentrations (as is the case inside the nucleus) the enzyme is complexed to DNA and thus is in the proper conformation for the interaction with aphidicolin whereas the free form of DNA polymerase α, predominant at low DNA concentrations, does not react or reacts less efficiently with aphidicolin. If this hypothesis were true, it would also explain our failure in the attempt to bind DNA polymerase α to a Sepharose-aphidicolin column (unpublished results). In this case the aphidicolin covalently bound to Sepharose regained the ability to interact with and therefore inhibit DNA polymerase α when DNA was added to the assay (unpublished results).

Table 1: Effect of aphidicolin on the utilization of natural and synthetic DNA-primed or RNA-primed DNA templates by HeLa cell DNA polymerase α

Template-primer	(^{3}H) dNTP	divalent cation	Aphidicolin µg/ml	Inhibition %
NATURAL				
Activated DNA	dCTP (25 µM)	Mg	2	72
Activated DNA	dCTP (25 µM)	Mg	20	96
Activated DNA	dCTP (25 µM)	Mn	2	10
Activated DNA	dCTP (25 µM)	Mn	20	60
Activated DNA*	dCTP (2.4 µM)	Mg	0.1	50
Activated DNA	dCTP (2.4 µM)	Mg	0.4	75
RNA-primed-fd DNA*	dCTP (2.4 µM)	Mg	0.1	50
RNA-primed-fd DNA	dCTP (2.4 µM)	Mg	0.4	75
SYNTHETIC				
polydT-oligodA	dATP	Mn	2	0
polydT-oligorA	dATP	Mg	2	0
polydC-oligodG	dGTP	Mg	2	0
polydC-oligodG	dGTP	Mg	50	5
polydC-oligodG	dGTP	Mn	2	0
polydC-oligodG	dGTP	Mn	50	5
polydI-oligodC	dCTP	Mn	2	0

*DNA polymerase α was present at low concentration to allow the addition of only an average of one nucleotide to each 3'OH primer, thus permitting to measure the inhibition of the initiation process. In such conditions the specific acitivity of ^{3}H-dCTP was 10,000 cpm/pmol and 100 was 5 pmol.

Aphidicolin inhibits nuclear DNA replication but has no effect on
mitochondrial and chloroplast DNA synthesis

Direct demonstration of the specific effect of aphidicolin in
vivo has been obtained by examining autoradiographs of animal spleen
lymphocytes and plant rice cells upon ^{3}H-thymidine incorporation in
the presence of aphidicolin. The autoradiographs at the ligth microscope (panels C and G of Figure 4) show that the drug completely inhibits the synthesis of DNA occuring in the nuclei of animal and
plant cells (panel A and E).

To distinguish between nuclear and organellar DNA synthesis,
we have also examined autoradiographs at the electron microscope of
sectioned spleen lymphocytes (panel B and D of Figure 4) and rice
cells (panel F and H). By counting the percentage of labelled mitochondria (in animal cells) or of both mitochondria and chloroplasts
in plant cells, we have found no significant difference in the presence or absence of the drug.

These results correlate well with those obtained with the extracted enzymes and indicated that (i) DNA polymerase α is essential
for nuclear DNA replication. Minor contribution of DNA polymerase β
and γ to this process is not absolutely excluded; however these polymerases are unable to carry out significant nuclear DNA synthesis in
the absence of an active DNA polymerase α; (ii) the effect of aphidicolin is specific for nuclear DNA replication whereas the incorporation of the precursor into mitochondrial and plastid DNAs is not
affected; (iii) the elongation of mitochondrial and plastid DNA does
not depend on the activity of DNA polymerase α.

The effect of aphidicolin on DNA repair synthesis depends on physiological conditions

Conflicting results have appeared on the effect of aphidicolin
on DNA repair synthesis (10). This may be due to the fact that (i)
cells at different phases of the cell-cycle have been used in different experiments; (ii) DNA repair was assayed by different techniques which measure specific and different stages of the multi-step
DNA repair synthesis process; (iii) different types of damage have
been studied.

In the experiments where either exponentially growing HeLa
cells (panel C of Figure 5) or protoplasts isolated from leaves of
Nicotiana sylvestris (panel D) were UV-irradiated and DNA repair was
measured by autoradiography, we could not detect a significant inhibitory effect of the drug, as compared with unirradiated cells (panel
A) or protoplasts (panel B), respectively. Thus, at least in the
tested experimental conditions, DNA repair synthesis was not ascribable to DNA polymerase α and was most probably due to DNA polymerase β.

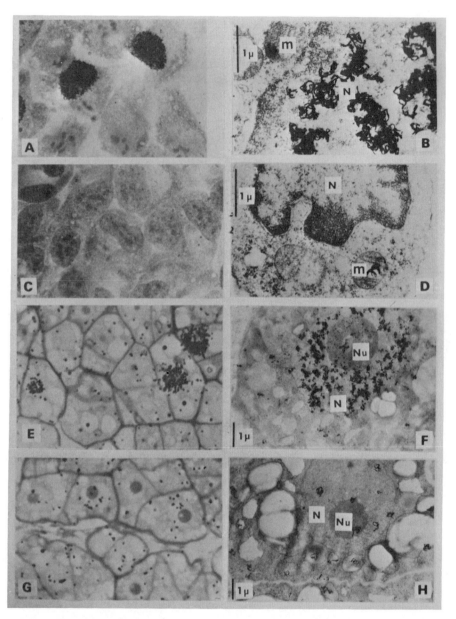

Figure 4: Effect of aphidicolin on nuclear and organellar DNA synthesis in eukaryotic cells. Autoradiographs of HeLa (A,C) and O.sativa
(E,G) cells incubated without (A,E) or with (C,G) aphidicolin. Electron microscope autoradiographs of sectioned cells of ConA stimulated
rabbit spleen lymphocytes (B,D) and of O.sativa (F,H) cells incubated
without (B,F) or with (D,H) aphidicolin. N = nucleus; Nu = nucleolus;
m = mitochondria. Autoradiographs were reproduced from Ref. 23 (A,C);
12 &E,G); 11 (B,D); 12 (F,H).

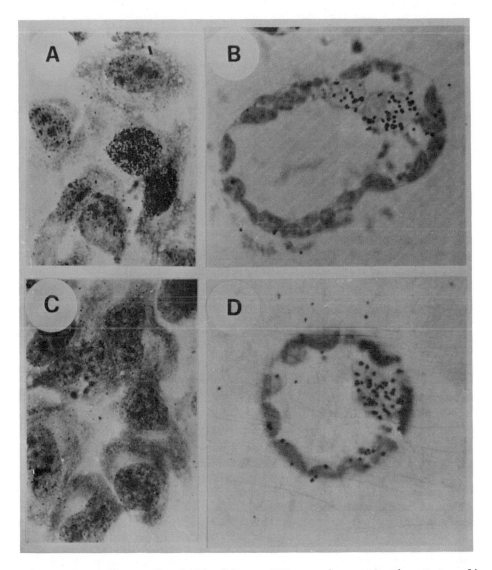

Figure 5: *Effect of aphidicolin on DNA repair synthesis. Autoradio-graphs of HeLa (A,C) cells and* <u>Nicotiana sylvestris</u> *(B,D) protoplasts incubated with (^{3}H)thymidine in the presence of aphidicolin. A,B: no irradiation; C,D: UV-irradiated. Autoradiographs A,C are reproduced from Ref. 23; autoradiographs B,D are reproduced from Ref. 24.*

On the other hand, significant inhibition of DNA repair synthesis by the drug was observed, in different experimental conditions, by other investigators, particularly when damaged quiescent cells (such as fibroblasts at confluency or lymphocytes from periferal

blood) were used (10). This suggested the involvement of DNA polymerase α in the assayed repair process. In fact sensitivity of DNA polymerase β to aphidicolin in these cells has been ruled out by the
insensitivity of the enzyme purified from HeLa cells, also when tested
at those low levels (down to 0.05 μM) of dNTPs that are physiologically found in quiescent cells.

Thus, it seems reasonable to postulate that both nuclear DNA
polymerases α and β are involved in DNA repair synthesis and that
physiological conditions, type of damage, cellular differentiation
and possibly other factors determine the extent of their involvement.

Mutants cells resistant to aphidicolin confirm the mechanism of action of the drug

Confirmation of the hypothesis that DNA polymerase α is the
main target for aphidicolin and that the drug competes with dCTP came
from the isolation of mutants containing an altered DNA polymerase α
either resistant to aphidicolin (18) or with an apparent K_m for dCTP
much lower than the parental enzyme (19). However resistant mutants
with other molecular alterations such as a increased level of DNA
polymerase α, ribonucleotide reductase or of dNTPs have also been
isolated (17).

Synchronization of the cell cycle of eukaryotic cells can be obtained following a treatment of the cells with aphidicolin

The action of the drug is reversible: if cells (or tissues) are
treated with aphidicolin for periods of times up to two generations
and then washed, growth resumes without apparent delay. This and the
fact that other important cellular functions, such as RNA, proteins
and nucleotides synthesis are not disturbed by the drug, have suggested the use of aphidicolin to synchronize cells both in culture
and in intact organisms. In fact, in the presence of the drug, cells
are expected to accumulate at the G_1/S boundary of the cell cycle.

Indeed, both animal (13) and plant cells (20, 21) have been efficiently synchronized by an appropriate treatment with aphidicolin
(Figure 6). An even higher degree of synchronization can be obtained
by a double block with the drug (17).

Synchronization of plant cells with aphidicolin is of particular interest owing to the lack of methods to obtain healthy and highly
synchronized cells: following the aphidicolin treatment all cycling
carrot cells (96%) enter synchronously the S phase (20) while in the
case of tobacco cells (22) or root tip meristems (21) up to 80% synchronous mitoses have been observed.

Thus aphidicolin is particularly suitable for the preparation
of large quantities of cells synchronized at the end of the G_1 phase.

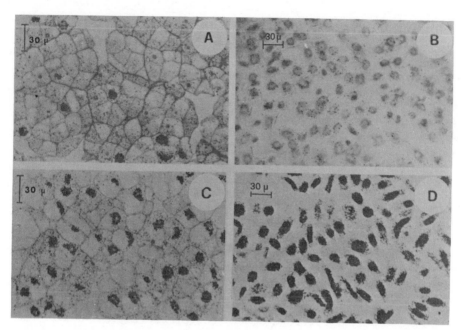

*Figure 6: Synchronization of plant cells with aphidicolin. Autora-
diographs at light microscope of sectioned Daucus carota (A,C) cells
or of root tips of embryos of Haplopappus gracilis (B,D). A: asyn-
chronous cell population labelled with (³H)thymidine for 3 hours.
C: aphidicolin-synchronized cells labelled with (³H)thymidine for
3 hours after the removal of the drug. B,D: embryos of Haplopappus
gracilis after incubation for 1 hour in the presence of (³H)thymidine
during the last hour of the aphidicolin treatment (B) or during the
first hour after the removal of the drug (D). Autoradiographs A,C
are reproduced from Ref. 20; autoradiographs B,D are reproduced from
Ref. 21.*

The apparent no adverse effect on major metabolic processes makes
them a most suitable material for the study of many cell-cycle related events.

CONCLUDING REMARKS

Aphidicolin has proved extremely useful in elucidating the
functional role of eukaryotic DNA polymerases, for cell synchronization and for the isolation of mutants with altered DNA polymerase α,
nucleotide reductase and dNTPs levels.

ACKNOWLEDGEMENTS

    We thank Mr. A. Rebuzzini and M.T. Chiesa, for technical assistance and Miss D. Tavarnè for typing the manuscript. This research was partially supported by C.N.R. grants from the Progetti Finalizzati "Controllo Malattie da Infezione" and "Oncologia"; by C.N.R., Italy, special grant IPRA-Sub-project 1 and by the "Biomolecular Engineering Programme" of the Commission of the European Communities, Contract GB1-6-031-I(S).

REFERENCES

1.   Kornberg,A. (1980) "DNA replication", Freeman W.H. Company, San Francisco.
2.   Kornberg,A. (1982) "1982 supplement to DNA replication" Freeman W.H. and Company, San Francisco.
3.   Amileni,A., Sala,F., Cella,R. and Spadari,S. (1979) Planta 146, 521.
4.   Sala,F., Amileni,A., Parisi,B. and Spadari,S. (1980) Eur. J. Biochem. 112, 211.
5.   Christophe,L., Tarrago-Litvak,L., Castroviejo,M. and Litvak,S. (1981) Plant Sci. Lett. 21, 181.
6.   Amileni,A., Sala,F., Alti,P., Lanzarini,P., Pedrali-Noy,G. and Spadari,S. (1981) Giorn. Bot. Ital. 115, 230.
7.   Bryant,J.A. (1980) Biol. Rev. 55, 237.
8.   Brundret,K.M., Dalziel,W., Hesp,B., Jarvis,J.A.J. and Niedle,S. (1972) J. Chem. Soc. Ser. D. Commun. 1027.
9.   Ikegami,S., Taguchi,T., Ohashi,M., Oguro,M., Nagano,H. and Mano,Y. (1978) Nature 275, 458.
10. Spadari,S., Sala,F. and Pedrali-Noy,G. (1982) TIBS 7, 29.
11. Geuskens,M., Hardt,N., Pedrali-Noy,G. and Spadari,S. (1981) Nucleic Acids Res. 9, 1599.
12. Sala,F., Galli,M.G., Levi,M., Burroni,D., Parisi,B., Pedrali-Noy,G. and Spadari,S. (1981) FEBS Lett. 124, 112.
13. Pedrali-Noy,G., Spadari,S., Miller-Faures,A., Miller,A.O.A., Kruppa,J. and Koch,G. (1980) Nucleic Acids Res. 9, 1599.
14. Oguro,M., Suzuki-Hori,C., Nagano,M., Mano,Y. and Ikegami,S. (1979) Eur. J. Biochem. 97, 603.
15. Pedrali-Noy,G. and Spadari,S. (1980) J. Virology 36, 457.
16. Sala,F., Parisi,B., Burroni,D., Amileni,A.R., Pedrali-Noy,G. and Spadari,S. (1980) FEBS Lett. 117, 93.
17. Spadari,S., Pedrali-Noy,G., Ciomei,M., Falaschi,A. and Ciarrocchi,G. (1984) Toxicologic Pathology 12, 2 issue.
18. Sugino,A. and Nakayama,K. (1980) Proc. Natl. Acad. Sci. USA 77, 7049.
19. Liu,P.K., Chang,C-C., Trosko,J.E., Dube,D.K., Martin,G.M. and Loeb,L.A. (1983) Proc. Natl. Acad. Sci. USA 80, 797.
20. Sala,F., Galli,M.G., Nielsen,E., Magnien,E., Devreux,M. Pedrali-Noy,G. and Spadari,S. (1983) FEBS Lett. 153, 204.

21.   Galli,M.G. and Sala,F. (1983) Plant Cell Reports 2, 156.
22.   Nagata,T., Okada,K. and Takebe,I. (1982) Plant Cell Reports
           1, 250.
23.   Hardt,N., Pedrali-Noy,G., Focher,F. and Spadari,S. (1981)
           Biochem. J. 199, 453.
24.   Sala,F., Magnien,E., Galli,M.G., Dalschaert,X, Pedrali-Noy,G.
           and Spadari,S. (1982) FEBS Lett. 138, 213.

III
Proteins Acting at the Origin of DNA Replication

THE ORIGIN OF DNA REPLICATION OF BACTERIOPHAGE f1 AND ITS

INTERACTION WITH THE PHAGE GENE II PROTEIN

Gian Paolo Dotto, Kensuke Horiuchi and
Norton D. Zinder

The Rockefeller University
New York, N.Y. 10021, USA

ABSTRACT

The origin of DNA replication of bacteriophage f1 consists of
two functional domains: 1) a "core region", about 40 nucleotides long,
that is absolutely required for viral (plus) strand replication and
contains three distinct but partially overlapping signals, a) the
recognition sequence for the viral gene II protein, which is necessary
for both initiation and termination of viral strand synthesis, b) the
termination signal, which extends for 8 more nucleotides on the 5'
side of the gene II protein recognition sequence, c) the initiation
signal that extends for about 10 more nucleotides on the 3' side of
the gene II protein recognition sequence; 2) a "secondary region",
100 nucleotides long, required exclusively for plus strand initiation.
Disruption of the "secondary region" does not completely abolish the
functionality of the f1 origin but does drastically reduce it (1%
residual biological activity). This region, however, can be made
entirely dispensable by mutations elsewhere in the phage genome.

RESULTS AND DISCUSSION

The small icosahedral or filamentous single-stranded DNA phages
such as ϕX174 or f1 (fd, M13) have served as useful model systems for
studying DNA replication (1). A wealth of information is now available
concerning both the enzymes involved and their mechanisms of action
(2). What is still missing is detailed knowledge of the signals present in the phage genome which are required for DNA replication to
be specifically initiated and terminated, and how these signals are
recognized and function. The present report summarizes our efforts

185

to characterize the region around the origin of replication of bacteriophage fl that is required for efficient viral strand synthesis
to occur.

     After entering the bacterial cell, the fl viral (plus) strand
serves as template for the synthesis by host enzymes of the complementary (minus) strand. In particular, the host RNA polymerase recognizes a specific signal on the fl plus strand and, beginning at the
"minus origin", synthesizes an RNA primer of about thirty nucleotides
(3). The primer is subsequently elongated by DNA polymerase III and
eventually a double-stranded, circular, superhelical molecule (RFI)
is formed. Viral (plus) strand synthesis is then initiated by the
viral gene II protein which introduces a nick at a specific site
(plus origin) on the plus strand of the RFI molecule (4). Elongation
of the 3' end of the nick by DNA polymerase III is accompanied by
displacement of the old viral strand (5) via a rolling circle mechanism (6). In addition to its role in initiation, gene II protein
is also able, after one round of plus strand synthesis, to cleave
the nascent single-stranded tail from the replicative intermediate
and seal it to form a covalently closed circle (7). The final products, in vivo, are a closed, single-stranded molecule and a closed,
double-stranded replicative form (RF) (5). In the early stages of
phage infection, the single-stranded circular DNA formed serves as
template for minus strand synthesis to yield more RF molecules. In
later stages, the single strands interact with the viral gene V
protein (8, 9) and are subsequently packaged into viral particles.
Gene V protein, in addition to its role as a single-stranded DNA
binding protein, regulates the amount of gene II protein present by
inhibiting translation of gene II mRNA (10, 11). In this way gene II
protein is normally maintained inside the infected cell at very low
levels.

     The phage "functional origin" of replication has been previously
defined as the minimal fl sequence that, when harbored in a plasmid,
causes it to enter the fl mode of replication provided that helper
phage is present (12, 13). This is manifested in three ways: 1) stimulation of plasmid RF synthesis; 2) ability of the plasmid to interfere
with fl DNA replication; and 3) ability of the plasmid to yield
virion-like particles that exclusively contain plasmid single-stranded
DNA and transduce resistance to antibiotics (transducing particles).
Such a sequence includes the fl plus strand nicking site and extends
for only 12 nucleotides on its 5' side but for more than 100 on its
3' side (12, 14) (Figure 1). (The fl minus origin, which lies about
30 nucleotides on the 5' side of the plus origin, is dispensable in
this system because other sequences elsewhere in the plasmid DNA can
serve for minus strand initiation (14, 15)). The fl plus origin can
be divided into two domains: A, 40 nucleotides long, and B, 100 nucleotides long. Disruption of domain B by deletions or insertions
causes a 100-fold drop in biological anctivity of the origin (as
measured in the chimeric plasmid system), whereas disruption of

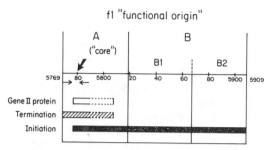

Figure 1: *The fl "functional origin", its signals and its domains.*
Numbers indicate nucleotide positions on the fl map (23, 24). ⟵ ,
gene II protein nicking site;⟶ ⟵, *palindromic region around the*
gene II protein nicking site. The locations of the gene II protein
recognition sequence and of the signals for initiation and termina-
tion of plus strand synthesis are indicated. Also shown are the
domains into which the fl functional origin can be divided. The site
of the four nucleotides insertion in the RS18 (or R209) origin is
also indicated (interrupted line). It separates domain B into two
"subdomains": B1 and B2. For more details see text.

domain A causes a drop of 10,000-fold or more. For this reason,
domain A can be considered as the "core" of the functional origin.

Domain A contains three partially overlapping sequences essential for viral (plus) strand replication: 1) the gene II protein in
vitro recognition sequence; 2) a sequence required for initiation of
viral strand synthesis; and 3) a sequence required for its termination. Domain B contains sequences required exclusively for initiation of viral strand synthesis.

The gene II protein recognition sequence was determined in
vitro by measuring the ability of purified gene II protein to nick
DNA of chimeric plasmids containing various deletions around the fl
origin. It extends from not more than four nucleotides on the 5'
side of the gene II protein nicking site (16) to 11 to 29 nucleotides
on its 3' side (Fig. 1) (17). It is strongly asymmetric. It does
not include the palindrome around the gene II protein nicking site,
once thought to be important for gene II protein recognition. Moreover, there are no obvious features that might explain why gene II
protein only reacts with superhelical fl DNA (18).

The sequence required for nicking by gene II protein in vivo
should be the same as that recognized in vitro. This is suggested
by the fact that the 5' and 3' boundaries of the gene II protein in
vitro recognition sequence seem to coincide, respectively, with the
5' boundary of the signal for initiation of in vivo plus strand syn-

thesis and the 3' boundary of the signal for termination (see below). This fact, unlikely to be coincidental, suggests that the gene II protein recognition sequence is an important if not the only determinant in both initiation and termination of plus strand synthesis.

The signals for initiation and termination of plus strand synthesis were identified by using a system of chimeric plasmids containing two fl functional origins inserted in the same orientation (Figure 2). With such cells, following fl infection, synthesis of plus strand DNA is initiated at either one of the two fl origins and

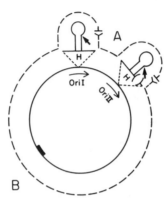

Figure 2: Diagram of the experimental system used to determine the effects of mutations on the signals for initiation and termination of fl plus strand synthesis. OriI and OriII: fl origins inserted in a plasmid. OriI in all cases is fully proficient for both initiation and termination. OriII contains various deletions or insertions (indicated by an open space between brackets) and might be either 1) proficient for both initiation and termination, or 2) only for termination, or 3) only for initiation, or 4) for neither initiation nor termination. fl-dependent plasmid DNA synthesis (interrupted lines) is predicted to vary according to these possibilities (see below). A and B: lesser and greater arc of the chimera. Thin arrows indicate the direction of plus strand synthesis. Thick arrow: site of gene II protein nicking (plus origin). (>——), and (——⊣), the signals for initiation and termination of plus strand, respectively. In this system, if oriII lacks the signal for initiation but retains that for termination, then, upon fl infection, DNA synthesis should occur only from oriI to oriIII, and of the two components of the chimera, only A should be found inside the cell as an independent replicon. On the other hand, if oriII lacks the signal for termination but not for initiation, DNA synthesis should start from both oriI and oriII but should terminate only at oriI, and only component B and the starting chimera should be synthesized inside the cell. If oriII lacks the signals for both initiation and termination, the only kind of plasmid DNA found inside the cell should be that of the starting chimera.

is terminated at the other. Hence, the chimeric plasmids segregate into two replicons, each of them containing only one fl origin (19). The efficiency of fl origins containing various deletions to function as signals for either initiation or termination of plus strand synthesis could thus be assessed directly as illustrated in Fig. 2. In this manner the 5' and 3' boundaries of the signals for plus strand initiation and termination (17, 20, 21) were mapped.

The signals for termination of fl plus strand synthesis extends from 12 nucleotides on the 5' side of the gene II protein nicking site to 11 to 29 nucleotides on its 3' side (Fig. 1). It is likely to include only two elements: 1) the gene II protein recognition sequence, discussed above, and 2) the palindrome located around the gene II protein nicking site which extends for a further 8 nucleotides on the 5' side of the gene II protein recognition site. The involvement of gene II protein in the termination of plus strand synthesis has been studied in vitro (7). After one round of replication, gene II protein is able to cleave the displaced single-stranded tail from the rolling circle intermediate and then seal it to form a covalently closed circle (22). It is reasonable to assume that gene II protein recognizes the same sequence for nicking the RFI molecule at the onset of replication and for cleaving the nascent strand after one round of synthesis. The DNA conformation, however, could be quite different in the two situations. The palindrome around the gene II protein nicking site might be required for the gene II protein recognition sequence to assume the proper conformation after one round of synthesis. Alternatively, the palindrome might be required only to bring together the 5' and 3' ends of the single-stranded molecule for circularization. Preliminary results, using an in vitro Escherichia coli replicating system, seem to support this second possibility (Dotto, unpublished observations).

The signal for initiation includes the gene II protein recognition sequence and extends for about 100 nucleotides downstream (Fig. 1). Therefore, it consists of the entire fl functional origin, (i.e. both domain A and B), excluding only the 5' half of the palindrome around the gene II protein nicking site. The filamentous phages fl, fd, and M13 are very often used as single-stranded cloning vectors (25). One of the most common sites for cloning lies within domain B, in the phage "functional origin". To further analyze the effect of foreign DNA in this crucial region, we have inserted only four nucleotides - one of our cloning vectors, R209, - and found it to be extremely detrimental to the origin function. To restore the replicative function a second, trans-acting mutation had to occur in the phage genome (R218). The effect of this mutation is quite surprising in that it is not only able to overcome the negative effect of the four-nucleotide insertion in the R218 origin, but it renders altogether dispensable a large part of the origin itself. Of the 140 nucleotides that constitute the fl functional origin, only 40 (domain A) remain essential while the rest (domain B) are no longer required (Dotto

G. P. DOTTO ET AL.

and Zinder, unpublished result). The site of the 4-nucleotide insertion was the same as that employed in the construction of the M13-
derived Mpl vector (26). Mpl also contains a mutation able to reduce
the minimal sequence required as a functional origin to a stretch
of only 40 nucleotides (domain A). This effect of the Mpl mutation
is similar to that in R218 and suggests that the two mutations, although selected independently, might be of an analogous nature.

The R218 mutation has been fully characterized (Dotto and
Zinder, unpublished result). It consists of a nucleotide substitution within gene V, leading to a single amino acid change in the
gene V protein. The mutated protein presumably retains its ssDNA
binding activity (R218 gives normal yields of phage) but appears to
lose its ability to repress translation of the viral gene II mRNA.
A markedly increased level of gene II protein is found inside the
R218-infected cell which is likely to account for the drastic reduction in the minimal sequence required as a functional origin.
This conclusion needs explanation. The role hitherto attributed to
gene II protein in the initiation of plus strand synthesis was the
introduction of a specific nick at the plus origin. It is now probable that this protein is otherwise involved in the initiation process, perhaps in the formation of the new replication fork. Whatever
the mechanism, it is clear that the nature of the DNA-protein interactions required for the initiation of viral strand replication can
be drastically altered by a simple increase in the intracellular
concentration of one of the proteins involved. It will be of great
interest to see whether the specific interaction suggested by the
in vivo studies, can be studied in vitro.

ACKNOWLEDGMENT

This work was supported by grants from the National Science
Foundation and the National Institutes of Health.

REFERENCES

1. Denhardt,D.T., Dressler,D. and Ray,D.S. eds. The Single-
        Stranded DNA Phages, (1978) Cold Spring Harbor Laboratory,
        Cold Spring Harbor, New York.
2. Kornberg,A. (1980) DNA replication. Freeman, San Francisco.
3. Geider,K., Beck,E. and Schaller,H. (1978) Proc. Natl. Acad.
        Sci. USA 75, 645.
4. Meyer,T.F., Geider,K., Kurz,C. and Schaller,H. (1979)
        Nature 278, 356.
5. Horiuchi,K., Ravetch,J.V. and Zinder,N.D. (1978) Cold Spring
        Harbor Symp. Quant. Biol. 43, 389.
6. Gilbert,W. and Dressler,D. (1968) Cold Spring Harbor Symp.
        Quant. Biol. 33, 473.

7.   Meyer,T.F. and Geider,K. (1982) Nature 296, 828.
8.   Mazur,B.J. and Model,P. (1973) J. Mol. Biol. 78, 285.
9.   Mazur,B.J. and Zinder,N.D. (1975) Virology 68, 490.
10. Yen,T.S.B. and Webster,R.E. (1982) Cell 29, 337.
11. Model,P., McGill,C., Mazur,B. and Fulford,W.D. (1982)
       Cell 29, 329.
12. Dotto,G.P., Enea,V. and Zinder,N.D. (1981) Virology 114, 463.
13. Dotto,G.P., Enea,V. and Zinder,N.D. (1981) Proc. Natl. Acad.
       Sci. USA 78, 5421.
14. Cleary,J.M. and Ray,D.S. (1981) J. Virol. 40, 197.
15. Dotto,G.P. and Zinder,N.D. (1983) Virology 130, 252.
16. Dotto,G.P., Horiuchi,K., Jakes,K.S. and Zinder,N.D. (1982)
       J. Mol. Biol. 162, 335.
17. Dotto,G.P., Horiuchi,K. and Zinder,N.D. (1983) J. Mol. Biol.
       in press.
18. Meyer,T.F. and Geider,K. (1979) J. Biol. Chem. 254, 12642.
19. Dotto,G.P. and Horiuchi,K. (1981) J. Mol. Biol. 153, 169.
20. Dotto,G.P., Horiuchi,K. and Zinder,N.D. (1982) Proc. Natl.
       Acad. Sci. USA 79, 7122.
21. Dotto,G.P., Horiuchi,K. Jakes,K.S. and Zinder,N.D. (1983)
       Cold Spring Harbor Symp. Quant. Biol. 47, 717.
22. Meyer,R.F., Baumel,I., Geider,K. and Bedinger,P. (1981)
       J. Biol. Chem. 256, 5810.
23. Beck,E. and Zink,B. (1981) Gene 16, 35.
24. Hill,D.F. and Petersen,G.B. (1982) J. Virol. 44, 32.
25. Zinder,N.D. and Boeke,J.D. (1982) Gene 19, 1.
26. Messing,J., Gronenborn,B., Muller-Hill,B. and Hofschneider,
       P.H. (1977) Proc. Natl. Acad. Sci. USA 74, 3642.

OVERPRODUCTION AND PURIFICATION OF THE GENE 2 PRODUCT INVOLVED

IN THE INITIATION OF PHAGE φ29 REPLICATION

Luis Blanco, Juan A. García, José M. Lázaro
and Margarita Salas

Centro de Biología Molecular (SDIC-UAM)
Universidad Autónoma Canto Blanco
Madrid - 34, Spain

INTRODUCTION

The <u>Bacillus subtilis</u> phage φ29 has a linear, double-stranded
DNA with the protein product of gene 3, p3, covalently linked to the
two 5' ends (1). φ29 replication starts at either DNA end by a novel
mechanism by which the terminal protein p3 primes replication by
reaction with dATP and formation of a protein p3-dAMP covalent complex that provides the 3'OH group needed for elongation by the DNA
polymerase. The initiation reaction, i.e. the formation of the protein p3-dAMP covalent complex, requires φ29 DNA-protein p3 as a template, free protein p3 and the gene 2 product (2, 3). We present here
the cloning of gene 2 and the overproduction and partial purification
of protein p2. A DNA polymerase activity has been found associated
with protein p2.

RESULTS AND DISCUSSION

<u>Cloning and expression of gene 2 in Escherichia coli: Partial puri-</u>
<u>fication of protein p2</u>

The φ29 DNA Hind III B fragment, containing gene 2, was cut
with the nuclease Bcl I (see Fig. 1) and cloned first in the Hind III
and Bam HI sites of plasmid pBR322 and then, to achieve a high level
of synthesis of protein p2, the Eco RI-Pst I fragment of the gene
2-containing recombinant plasmid was cloned in the corresponding sites
of plasmid pPLc28 (4) to put protein p2 under the control of the
lambda P_L promoter. After heat inactivation of the thermosensitive

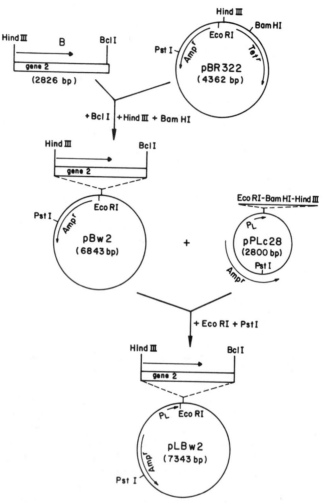

Figure 1: *Cloning of gene 2 in plasmid pPLc28 under the control of the P_L promoter of phage lambda.*

lambda repressor carried in the <u>Escherichia coli</u> lysogen host, a protein of Mr 68,000, the size expected for protein p2 (5), was produced accounting for about 2% of the de novo synthesized protein. The protein p2 made in <u>Escherichia coli</u> was active in the in vitro formation of the p3-dAMP initiation complex when supplemented with extracts from <u>sus2</u>-infected <u>Bacillus subtilis</u> or with extracts from <u>Escherichia coli</u> transformed with the gene 3-containing recombinant plasmid pKC30 Al (6) (see Fig. 2).

    Protein p2 has been partially purified from the <u>Escherichia coli</u> cells transformed with the gene 2-containing recombinant plas-

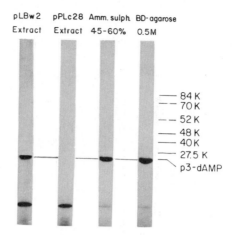

*Figure 2:   Formation of the protein p3-dAMP complex with protein p2
from different purification steps.* Extracts from cells harbouring
the gene 2-containing recombinant plasmid pLBw2 (32 μg) or the control plasmid pPLc28 (32 μg), the 45-60% ammonium sulfate precipitate
(3.5 μg) and the fraction eluted from a blue dextran-agarose column
at 0.5 M NaCl (74 ng), both from the gene 2-containing extracts, were
used for the in vitro formation of the p3-dAMP complex (Peñalva and
Salas, 1982) in the presence of extracts from Escherichia coli transformed with the gene 3-containing recombinant plasmid pKC30 A1 (15
μg).

mid pLBw2 by ammonium sulfate precipitation and affinity chromatography on a blue dextran-agarose column. Protein p2, eluted from the
latter at 0.5 M NaCl, was active in the formation of the initiation
complex in vitro when supplemented with protein p3-donor extracts
as indicated above (Figure 2).

DNA polymerase activity associated with protein p2

Table 1 shows the DNA polymerase activity, assayed with polydA
(dT)$_{12-18}$ as template, of the fractions from the purification of
protein p2 starting with Escherichia coli harbouring the gene 2-con-
taining plasmid pLBw2 or, as a control, with Escherichia coli transformed with plasmid pPLc28. It can be seen that, after the blue
dextran-agarose chromatography, the DNA polymerase activity in the
gene 2-containing cells is about 200 fold higher than in the control
bacteria, suggesting that protein p2 has DNA polymerase activity.

The DNA polymerase present in the protein p2-containing fraction could also elongate the φ29 DNA-protein p3 complex when supplemented with purified protein p3 (7), suggesting that protein p2, not
only is involved in the initiation reaction, but also may be the DNA
polymerase that elongates the φ29 DNA chain.

Table 1: DNA polymerase activity associated with protein p2

| | DNA polymerase activity, nmol/mg[a] | |
	Cells harbouring gene 2- containing plasmid pLBw2	Cells harbouring control plasmid pPLc28
Extract	3	3
Ammonium sulfate, 45-60%	9	6
Blue dextran-agarose, 0.5 M NaCl	1142	6

[a] The DNA polymerase activity was assayed with polydA-(dT)$_{12-18}$ as template as described (7).

ACKNOWLEDGEMENTS

   This investigation has been aided by Grant 2 RO1 GM27242-04
from the National Institutes of Health and by Grants from the
Comisión Asesora para el Desarrollo de la Investigación Científica
y Técnica and Fondo de Investigaciones Sanitarias. L.B. and J.A.G.
were Fellows of the Plan de Formación de Personal Investigador and
the Spanish Research Council, respectively.

REFERENCES

1.   Salas,M., Mellado,R.P., Viñuela,E. and Sogo,J.M. (1978)
           J. Mol. Biol. 119, 269.
2.   Peñalva,M.A. and Salas,M. (1982) Proc. Natl. Acad. Sci. USA
           79, 5522.
3.   Blanco,L., García,J.A., Peñalva,M.A. and Salas,M. (1983)
           Nucl. Acids Res. 11, 1309.
4.   Remaut,E., Stanssens,P. and Fiers,W. (1981) Gene 15, 81.
5.   Yoshikawa,H. and Ito,J. (1982) Gene 17, 323.
6.   García,J.A., Pastrana,R., Prieto,I. and Salas,M. (1983)
           Gene 21, 65.
7.   Salas,M., Blanco,L., Prieto,I., García,J.A., Mellado,R.P.,
           Lázaro,J.M. and Hermoso,J.M. (1983), this volume.

AUTOREPRESSION OF THE dnaA GENE OF ESCHERICHIA COLI

Tove Atlung, Erik Clausen and Flemming G. Hansen[1]

University Institute of Microbiology
Øster Farimagsgade 2A
DK-1353 Copenhagen K, Denmark

[1]Department of Microbiology
The Technical University of Denmark
DK-2800 Lyngby-Copenhagen, Denmark

INTRODUCTION

The dnaA protein is essential for initiation of replication from the chromosomal origin, oriC (1), in Escherichia coli. It interacts with the RNA polymerase (2) in an early step in the initiation process (3). The dnaA protein is also involved in termination of transcription - dnaA(Ts) mutants show increased readthrough at the trp operon attenuator at nonpermissive temperature (4). The dnaA gene has been cloned (5) and the nucleotide sequence of both the promoter region and the structural gene has been determined (6, 7); the gene codes for basic polypeptide of 52.5 kD molecular weight. Transcript start sites corresponding to two promoters were found in vivo (6). Between the two promoters we found an 11 bp nucleotide sequence which is present 4 times within the minimal origin (6). These sequences are amongst those that are highly conserved within the origins of different species of Enterobacteriaceae (8). It has been suggested that the dnaA protein is regulating its own synthesis (9). To study the regulation of the dnaA gene we have fused the dnaA promoter region to the tet and lacZ genes. We found that an increased concentration of wild type dnaA protein, supplied by a second plasmid, caused repression of transcription from the dnaA promoter region. By genetic analysis it was shown that the repression took place at dnaA2p and was dependent on the presence of the 11 bp oriC homology.

MATERIALS AND METHODS

Strain MC1000 (10) was used throughout this study. Plasmids pBR322 (11) and pACYC184 (12) were used as vectors. Growth media were NY and AB minimal medium (2) supplemented with 0.2% glycerol, 1 µg/ml of thiamine and 100 µg/ml of leucine. Ampicillin and chloramphenicol were added at 200 µg/ml and 25 µg/ml respectively. Assays for β-galactosidase activity were according to Miller (13) using sonicated cell extracts; 1 unit was defined as 1000 x OD_{420}/min x ml. Plasmid copy numbers were determined as described (14) with the modification that the DNA was digested with EcoRI prior to electrophoresis. DNA manipulations were essentially as described in Maniatis et al. (15). The promoter cloning vehicles, pTAC909 and pTAC911, were derived from pBR322 by removal of the tet gene promoter, in pTAC909 by S1 nuclease digestion of the HindIII site and in pTC911 by recloning of the BAL-31 deletion from pMLB508.101, which extends from bp 13 to bp 63 in the pBR322 sequence (M.L. Berman personal communication, 16). The plasmid pKK2254 was obtained from K. Clemmensen and C. Petersen, it is a derivative of pMLB1034 (17) carrying an EcoRI site that allowed a fusion of the first 22 aminoacids of the dnaA protein in frame with β-galactosidase.

RESULTS

Transcriptional fusions between the dnaA promoters and the tet gene

were obtained by insertion of the EcoRI fragment carrying the dnaA promoter region and the start of the structural gene (7) into the EcoRI site of pTAC909 and pTAC911. From the initial clones (pTAC926 and pTAC929) a number of deletion derivatives were constructed taking advantage of the unique restriction enzyme sites in and around the dnaA promoters. Mutations in dnaA2p were obtained by DNA polymerase and S1 nuclease treatment of the BglII site which is part of the -35 region of this promoter (7). Subsequently deletions extending from the unique ClaI site upstream from the dnaA promoters in pTAC929 were made by exonuclease BAL-31 treatment.

The transcriptional activity from the dnaA promoter clones was estimated by determination of the level of tetracycline resistance (Tc^R) conferred by these plasmids (Table 1). The plasmids were transformed into derivatives of MC1000 containing the indicated pACYC184 derived plasmids to test for a regulatory effect of the dnaA protein on dnaA gene transcription. Plasmid pFHC871 carries the dnaA promoters and the intact structural gene inserted as a ClaI XhoI fragment into the ClaI SalI sites of pACYC184; pTAC1302 and pTAC1303 are BssHII cutbacks of this plasmids having 121 and 123 bp deletions both inactivating the structural dnaA gene; the plasmid pNF2449, which has a fragment containing the rplJ and rplL genes inserted in the tet gene of pACYC184, served as a general control. In the presence of pFHC871

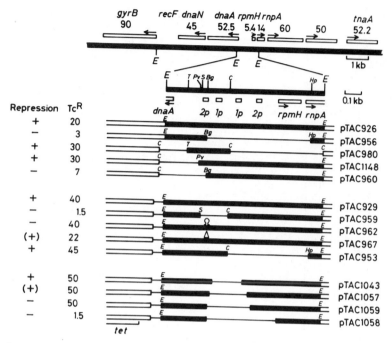

Figure 1: Structure of plasmids with transcriptional fusions between the <u>dnaA</u> promoter region and the <u>tet</u> gene. Top: map of the <u>dnaA</u> region of the <u>Escherichia coli</u> chromosome (5, Hansen et al. manuscript subm. Mol. Gen. Genet.) showing position of the genes, size of the gene products in kD, and direction of transcription. Below: enlargement of the <u>dnaA</u> promoter fragment showing position of the promoters and restriction enzyme sites (6). Restriction enzyme sites used in the construction are indicated above the different plasmids. Incompatible restriction enzyme ends were made flush with S1 nuclease before religation. The four plasmids at the bottom are representative examples of deletions obtained by BAL-31 exonuclease treatment of pTAC929 opened in the unique <u>Cla</u>I site. <u>Cla</u>I linkers were inserted before religation and the extent of the deletions was determined by restriction enzyme analysis. The endpoints in the <u>dnaA</u> promoter region were determined with a precision of ±5 bp. The level of Tc^R (μg/ml) conferred by the plasmids in strain MC1000 is indicated to the left together with a summary of the data on repression by <u>dnaA</u> protein from Table 1. Bg: <u>Bgl</u>II; C: <u>Cla</u>I; E: <u>Eco</u>RI; Hp: <u>Hpa</u>I; Pv: <u>Pvu</u>I; S: <u>Sac</u>II; T: <u>Taq</u>I. Ω : 4 bp insertion and Δ : 4 bp deletion in the <u>Bgl</u>II site.
━━━━ chromosomal DNA, ──────── deleted DNA segment.
▭ pBR322 DNA, pTAC909 was used as vector;
▭─ pBR322 DNA, pTAC911 was used as vector.

Table 1:   Transcription from the dnaA promoters measured as the level of tetracycline resistance[a]

pACYC184 plasmid pBR322 plasmid	none	pNF2449	pFHC871 dnaA+	pTAC1302 dnaAΔ121	pTAC1303 dnaAΔ123
pBR322	60-75	75	75	75	75
pTAC926	20	20	10	20	20
pTAC956	3	<5	<5	<5	–
pTAC980	30	35	20	35	–
pTAC1148	30	30	15	30	–
pTAC960	7	10	7	7-10	–
pTAC929	40	45	25	45	45
pTAC959	1.5	<5	<5	<5	–
pTAC953	45	55	35	55	–
pTAC962	40	40	40	40	–
pTAC967	22	25	17	25	–
pTAC1043	50	–	30	55	55
pTAC1057	50	–	45	60	60
pTAC1059	50	–	60	60	–
pTAC1058	1.5	–	<5	<5	–

[a] Measured as μg/ml of tetracycline on NY plates as described (18).

–:Not determined.

a two fold reduction in Tc^R was observed with all promoter clones
carrying an intact dnaA2p region. The effect was independent of
sequences downstream from the transcript start site (compare pTC926
and pTC1148) and independent of the presence of dnaAlp (compare
pTAC929 and pTAC1043). Transcription from dnaAlp seemed to be unaffected by pFHC871 (pTAC960). The 4 bp insertion in the -35 region
of dnaA2p (pTAC962) abolished the effect without apparent change in
the promoter strength, indicating that initiation of transcription
from this promoter is the target for the autorepression by the
surplus of dnaA protein supplied by pFHC871. The BAL-31 deletions
showed that also sequences located upstream from dnaA2p are required
for repression. This site is located around position -55 in the
promoter where the 11 bp homology to oriC was found (6). The promoter
clones gave the same level of Tc^R in the control strain and in the
strains carrying the derivatives of pFHC871 with the internal deletions in the structural dnaA gene. This showed that the repression
was due to the dnaA protein and not to the presence of extra dnaA
promoters or a product made from the promoter region.

Translational fusions between the dnaA gene and the lacZ gene

were constructed by insertion of the EcoRI fragment from
pFHC539, a pBR322 analog of pFHC871, into the EcoRI site of pKK2254.
In plasmid pTAC1377 the fused gene is transcribed from a wild type
dnaA promoter region, while the plasmid pTAC1448 carries the 4 bp
insertion in the -35 region (analogous to pTAC962). A control plasmid
lacking the dnaA promoters, pTAC1382 was made by insertion of the
EcoRI fragment from pTAC956 (Fig. 1).

These plasmids were transformed into the Δlac strain MC1000
containing the pACYC184 derived plasmids (Table 2). The translational
fusions were made to get a better measurement of the dnaA autorepression. We found that the dnaA´´lacZ gene expression was reduced to
60% in rich medium and to about 30% in minimal medium by the presence of pFHC871. The contribution of transcription from dnaAlp,
which is present on both fusion plasmids, is in the order of 20-30%
of the unrepressed level (Table 1, and data not shown) therefore
the repression exerted on danA2p at the low growth rate must be
85-95%. We found that the relative copy number of pBR322 and pACYC184
derived plasmids (1.5) was independent of growth rate and the genetic
constitution of the plasmids. The copy numbers per ml of culture at
OD$_{450}$ = 1 varied little if at all between the different strains but
considerably with growth rate, being 4-6 fold higher in minimal than
in rich medium. The higher levels of β-galactosidase activity found
in the slowly growing cultures is probably due to this difference
in copy number. The higher degree of repression might also be caused
by the increased copy number of the plasmid carrying the dnaA$^+$ gene
giving rize to a higher concentration of dnaA protein.

Table 2: Repression of dnaA''β-galactosidase by dnaA protein.[a]

pBR322 plasmid dnaA''lacZ	pACYC184 plasmid	Growth medium		
		NY	glycerol	minimal
pTAC1377 dnaA+	pFHC871 dnaA+	460 600	1000	800
	pTAC1302 dnaAΔ121	680 950	2150	2500
pTAC1448 dnaA2pΩ4bp	pFHC871 dnaA+	770 —	1900	2900
	pTAC1302 dnaAΔ121	720 —	2300	3200
pTAC1382 ΔdnaAp	none	12 —	130	—

[a] The strains were grown exponentially at 37°C. 6-8 samples for determination of β-galactosidase were taken during 2-4 generations. β-galactosidase activity is given as units/mlxOD$_{450}$. Values from two experiments are shown.

DISCUSSION

In this study we have shown that a surplus of dnaA protein caused repression of transcription from the dnaA promoter region.This repression took place at the promoter designated dnaA2p (6) and was dependent on sequences around position -55 where an 11bp homology with oriC is located (6; see Table 3). This sequence is also found in the promoter for the 16 kD protein gene which is located clockwise of oriC (18). In minichromosome pBR322 chimeras transcription from the 16 kD promoter leads to out-titration of dnaA protein in certain dnaA(Ts) mutants (F.G. Hansen unpublished results). The 11 bp sequence is also found in the origin of the plasmid pSC101 (20) which depends on dnaA protein for replication. This suggests that the sequence is a dnaA operator site also in initiation of replication. The same sequence is, however, present in the vicinity of some plasmid origins from which initiation takes place in the absence of active dnaA protein (see Table 3).

Table 3:   Occurrence of the 11bp sequence

GTTATCCACAG	"consensus sequence"
TTTATCCACAG	in dnaA2p in position -61[a]
CTTTTCCACAG	in 16kD promoter in position -47
GTTATCCACAG	in P_{ori-r}[b] in position -54
GTTATCCAAAG	in P_{ori-1}[b] in position -56
GTTATCCACAG	in oriC
GTTATACACAA	in oriC
GTTATACACAG	in pSC101 origin
GTTATCCACAG	near origins of ColE1[c], pBR322[d], RSF1030[d]
GTTACCCACAG	near origin of ClodF13[d]
CTTATCCACAT	near origin of R100[e]

a)   Position relative to transcript start site.
b)   The two promoters found in vitro by Lother et al. (19).
c)   At position +85 from the origin (20).
d)   From the compiled sequence data in Selzer et al. (21).
     The sequence is not present in p15A.
e)   Between the repA gene and the origin at base number -57 in the sequence of Rosen et al. (22).

We showed that the repression is caused by overproduction of dnaA protein. It cannot, at present, be excluded that the effect is mediated through the product of another gene regulated by the dnaA protein. The synthesis of at least two chromosomally encoded proteins was stimulated when dnaA protein was overproduced 200-1000 fold upon inducation of a plasmid carrying the λp_L in front of the dnaA gene (T. Atlung unpublished results). These proteins could also be over-produced in the presence of pFHC871 used in this study. Thus, whether the dnaA protein itself acts at the 11 bp operator sequence has to be verified by in vitro experiments.

The autorepression of the dnaA gene, probably, only plays a minor role in regulation of dnaA protein synthesis under normal growth conditions. We have not yet determined the actual increase in dnaA protein concentration in the strain carrying pFHC871, but we estimate that it is in the order of 2-4 fold in rich medium and 5-10 fold in minimal medium where 50% and 90% repression of dnaA2p was observed respectively.

The dnaA protein is required for initiation of replication both in vivo and in vitro (23) and acts together with RNA polymerase as shown by the allele specificity of suppression of dnaA(Ts) mutants by rpoB mutations and the observation that all of 12 Ts mutants tested were suppressible by rpoB mutations (2, T. Atlung unpublished results). There are no indications yet that the dnaA protein acts as a repressor for transcription in the oriC region. The interaction with the RNA polymerase and the 11 bp sequence might be of a differ-ent kind in initiation of replication than that in dnaA gene regula-tion. The dnaA protein might affect transcription termination in oriC as inactivation of dnaA protein in Ts mutants was found to increase readthrough at the trp operon attenuator (4).

REFERENCES

1.   Meijer,M., Beck,E., Hansen,F.G., Bergmanns,H.E.N., Messer,W., von Meyenburg,K. and Schaller,H. (1979) Proc. Natl. Acad. Sci. USA 76, 580.
2.   Atlung,T. (1981) ICN-UCLA Symp. Mol. and Cell. Biol. 21, 297.
3.   Lark,K.G. (1972) J. Mol. Biol. 64, 47.
4.   Atlung,T. and Hansen,F.G. (1983) J. Bacteriol. 156, 985.
5.   Hansen,F.G. and von Meyenburg,K. (1979) Mol. Gen. Genet. 175, 135.
6.   Hansen,F.G., Hansen,E.B. and Atlung,T. (1982) EMBO J. 1, 1043.
7.   Hansen,E.B., Hansen,F.G. and von Meyenburg,K. (1982) Nucleic Acids Res 10, 7373.
8.   Zyskind,W., Harding,N.E., Takeda,Y., Cleary,J.M. and Smith,D.W. (1981) ICN-UCLA Symp. Mol. and Cell. Biol. 21,13.
9.   Hansen,F.G. and Rasmussen,K.V. (1977) Mol. Gen. Genet. 155, 219.

10. Casadaban,M.J. and Cohen,S.N. (1980) J. Mol. Biol. 138, 179.
11. Bolivar,F., Rodrigues,R.L., Green,P.J., Betlach,M.C., Heyneker,H.L., Boyer,H.W., Crosa,J.H. and Falkow,S. (1977) Gene 2, 95.
12. Chang,A.C.Y. and Cohen,S. (1978) J. Bacteriol. 134, 1141.
13. Miller,J.H. (1972) "Experiments in molecular genetics". Cold Spring Harbor Laboratory, New York.
14. Stuber,D. and Bujard,H. (1982) EMBO J. 1, 1399.
15. Maniatis,T., Fritsch,E.F. and Sambrook,J. (1982) "Molecular cloning". Cold Spring Harbour Laboratories.
16. Sutcliffe,J.G. (1979) Cold Spring Harbor Symp. Quant. Biol. 43, 77.
17. Weinstock,G.M., ApRhys,C., Berman,M.L., Hampar,B., Jackson,D., Silhavy,T.J., Weiseman,J. and Zweig,M. (1983) Proc. Natl. Acad. Sci. USA 80, 4432.
18. Hansen,F.G., Koefoed,S., von Meyenburg,K. and Atlung,T. (1981) ICN-UCLA Symp. Mol. and Cell. Biol. 22, 37.
19. Lother,H., Bukh,H-J., Morelli,G., Heimann,B., Chakraborty,T. and Messer,W. (1981) ICN-UCLA Symp. Mol. and Cell. Biol. 22, 57.
20. Churchward,C., Linder,P. and Caro,L. (1983) Nucleic Acids Res. 11, 5645.
21. Selzer,G., Som,T., Itoh,T. and Tomizawa,J.-I. (1983) Cell 32, 119.
22. Rosen,J., Ohtsubo,H. and Ohtsubo,E. (1979) Mol. Gen. Genet. 171, 287.
23. Fuller,R.S., Kaguni,J.M. and Kornberg,A. (1981) Proc. Natl. Acad. Sci. USA 78, 7370.

REPLICATION FUNCTIONS ENCODED BY THE PLASMID pSC101

G. Churchward, P. Linder and L. Caro

Département de Biologie Moléculaire
Université de Genève
30, Quai Ernest-Ansermet
CH-1211 Genève 4, Switzerland

SUMMARY

We describe the mapping of several genetic loci involved in
the replication of the pSC101 plasmid. These include the origin of
replication and a short segment of DNA that encodes a pSC101 incompatibility function. This short segment lies within the origin region.
Flanking the incompatibility segment are two loci, repA and repB,
which are required for replication. The product of the repA locus is
shown to be trans-acting.

INTRODUCTION

The pSC101 plasmid (1) was probably isolated from Salmonella
panama (2) and specifies resistance to the antibiotic tetracycline.
In Escherichia coli it is characterized by a rather low copy number,
5-6 copies per cell (3, 4). Replication of pSC101 requires the DnaA
protein of Escherichia coli (4, 5). This protein has been shown to
be required for the initiation of chromosomal DNA replication both
in vivo (6) and in vitro (7). To date, pSC101 is the only replicon,
apart from the origin of replication of the chromosome, oriC, where
a requirement for DnaA protein has been demonstrated. These two
properties, low copy number and requirement for DnaA protein, have
made pSC101 an attractive model for the investigation of the mechanism of initiation of DNA replication, and its regulation, in
Escherichia coli.

We present here a summary of the results obtained in our experiments to identify and map pSC101 encoded replication functions
(8, 9). To produce mutations which could be easily mapped we used
mutagenesis by insertion of the transposon gamma-delta (Tn1000; see
ref. 10). Insertion mutations were isolated into a hybrid plasmid,
pLC709, consisting of a HincII fragment of pSC101 ligated to a colE1
plasmid bearing a selectable marker. The properties of the insertion
plasmids and of DNA fragments cloned from them onto pBR322, a colE1-
type vector, allowed us to construct a map of the pSC101 replication
region. The insertion plasmids also facilitated the sequencing of
this region.

RESULTS

        A restriction map of pSC101, and of the HincIIA fragment bear-
ing the replication origin of pSC101, is shown in Figure 1. The
segment of pSC101 from the leftward HincII cleavage site to the RsaI
cleavage site, when cloned on pBR322, permits transformation of a
polA⁻ strain. A HaeII fragment contained within the HincII fragment
has also been shown to function as an autonomous replicon (11).

        The positions of several insertions of Tn1000 into the replica-
tion region of pSC101 are shown in Fig. 1. Insertion 22 completely
abolishes, and insertion 36 partially abolishes, incompatibility
towards a pSC101 plasmid. A DNA fragment carrying the left end of
Tn1000 and the pSC101 segment leftward of the site of insertion 38
(as drawn in Fig. 1) expresses incompatibility towards pSC101, when
cloned on pBR322, but a similar fragment cloned from insertion 14
does not. Conversely, a DNA fragment carrying the right end of Tn1000
and sequences rightwards of insertion 14 expresses incompatibility
towards pSC101, but a similar fragment from insertion 38 does not.

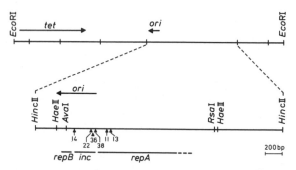

Figure 1: *Restriction map of pSC101 and the HincIIA replication
fragment. The map of the HincIIA fragment shows the position of the
Tn1000 insertions and the loci referred to in the text. Only the
relevant HaeII, AvaI and RsaI sites within the HincIIA fragment are
shown.*

These results show that the locus responsible for pSC101 incompatibility lies between the sites of insertions 14 and 38, a segment of
270 base pairs.

    Insertion 22, in addition to its effect on the expression of
incompatibility, is absolutely defective in DNA replication. The
defect cannot be complemented by pSC101. Examination of replicating
molecules by electron microscopy shows that the majority of the
molecules initiate replication in the vicinity of the site of insertion 22. This replication procedes leftwards as shown in Figure 2.
Insertion 22 therefore probably inactivates the replication origin
of pSC101. The extent of the <u>ori</u> locus is not known, but a DNA fragment extending leftwards from the site of insertion 38, when cloned
on pBR322, allows transformation of a <u>polA</u>⁻ strain provided that a
pSC101 plasmid is already present in the strain. Taken together with
the observation that a <u>Hae</u>II fragment is capable of replication,
this suggests that <u>ori</u> lies within 450 basepairs leftwards of insertion 38. The site of this insertion probably lies just within
the origin since transformation of a <u>polA</u>⁻ strain by a pLC709::
Tn<u>1000</u>-38 plasmid is reduced but not abolished.

    We have determined the nucleotide sequence around the origin
of replication of pSC101. The sequence shows that there is very
little homology between the origin of replication of pSC101 and <u>ori</u>C,
despite the requirement for DnaA protein for replication. However,
the sequence does reveal several features found in other replication origins. Adjacent to the site of insertion 22 is an 80 base
segment containing 85% A-T base pairs followed by three direct repetitions of an 18 base pair sequence (Fig. 2). Insertion 22 is located
in the rightmost repeat (as shown in Fig. 2).

    Insertions 13 and 11 completely abolish replication, as detec-
ted by the ability of a pLC709::Tn<u>1000</u> plasmid to transform a <u>polA</u>⁻
strain. However, provided that pSC101 is present as a resident
plasmid, a plasmid carrying insertion 11 or 13 will transform such
a strain. This suggests that insertions 11 and 13 inactivate a function required for replication but that this function can be provided

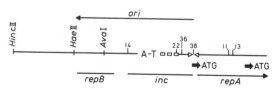

*Figure 2: Features of the DNA sequence of the origin region. The
position of direct (□□□) and inverted (▷ ◁) repeated sequences is
shown. Bold arrows (➡) denote promotor-like sequences.*

in trans by pSC101. This interpretation is confirmed by measurements
of the rate of replication of a pLC709::Tn1000-13 plasmid after a
temperature shift in a polA4113 strain where DNA polymerase I activity is temperature-sensitive. We therefore designate this function
repA.

Nucleotide sequence data show that to the right of insertion
22 are inverted repeats of a 22 base pair sequence. The site of insertion 38 is within one of these repeats, and interrupts a promoter-
like sequence that could direct transcription to the right. This
putative promoter is followed by an ATG initiation codon and a
reading frame open to the end of the sequenced segment. A fragment
of DNA containing the putative transcription and translation initiation signals has been cloned into the pMC1403 plasmid (12) so that
an in-phase fusion was produced between the open reading frame and
the β-galactosidase gene segment carried by the plasmid. The resulting
plasmid synthesised high levels of β-galactosidase indicating that
both the transcription and translation signals function in vivo.
An obvious conclusion would be that these are the promoter and initiation codon of the repA gene, and that the inverted repeats in the
promoter have some role in the regulation of repA expression. However, the situation is complicated by the observation that a sequence
downstream also contains a functional promoter (13) and that this
is followed by possible translation initiation signals. Cloning of
these signals into pMC1403 also results in production of β-galactosidase implying that two proteins may be synthesized from the
repA region. Since the two initiation codons are in the same phase,
the carboxy-terminal end of the longer protein would be identical
to the shorter protein.

Evidence for another function required for replication comes
from the properties of a pBR322 plasmid carrying a HincII-RsaI fragment bearing the inc and repA loci (Fig. 1), and a derivative where
pSC101 sequences between the HincII and AvaI sites have been deleted.
The deletion derivative is no longer capable of transforming a polA⁻
strain and, in contrast to the parent plasmid, no longer replicates
at high temperature in the polA4113 strain. Cleavage with AvaI interrupts a short open reading frame that could encode a protein of
3500 D. Since this reading frame is preceded by a promoter-like
sequence it is a candidate for the replication function and we
provisionally designate the locus repB.

DISCUSSION

The results presented here describe the identification, mapping
and structural analysis of several pSC101 replication functions. In
particular, the inc locus, responsible for pSC101 incompatibility
and implicated in the regulation of pSC101 replication (3), has
been shown to be a short segment of DNA containing three direct

repeats of an 18 base pair sequence. A similar structure has been found for the inc loci of several other plasmids (14-16). As yet we do not know the nature of the product responsible for incompatibility or its mechanism of action. This same segment of DNA has been shown to be an essential part of the origin of replication of pSC101. A similar relationship between inc and ori loci has been demonstrated for the plasmids RK2 and RK6 (17, 18).

We have also identified two other loci, repA and repB. The products of these loci remain to be identified, and, in particular, the question of whether or not the repA locus encodes one or two proteins in vivo remains open. However, the identification of these loci represents the first step in the analysis of the mechanism of pSC101 replication, and of the role of the DnaA protein in this process.

ACKNOWLEDGEMENTS

This work was supported by grant no. 3.169.0.81 from the Swiss National Foundation.

REFERENCES

1. Cohen,S.N. and Chang,A.C.Y. (1973) Proc. Natl. Acad. Sci. USA 70, 1293.
2. Cohen,S.N. and Chang,A.C.Y. (1977) J. Bacteriol. 132, 734.
3. Cabello,F., Timmis,K. and Cohen,S.N. (1976) Nature 259, 285.
4. Frey,J., Chandler,M. and Caro,L. (1979) Mol. Gen. Genet. 174, 117.
5. Hasunuma,K. and Sekiguchi,M. (1977) Mol. Gen. Genet. 154, 225.
6. Hirota,Y, Ryter,A. and Jacob,F. (1968) Cold Spring Harbor Symp. Quant. Biol. XXXIII, 677.
7. Fuller,R.S., Kaguni,J.M. and Kornberg,A. (1981) Proc. Natl. Acad. Sci. USA 78, 7370.
8. Linder,P., Churchward,G. and Caro,L. (1983) J. Mol. Biol. 170, 287.
9. Churchward,G., Linder,P. and Caro,L. (1983) Nucl.Acids Res. 11, 5645.
10. Guyer,M. (1978) J. Mol. Biol. 126, 347.
11. Meacock,P.A. and Cohen,S.N. (1980) Cell 20, 529.
12. Casadaban,M.J., Chou,J. and Cohen,S.N. (1980) J. Bacteriol. 143, 971.
13. Pannekoek,H., Maat,J., Berg,E.rd. and Nordermeer,I.(1980) Nucl. Acids Res. 8, 1535.
14. Stalker,D.M., Thomas,C.M. and Helinski,D.R. (1981) Mol. Gen. Genet. 181, 8.
15. Stalker,P.M., Kolter,R. and Helinski,D.R. (1982) J. Mol. Biol. 161, 33.

16. Murotsu,T., Matsubara,K., Sugisachi,H. and Takanami,M. (1981) Gene 15, 257.
17. Kolter,R. and Helinski,D.R. (1982) J. Mol. Biol. 161, 45.
18. Thomas,C.M., Meyer,R. and Helinski,D.R. (1980) J. Bacteriol. 141, 213.

HOW DOES ROP WORK ?

D.W.Banner, R.M.Lacatena, L.Castagnoli,
M.Cornelissen and G.Cesareni

European Molecular Biology Laboratory
Postfach 10.2209
D-6900 Heidelberg, Fed. Rep. of Germany

Plasmids of the ColEl family provide a convenient model system
for the investigation of the molecular mechanisms which regulate initiation of DNA replication in Escherichia coli. ColEl replicates in
bacteria at a copy number of approximately 15. The observations that
this copy number could be increased by mutagenesis and that some of
these mutants are complemented in trans by plasmids of the same incompatibility group suggest that ColEl replication is negatively regulated (1-6). An attractive model of the processes of initiation
of DNA replication was proposed by Tomizawa and coworkers who were
able to reassemble from purified enzymes an in vitro system which
can carry out the biochemical steps involved in primer formation and
initial deoxyribonucleotide polymerization (3, 7). When origin sequences are incubated in the presence of RNA polymerase, two main transcripts can be identified (Figure 1). The longer one (RNA2) starts
from a promoter 550 nucleotides upstream from the replication origin
and is believed to form a hybrid with the DNA template in the origin
region. This hybrid is processed by RNAseH to yield an RNA molecule
of 550 nucleotides which can serve as primer for deoxyribonucleotide
polymerization by DNA polymerase I.

The second transcript, RNAl, is complementary to 108 nucleotides in the 5' portion of RNA2. Genetic and biochemical evidence suggest that RNAl is an inhibitor of ColEl replication. Secondary structure prediction and enzymatic digestion kinetics (8) indicate a
clover-leaf folding of RNAl. The isolation and characterization of
a large number of mutants in the target of RNAl indicates that seven
bases in two of the predicted loops are critical for the interaction
of RNAl with RNA2 (5). It is presumed that RNAl stabilizes a comple-

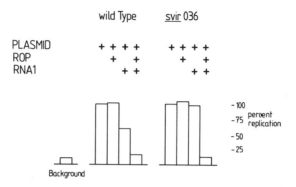

Figure 1: *Transcription pattern and regulatory circuits in the*
replication origin of ColEl. Numbers refer to the distance in base-
pairs from the origin of DNA replication. Transcripts are represen-
ted as wavy lines. Bom is a region required in cis for mobilization
of ColEl and pMBl derivatives by conjugative plasmids.

mentary structure in RNA2 and that this interferes in some way with
RNAseH processing at the origin.

      Deletion analysis has identified a second region, 500 nucleo-
tides downstream from the replication origin which is also involved
in the control of plasmid copy number (6). Cesareni et al. (9) located in this region the rop gene which represses in trans the expression of genes transcribed under the control of the primer promoter.
We have proposed that a small protein of 63 aminoacids is the product of the rop gene and is a second negative regulator of plasmid
replication. This 63 aminoacid protein has been overproduced under
the control of λp_L promoter and purified to homogeneity (10). The
observation that purified protein could inhibit plasmid replication
in a crude in vitro system confirms to us that the 63 aminoacid protein is the product of the rop gene. It has, however, never been
possible to demonstrate the activity of purified Rop in an in vitro
system reconstituted from purified RNA polymerase, RNAseH and DNA
polymerase. This suggests that one or more further factors are essential for inhibition by Rop.

      The experiments in Figure 2 show that purified Rop is not suffi-
cient to inhibit plasmid replication in an extract prepared from cells
that did not contain any ColEl plasmid. Inhibition by Rop must then
require a plasmid encoded element. Genetic and biochemical work (manuscript in preparation) pointed to RNAl as the candidate factor.
When RNAl is purified from a denaturing polyacrylamide gel and added
to the system, Rop inhibition may be shown. This is more clear if the
target plasmid is a mutant such as svir036 whose RNA2 is less sensitive to wild type RNAl inhibition (5). In this case, when RNAl is

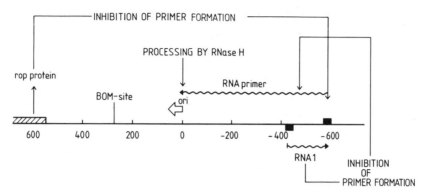

Figure 2: *Control of plasmid replication by purified Rop in a cell extract. (^{3}H) dTTP incorporation into acid insoluble material was measured after addition of supercoiled plasmid (0.1 μg) to a crude cell extract prepared according to Staudenbauer (14). Columns represent percentage of replication. 100% replication corresponds to approximately 10 pmoles of deoxynucleotide incorporated into high molecular weight RNA. Rop was purified according to Lacatena et al. (10), RNA1 was purified after electrophoresis on a 6% urea gel from total RNA extracted from cells containing the plasmid pAT153. RNA1 was identified by comigration with labeled RNA1 synthesized in vitro on a plasmid template.*

added, little or no inhibition is observed. In the presence of Rop and RNA1, however, replication decreases to approximately one fifth of the control level. This result might suggest that the function of Rop is simply that of increasing the inhibitory function of RNA1 on the processing of the primer, possibly by increasing its affinity for the primer precursor. This model was recently proposed by Som and Tomizawa (11). If this were the case, the identification of the rop gene by its ability to inhibit in vivo β-galactosidase synthesis under the control of the primer must be considered as a fortuitous observation. The transcription fusion constructions that permitted the identification of the rop gene do not contain the origin, that is the site of processing by the enzyme RNAseH. We have, however, recent evidence that the inhibitory activity of Rop on β-galactosidase transcription in the fusion phages is also dependent on RNA1 (manuscript in preparation).

Table 1 shows that β-galactosidase synthesis in strains containing the phage φBG42 (9) is not affected by a plasmid unrelated to ColE1 in which the rop gene had been cloned. In a similar way a plasmid which synthesizes only RNA1 is not sufficient to inhibit β-galactosidase activity. The presence of both plasmids inside the cells however results in a decrease of β-galactosidase activity.

Table 1: lacZ expression under the control of the primer promoter

Fusion phage[a]	Plasmid[b]	Inhibitors synthesized	βgal activity[c]
132			20
φBG42			70
"	pBR322	Rop + RNA1	30
"	pGC8	Rop	60
"	pac163pII1	RNA1	70
"	pGC8 + pac163pII1	Rop + RNA1	30

[a] Phage 132 was described by Maurer et al. (12); φBG42 by Cesareni et al. (9).

[b] pGC8 contains the Pst I Pvu II fragment of pBR322 (encoding Rop) in a small R6K derivative. pac163pIIl is a P15A derivative in which a DNA fragment encoding RNA1 from pMB1 has been cloned (Cesareni et al., in preparation).

[c] β-galactosidase activity is measured according to Miller (13).

This result parallels the results obtained in the in vitro replication system and suggests that both ways of assaying for Rop activity are actually measuring the same function. Experiments by Som and Tomizawa (11), and observations from ourselves, prove that at least part of the target of Rop inhibitory activity lies more than 50 nucleotides downstream from the primer promoter. This excludes the possibility that Rop and RNA1 might affect the initiation of transcription of the primer precursors in a quasi-classical way. We have considered three hypothetical molecular mechanisms by which Rop might act.

a)   Rop increases the ability of RNA1 to inhibit the processing of the primer precursor at the origin.

b)   Rop and RNA1 together affect the primer formation by causing the premature termination of transcription of the primer precursor.

c)   The interaction of Rop and RNA1 with the primer precursor affects the stability of the latter against ribonuclease degradation.

Mechanism a) assumes that the primer precursor is transcribed past the origin. This is inconsistent with the in vivo results described above which show that genes under the control of the primer precursor are not expressed in the presence of Rop and RNA1. The possibility remains that the decrease in β-galactosidase activity observed is due to artefacts at the level of translation. This, however, seems improbable because it would be difficult to explain why such artefacts should also be observed in fusions with genes as different as lacZ (9), galK (11) and the tetracycline resistance gene (unpublished observations). Experiments are in progress to define precisely the mechanism of Rop activity at the molecular level.

REFERENCES

1.   Gelfand,D.H., Shepard,H.M., O'Farrell,P.H. and Polisky,B.
         (1978) Proc. Natl. Acad. Sci. USA 75, 5869.
2.   Conrad,S.E. and Campbell,J.L. (1979) Cell 18, 61.
3.   Tomizawa,J., Itoh,T., Selzer,G. and Som,T. (1981) Proc. Natl.
         Acad. Sci. USA 78, 1421.
4.   Lacatena,R.M. and Cesareni,G. (1981) Nature 294, 623.
5.   Lacatena,R.M. and Cesareni,G. (1983) J. Mol. Biol. 170, 635.
6.   Twigg,A.J. and Sherratt,D. (1980) Nature 283, 216.
7.   Itoh,T. and Tomizawa,J. (1980) Proc. Natl. Acad. Sci. USA
         77, 2450.
8.   Tamm,J. and Polisky,B. (1983) Nucleic Acid Res. 11, 6381.
9.   Cesareni,G., Muesing,M.A. and Polisky,B. (1982)
         Proc. Natl. Acad. Sci. USA 79, 6313.
10. Lacatena,R.M., Banner,D.W. and Cesareni,G. (1983)
         in: Mechanism of DNA Replication and Recombination,
         Cozzarelli,N.R., ed., UCLA Symposia on Molecular and
         Cellular Biology, Alan R. Liss, Inc., New Series,
         Vol. X, 327.
11. Som,T. and Tomizawa,J. (1983) Proc. Natl. Acad. Sci. USA
         80, 3232.
12. Maurer,R., Meyer,B. and Ptashne,M. (1980) J. Mol. Biol.
         139, 147.
13. Miller,J.H. (1972) Experiments in molecular genetics, Cold
         Spring Harbor Laboratory, New York.
14. Staudenbauer,W.L. (1976) Molec. Gen. Genet. 145, 273.

GENE A PROTEIN OF BACTERIOPHAGE φX174 IS A HIGHLY SPECIFIC
SINGLE-STRAND NUCLEASE AND BINDS VIA A TYROSYL RESIDUE TO
DNA AFTER CLEAVAGE

A.D.M. Van Mansfeld, P.D. Baas and H.S. Jansz

Institute of Molecular Biology and Laboratory for
Physiological Chemistry
State University of Utrecht
Padualaan 8
3584 CH Utrecht, The Netherlands

ABSTRACT

The sequence specificity of the endonuclease activity of gene
A protein and A* protein was studied using synthetic oligonucleotides
containing (part of) the sequence of the origin of φX RF DNA replication and single-stranded (ss) DNA fragments of φX and G4. From a
comparison of the sequences that are cleaved a consensus sequence
for cleavage of ssDNA by gene A protein has been deduced. This consensus sequence occurs in ssDNA of both φX and G4 at the origin and
at one additional site. This is surprising since the rolling circle
mechanism demands that gene A protein cleaves at the origin only.
However, it could be shown that in the presence of SSB protein the
ssDNAs of φX and G4 are only cleaved at the origin, which is probably
due to a strong gene A protein binding site, the key sequence, which
forms part of the 30 b.p. origin region of φX and related bacteriophages.

Gene A protein and A* protein bind covalently to the DNA at
the 5'-end of the cleavage site. Using a uniquely, internally [32]P-
labelled oligonucleotide as a substrate, it was shown that gene A
protein and A* protein are bound via a tyrosyl residue to the 5'-
phosphate of the phosphodiester bond which is cleaved.

INTRODUCTION

The interaction of gene A protein of bacteriophage φX174 and
the origin of φX RF DNA replication has been studied in our labora-

tory using various experimental approaches. Previous experiments
(1-6) have led to the following model (Fig. 1). The origin region
consists of a sequence of 30 b.p. which is strongly conserved in the
group of φX-related bacteriophages. In superhelical RFI DNA gene A
protein binds to a specific sequence in the origin region: the key
sequence. Binding causes local denaturation which is facilitated by
the A-T-rich region. As a consequence, the flanking region, the
recognition sequence, is exposed in a single-stranded form and can
be recognized and cleaved by gene A protein. The cleavage by gene
A protein creates a free 3'-OH (G), which acts as a primer for the
subsequent rolling circle DNA replication.

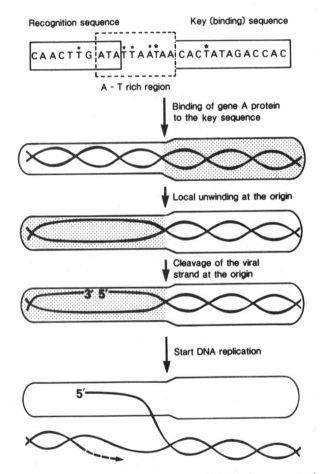

*Figure 1: Model of the interaction of φX gene A protein with the
origin. The sequence corresponds to nucleotides 4299-4328 of the
φX DNA sequence (16). The asterisks indicate nucleotides that can
be changed without loss of origin function.*

Using synthetic oligonucleotides with changes at different sites in the sequence and fragments of the ssDNAs of φX and G4 we have investigated which bases in the recognition sequence are critical for cleavage by gene A protein. The communication by Fluit et al. (this volume) focuses on another domain of the origin region, the key sequence.

The DNA at the 5'-end of the cleavage site is linked to gene A protein in a covalent bond (7, 8). At the end of one replication-round the genome length tail of the rolling circle is converted into circular viral DNA. This conversion is supposed to take place by the concerted cleavage-ligation-transfer reaction of the covalently bound gene A protein with the regenerated origin in the ss tail of the rolling circle (8, 9). As a first step to elucidate the mechanism of this reaction we have determined the nature of the protein-DNA bond in the gene A protein-DNA complex.

A* protein is a second product of gene A. It is synthesized from an internal translational start in the same reading frame as is used for gene A protein (10). A* protein has retained a number of enzymatic activities of gene A protein (11, 12). Most experiments with gene A protein were also performed with A* protein.

RESULTS AND DISCUSSION

The sequence specificity of gene A protein and A* protein

The recognition sequence of gene A protein extends to the 5'-end of the conserved region of 30 b.p.. This has been shown by changing the 5'-terminal C in the decamer CAACTTGATA to A. The changed decamer is not cleaved by gene A protein (6). To determine the sequence specificity at the 3'-end partial digestion of the decamer was performed and three oligonucleotides shown in Table 1 were isolated. These products were labelled at the 5'-end with ^{32}P and extended at the 3'-ends using terminal transferase and one of the four dNTP's at a time. The resulting mixtures (which contain the starting oligonucleotide, with one dNMP added, two dNMP's, etc.) were incubated with gene A protein and A* protein. The products were analyzed by electrophoresis on a 25% polyacrylamide gel containing 7 M urea. After autoradiography the lanes were scanned. The increase of intensity of the heptamer band indicates cleavage by gene A protein (or A* protein). The results (Table 1) show that length and sequence at the 3'-end determine whether an oligonucleotide is cleaved. The first nucleotide after G must be A, the second T or G, the third can be any nucleotide. Gene A protein and A* protein have the same sequence requirements, CAACTTGA$^{T}_{G}$N. Incubations with ssDNA might show whether other sequences are also cleaved by gene A protein. Therefore we digested φX ssDNA with HaeIII and labelled the fragments at the 5'-ends with ^{32}P. Analysis of the products after incubation with gene

Table 1:   Incubation of gene A protein or A* protein with
           oligonucleotides extended using terminal transferase

Oligonucleotide	Extended with	Production of heptamer	
		gene A protein	A* protein
C A A C T T G	dTTP	−	−
	dCTP	−	−
	dGTP	−	−
	dATP	−	−
C A A C T T G A	dTTP	+	+
	dCTP	−	−
	dGTP	+	+
	dATP	−	−
C A A C T T G A T	dTTP	+	+
	dCTP	+	+
	dGTP	+	+
	dATP	+	+

A protein (Fig. 2a, b) reveals that gene A protein cleaves the se-
quence CAACTTG$^{\downarrow}$ATA (at the origin) and at a second site TTACTCG$^{\downarrow}$AGG
(in fragment Z_8) which is indicated by the appearance of a new frag-
ment of 98 nucleotides (resulting from cleavage of fragment Z_{6B}) and
the appearance of a fragment of 11 nucleotides (resulting from cleavage of fragment Z_8), respectively. The fragment of 11 nucleotides was
detected using a 25% polyacrylamide gel (not shown). The second site
is confirmed by cleavage of the partial Z_5-Z_8 which yields a fragment
slightly longer than Z_5.

A similar experiment was performed with ssDNA fragments of G4
(Fig. 3a, b). Again gene A protein cleaves twice, the sequence
CAACTTG$^{\downarrow}$ATA (at the origin in fragment Z_{2A}) and the sequence
ATACTCG$^{\downarrow}$AGT (in fragment Z_{5A}), giving rise to fragments of 188 and
63 nucleotides, respectively. A* protein (Fig. 2d and Fig. 3d) produces a large number of fragments which reflect the relaxed sequence
specificity of A* protein (11). By comparing the cleavage sites in
ssDNAs and in the oligonucleotides a consensus sequence can be deduced
(Table 2). G at position 1 is indicated by the work of Brown et al.
(9). C at position 6 is consistent with the work of Baas et al. (3)
using site directed mutagenesis. Introduction of G at position 9 did
not yield viable phage (3). In ϕX and G4 viral DNA the consensus
sequence occurs only at the two sites that are cleaved by gene A
protein.

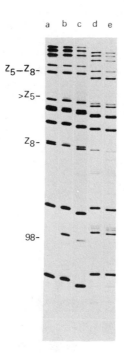

a b c d e

Z_5-Z_8-

$>Z_5-$

Z_8-

98-

Figure 2: Cleavage of φX ssDNA HaeIII fragments by gene A protein and A* protein. The ssDNA fragments were separated on a 6% polyacrylamide gel containing 7 M Urea using 40 V/cm for 3 hrs. The autoradiogram of the gel is shown. a: Control, 0.02 µg ssDNA fragments; b: + gene A protein, 0.5 ng, isolated according to Langeveld et al. (14); c: + SSB protein, 0.5 µg, purchased from P.L. Biochemicals, and gene A protein; d: A* protein, 1 ng; e: + SSB protein an A* protein.

Table 2:   Comparison of gene A protein cleavage sites

	1									10
Origin in φX and G4	C	A	A	C	T	T	G ↓	A	T	A
Second site in φX	T	T	A	C	T	C	G ↓	A	G	G
Second site in G4	A	T	A	C	T	C	G ↓	A	G	T
Oligonucleotides	C	A	A	C	T	T	G ↓	A	T/G	N

Consensus:

$$\begin{array}{ccccccccc}
C & & & & & & & & T & N \\
G & A & & & & T & & & & \\
  &   & A & C & T & & G ↓ & A & & G \\
A & T & & & & C & & & G & \\
T & & & & & & & & & T
\end{array}$$

    The conclusion that gene A protein can cleave ss φX viral DNA
not only at the origin, but also at a second site raises the following
question: Why is the ss tail of the looped rolling circle (8, 13) not
cleaved at the second cleavage site? This question also retains to
the fact that in a recombinant plasmid which contains the potential
cleavage site CAACTTGATATTAATA besides a complete φX origin (plasmid
pPR903) the former sequence is not cleaved during rolling circle DNA
synthesis (13). One possibility is that in the replication systems
(8, 13) some of the potential cleavage sites are protected against
gene A protein by single-strand binding (SSB) protein. This was tested
in the following experiments.

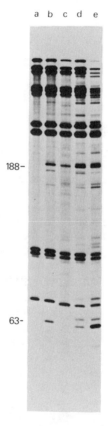

*Figure 3:   Cleavage of G4 ssDNA HaeIII fragments by gene A protein
and A* protein. The ssDNA fragments were separated on a 6% polyacrylamide gel containing 7 M Urea using 40 V/cm for 3 hrs. The autoradiogram of the gel is shown. a: Control, 0.02 μg ssDNA fragments;
b: + gene A protein, 0.5 ng; c: + SSB protein, 0.5 μg and gene A protein, 1 ng; d: + A* protein, 1 ng; e: + SSB protein and A* protein.*

Prior to incubation with gene A protein SSB protein was added to the mixture of ϕX ssDNA fragments. Analysis of the products obtained after incubation with gene A protein (Fig. 2c) shows the presence of the 98 nucleotide band, the absence of the cleavage product of the Z_5-Z_8 partial. The intensity of the Z_8 band has not decreased. So the cleavage by gene A protein at the second site is greatly reduced in the presence of SSB protein. Incubation of the mixture of G4 ssDNA fragments with SSB protein and gene A protein gave similar results (Fig. 3c). Cleavage at the second site is suppressed by SSB protein. The cleavage pattern obtained with A* protein is not much changed in the presence of SSB protein (Fig. 2e and Fig. 3e). In the incubation of G4 ssDNA fragments with A* protein, SSB protein even seems to favour cleavage at the second site since the intensity of the band of 63 nucleotides is increased.

The increased cleavage specificity of gene A protein in the presence of SSB protein may be explained in line with model presented in Fig. 1. The binding of SSB to ssDNA is stronger than that of gene A protein. This can be inferred from the fact that gene A protein is eluted from ssDNA cellulose columns at a lower salt concentration than SSB (14, 15). We assume that SSB protein prevents binding of gene A protein to ssDNA, which is therefore not cleaved. Only at the specific gene A protein binding site (the key sequence), at the origin, can the gene A protein displace SSB protein and bind and cleave the ss ϕX DNA. This explanation predicts that the complete origin sequence is required to cleave ssDNA in the presence of SSB protein which is in agreement with the results of Reinberg et al.(13).

The protein-DNA bond in the gene A protein-DNA complex

The protein-DNA bond was analyzed using the uniquely, internally ^{32}P-labelled oligonucleotide CAACTTG*ATATTAATAAC. This oligonucleotide was obtained by coupling CAACTTG to 5'-labelled ATATTAATAAC with T4 DNA ligase using a complementary oligonucleotide. The sequence of this oligonucleotide corresponds to the first 18 nucleotides of the origin region (Fig. 1). Gene A protein and A* protein were incubated with this radioactive substrate and analyzed on a SDS-polyacrylamide gel. Both proteins become labelled by this procedure (Fig. 4). This shows that gene A protein and A* protein bind covalently to the phosphate of the phosphodiester bond between G and A (nucleotide residues 4305 and 4306 in the complete ϕX DNA sequence (16). As expected, the presence of the covalently bound fragment of 11 nucleotides reduced the electrophoretic mobility of both proteins. In order to determine to which amino acid residue in gene A protein and A* protein the DNA is bound we hydrolyzed the ^{32}P-labelled protein in HCl. The hydrolysate was analyzed by paper electrophoresis. The radioactive material was detected by autoradiography. Phosphoserine, phosphothreonine and phosphotyrosine were electrophoresed as references and detected with ninhydrin. The position of the major radioactive spot coincides with inorganic phosphate (Fig. 5). The second spot coincides with phos-

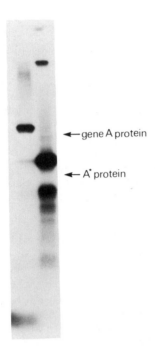

*Figure 4: Autoradiogram of the SDS-polyacrylamide gel of gene A
protein and A* protein after incubation with the internally ^{32}P-
labelled oligonucleotide CAACTTG*ATATTAATAAC. The arrows mark the
position of non-radioactive gene A protein and A* protein detected
by staining according to Boulikas and Hancock (20).*

photyrosine. The radioactive material present at the other positions
yielded phosphate and a radioactive spot at the position of phosphotyrosine after a second treatment with HCl. Thus, we conclude that
gene A protein and A* protein are linked to the 5'-phosphate of the
DNA via a tyrosyl residue. Recently, M. Roth at the Albert Einstein
College of Medicine (Bronx, N.Y.) also showed that gene A protein
is linked to DNA via a phosphotyrosyl bond (personal communication).
The topoisomerases I from Escherichia coli and M. luteus are also
bound to DNA via a tyrosyl residue (17). On the analogy of tyrosyl-
phosphate bonds of other proteins (18, 19), the tyrosyl-phosphate
in these complexes might be energy rich and thus supply the energy
required for joining of the protein bound 5' phosphoryl group of
the DNA to the 3' hydroxyl group of the acceptor DNA.

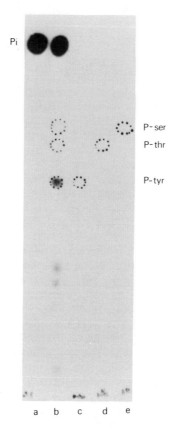

Figure 5: Autoradiogram after paper electrophoresis of the acid hydrolysates of ^{32}P-labelled gene A protein and A* protein. Hydrolysis was performed by incubation for 3 hrs. at 110°C in 3 N HCl. After lyophilization the samples were dissolved in H_2O and subjected to electrophoresis at 30 V/cm for 2 hrs. on Whatman 3M paper at pH 3.5. Radioactive material was detected by autoradiography, the non-radioactive phospho-amino acids were detected with ninhydrin. a: Hydrolysate of the radioactive oligonucleotide CAACTTG*ATATTAATAAC; b: hydrolysate of gene A protein and A* protein after incubation with the internally ^{32}P-labelled oligonucleotide; c: phosphotyrosine; d: phosphothreonine; e: phosphoserine.

ACKNOWLEDGEMENTS

We thank H.A.A.M. Van Teeffelen for skillfull technical assistance, G.H. Veeneman and J.H. Van Boom (Leiden) for the synthetic oligonucleotides and phosphotyrosine, P.J. Weisbeek (Utrecht) for T4 DNA ligase and The Netherlands Organization for the Advancement of Pure Research (ZWO) for financial support.

REFERENCES

1. Heidekamp,F., Langeveld,S.A., Baas,P.D. and Jansz,H.S. (1980)
       Nucl. Acids Res. 8, 2009.
2. Heidekamp,F., Baas,P.D. and Jansz,H.S. (1982) J. Virol. 42, 91.
3. Baas,P.D., Teertstra,W.R., Van Mansfeld,A.D.M., Jansz,H.S.,
       Van der Marel,G.A., Veeneman,G.H. and Van Boom,J.H.
       (1981) J. Mol. Biol. 152, 615.
4. Van Mansfeld,A.D.M., Langeveld,S.A., Baas,P.D., Jansz,H.S.,
       Van der Marel,G.A., Veeneman,G.H. and Van Boom,J.H.
       (1980) Nature (London) 283, 561.
5. Heidekamp,F., Baas,P.D., Van Boom,J.H., Veeneman,G.H.,
       Zipursky,S.L. and Jansz,H.S. (1981) Nucl. Acids Res.
       9, 3335.
6. Baas,P.D., Heidekamp,F., Van Mansfeld,A.D.M., Jansz,H.S.,
       Van der Marel,G.A., Veeneman,G.H. and Van Boom,J.H.
       (1981) in: "ICN-UCLA Symposia on Molecular and Cellular
       Biology: The Initiation of DNA Replication", Vol. XXI,
       D.S. Ray and C.F. Fox, eds., Academic Press, New York.
7. Ikeda,J.-E., Yudelevich,A., Shimamoto,N. and Hurwitz,J.
       (1979) J. Biol. Chem. 254, 9416.
8. Eisenberg,S. and Kornberg,A. (1979) J. Biol. Chem. 254, 5328.
9. Brown,D.R., Reinberg,D., Schmidt-Glenewinkel,D., Roth,M.,
       Zipursky,S.L. and Hurwitz,J. (1982) Cold Spring Harbor
       Symp. Quant. Biol. 47, 701.
10. Linney,E. and Hayashi,M. (1973) Nature New Biol. 245, 6.
11. Langeveld,S.A., Van Mansfeld,A.D.M., Van der Ende,A.,
       Van de Pol,J.H., Van Arkel,G.A. and Weisbeek,P.J. (1981)
       Nucl. Acids Res. 9, 545.
12. Brown,D.R., Hurwitz,J., Reinberg,D. and Zipursky,S.L. (1982)
       in: "Nucleases", S.M. Linn and R.J. Roberts, eds.,
       Cold Spring Harbor Labroatory, Cold Spring Harbor, p.187.
13. Reinberg,D., Zipursky,S.L., Weisbeek,P., Brown,D. and
       Hurwitz,J. (1983) J. Biol. Chem. 258, 529.
14. Langeveld,S.A., Van Arkel,G.A. and Weisbeek,P.J. (1980)
       FEBS Lett. 114, 269.
15. Kowalczykowski,S.C., Bear,D.G. and Von Hippel,P.H. (1981)
       in: "The Enzymes", Vol. XVI, S. Boyer, ed., Academic
       Press, New York, p. 373.
16. Sanger,F., Coulson,A.R., Friedmann,T., Air,G.M., Barrell,B.G.
       Brown,N.L., Fiddes,J.C., Hutchinson III,C.A.,
       Slocombe,P.M. and Smith,M. (1978) J. Mol. Biol. 125, 255.
17. Tse,Y.-C., Kirkegaard,K. and Wang,J.C. (1980) J. Biol. Chem.
       255, 5560.
18. Holzer,H. and Wohlhueter,R. (1972) Adv. Enzyme Reg. 10, 121.
19. Fukami,Y. and Lipman,F. (1983) Proc. Natl. Acad. Sci. USA
       80, 1872.
20. Boulikas,T. and Hancock,R. (1981) J. Biochem. Biophys. Methods
       5, 219.

GENE A PROTEIN INTERACTING WITH RECOMBINANT PLASMID DNAs

CONTAINING 25-30 b.p. OF THE φX174 REPLICATION ORIGIN

A.C. Fluit, P.D. Baas, H.S. Jansz, G.H. Veeneman[1]
and J.H. van Boom[1]

Institute of Molecular Biology and Laboratory for
Physiological Chemistry
State University of Utrecht
Utrecht, The Netherlands

[1]Department of Organic Chemistry
State University of Leiden
Leiden, The Netherlands

SUMMARY

Synthetic oligodeoxyribonucleotides, DNA ligase and DNA polymerase were used to construct double-stranded DNA fragments homologous
to the first 25, 27 or 30 b.p. of the 30 b.p. origin region of bacteriophage φX174 (nucleotides 4299-4328 of the φX174 DNA sequence).

The double-stranded DNA fragments were cloned into the kanamycin
resistance gene of pACYC177 (Amp[R], Km[R]). Transformants were picked
up by antibiotic selection and filter-hybridization using one of the
oligodeoxyribonucleotides as a probe. Approximate lengths of the inserts were determined by restriction enzyme analysis. Exact length
and orientation of each insert was determined by DNA sequencing.

Plasmid DNA with an insert homologous to the first 25 b.p. of
the φX174 origin is not nicked by the gene A protein. However, plasmid
DNA containing the 27 b.p. fragment in either orientation is nicked
by the gene A protein, as well as plasmid DNAs containing the first
28 b.p. or the complete 30 b.p. conserved origin region of the isometric phages.

231

INTRODUCTION

The bacteriophage ϕX174 gene A protein is necessary in various steps of the ϕX174 RF DNA replication cycle. The gene A protein initiates RF DNA replication by cleavage of the viral strand in superhelical DNA. The cleavage site is located in a 30 b.p. stretch within gene A, which is highly conserved among six isometric phages (1). The gene A protein is also involved during elongation and termination of viral strand DNA synthesis (2).

Several studies (3-7) in our laboratory have led to the following model for gene A protein action during initiation of RF DNA replication (see Figure 1 in the paper of Van Mansfeld et al. (10)). The 30 b.p. conserved origin region has two functional domains, a recognition sequence and a key (binding) sequence for gene A protein, which are separated by an AT-rich region.

Initiation of ϕX RF DNA replication takes place in three steps. First, the gene A protein binds to the key (binding) sequence. This binding causes local unwinding of the origin, which is driven by the available energy in the superhelical turns. The recongnition sequence becomes exposed in single-stranded form. Then the single-strand specific gene A protein can cleave the viral strand next to the G-residue at position 4305 of the ϕX174 DNA sequence (8, 9) and binds covalently to the 5'-phosphate of the A-residue at position 4306 (10). The resulting RFII-A protein complex can start rolling circle DNA replication when provided with additional factors.

The present study was undertaken to determine the minimum length of the ϕX174 origin region required for gene A protein action during initiation of replication. According to the model, the left part of the origin is formed by the recognition sequence of the gene A protein, which has been characterized extensively (3, 10). The right part of the origin is formed by the postulated key (binding) sequence, which boundaries are still undefined.

A previous study of Heidekamp et al. (6) had shown that the first 20 b.p. of the ϕX174 origin, which includes the recognition sequence, the complete AT-rich region and possibly part of the postulated key (binding) sequence, are insufficient for gene A protein interaction. However, in a recombinant plasmid DNA with a 49 b.p. insert of the isometric phage G4, including the 30 b.p. conserved origin sequence, the viral strand is cleaved by gene A protein and the plasmid DNA is successfully replicated in the ϕX174 in vitro replication system (11).

RESULTS AND DISCUSSION

Construction of recombinant plasmid DNAs containing the first 25, 27 or 30 b.p. of the φX174 replication origin

Double-stranded DNA fragments homologous to the first 25, 27 or 30 b.p. of the φX174 origin (nucleotides 4299-4328 of the φX174 DNA sequence (9)) were constructed following the strategy shown in Figure 1.

The oligodeoxyribonucleotide 1-16 of the (+) strand was hybridized for 2 hrs at 4°C to oligodeoxyribonucleotide 1-10 of the (-) strand. Twenty-fold excess of the latter oligodeoxyribonucleotide was used to minimize possible self-hybridization of the (+) strand oligodeoxyribonucleotide. Then a 25-fold excess of oligodeoxyribo-

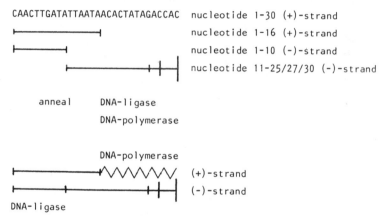

Figure 1: *Strategy for the construction of double-stranded φX174 origin fragments. The different oligodeoxyribonucleotides were labelled to a low specific activity using (γ-^{32}P)-ATP and T4 polynucleotide kinase purchased from PL Biochemicals Inc.). Sixty pmol of the (+) strand oligodeoxyribonucleotide was hybridized to 1200 pmol of the oligodeoxyribonucleotide 1-10 of the (-) strand for 2 hrs at 4°C in 71 mM KPO_4 (pH 7.4), 11.5 mM DTT in a volume of 37 μl. Then 13 μl containing -500 pmol 11-25, 11-27 or 11-30 of the (-) strand was added and hybridization continued for 1½ hrs at 4°C. Then, $MgCl_2$, ATP and an appropriate amount of T4 DNA ligase were added resulting in a buffer concentration of 31 mM KPO_4 (pH 7.4), 5 mM DTT, 25 mM $MgCl_2$ and 1 mM ATP. The ligation reaction was carried out for 16 hrs at 4°C. After ligation the mixture was diluted ten-fold and the incubation mixture was adjusted to 31 mM KPO_4 (pH 7.4), 5 mM DTT, 10 mM $MgCl_2$ and 50 μM of each of the four dNTPs. DNA polymerase reaction was carried out at 4°C during 2 hrs using 2.4 U Klenow DNA polymerase (purchased from PL Biochemicals Inc.).*

nucleotide 11-25, 11-27 or 11-30 of the (-) strand was added (depending on the fragment to be constructed) and hybridization was continued for 1½ hrs. For 16 hrs the (-) strand oligodeoxyribonucleotides
were ligated using T4 DNA ligase. The fragments were made double-
stranded using Klenow DNA polymerase. The different oligodeoxyribonucleotides were labelled with ^{32}P at their 5'-ends.

Double-stranded fragments were partially purified by applying
the total mixture to a BND-cellulose column (12). Fractions containing
double-stranded DNA fragments were concentrated after which salt and
smaller DNA fragments were removed by Sephadex G-50 chromatography.
Product formation and purification were analyzed by polyacrylamide
gel electrophoresis (Figure 2).

The double-stranded fragments were ligated into the unique
SmaI site of pACYC177, which is located in the kanamycin resistance
gene. Escherichia coli HF4704 (Thy⁻, uvrA, ϕX^S, sup⁻) was used as
the recipient strain in the transformation. Transformants were selected by plating on kanamycin and ampenicillin containing Giston-
agar plates.

The possibility existed that clones with 27 or 30 b.p. inserts
could not be detected by antibiotic selection, because it was not
known if an insertion of 9 or 10 amino acids renders the kanamycin
phosphotransferase inactive. However, the results showed that insertion of 9 or 10 codons at the SmaI site in the gene resulted in a
phenotype which is kanamycin sensitive. Further selection was carried
out using filter-hybridization of plasmid DNA isolated according to
the method of Birnboim and Doly (13) and by hybridization of colonies
grown on nitro-cellulose filters using plasmid copy number stimulation by chloramphenicol. The oligodeoxyribonucleotide 11-25 of the
(-) strand, labelled at the 5'-end with (^{32}P), was used as a probe
for the assay.

Colonies and DNA isolated from kanamycin resistant transformants
were negative. Approximately 70% of the colonies and plasmid DNAs of
kanamycin sensitive transformants were positive (Figure 3).

Characterization and DNA sequence analysis of recombinant plasmid
DNAs containing 25, 27 or 30 b.p. of the ϕX174 replication origin

Four colonies from each series of transformation experiments
were chosen for further study. A TaqI restriction enzyme digestion
was carried out to determine the presence and approximate length of
the insert. Two transformants of the 25 and 27 b.p. series had an
insert of the expected length. The remaining plasmids in these series
showed an insert longer than 25 or 27 b.p.. The 4 transformants from
the 30 b.p. series showed an insert of approximately 30 b.p. (see
also Figure 4). A larger quantity of plasmid DNA was prepared from
2 clones of each series, which showed a single insert.

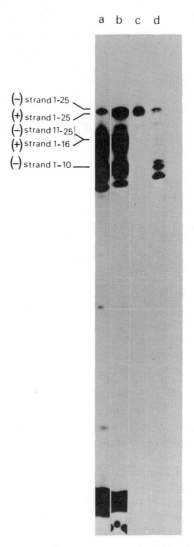

Figure 2: Analysis of the production and purification of double-stranded DNA fragments homologous to the first 25 b.p. of the $\phi X174$ origin on a denaturing polyacrylamide gel. Oligodeoxyribonucleotides were labelled with ^{32}P at their 5'-ends. Lane a: formation of oligodeoxyribonucleotide 1-25 of the (-) strand after incubation with T4 DNA ligase; lane b: formation of oligdeoxyribonucleotide 1-25 of the (+) strand after the DNA polymerase reaction; lane c: 25 b.p. double-stranded DNA fragments purified over BND-cellulose and Sephadex G-50; lane d: material in discarded fractions after the final Sephadex G-50 column. (Lengths are indicated in nucleotides).

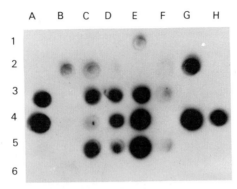

*Figure 3: Filter hybridization of DNA isolated according to Birnboim
and Doly (13) from kanamycin sensitive transformants of the 27 b.p.
series. Twenty-seven pmol of ^{32}P end-labelled oligodeoxyribonucleotide 11-25 of the (-) strand (specific activity: 1 x 10^6 cpm/pmol)
was hybridized to plasmid DNA prepared from 1 ml cells on filters
for 18 hrs at 23°C in buffer containing: 0.2% polyvinylpyrolidone,
0.2% Ficoll, 6 x SSC (1 x SSC equals 0.15 M sodium chloride, 0.015 M
sodium citrate, pH 7.0), 2 mg/ml BSA in a volume of 25 ml. Filters
were washed 6 times for 1½ hrs in 2 x SSC at 23°C. pACYC177, as a
negative control, is presented in B3 and G5. Only transformants A3,
E3, G4 and E5 were analyzed further. Restriction enzyme analysis
showed that plasmid DNA obtained from transformants A3 and E3 contained a single insert. Plasmid DNA obtained from the transformants
G4 and E5 showed a larger insert.*

Nucleotide sequence and orientation of the inserts was determined using the chemical degradation method of Maxam and Gilbert (14).
The following strategy was used. Plasmid DNA was linearized by restriction enzyme digestion with ClaI, followed by 5'-end labelling
with (γ-^{32}P)-ATP. A second restriction endonuclease digestion was
carried out with FnuDII, which gives rise to 9 fragments. Two fragments are labelled; one 34 b.p. fragment and a larger fragment of
approximately 340 b.p., carrying the insert. The total mixture was
used in chemical modification, because the small fragment does not
interfere with the reading of the nucleotide sequence of the insert.
Results are shown in Table 1.

Gene A protein interaction with recombinant plasmid DNAs containing
25, 28 or 30 b.p. of the φX174 replication origin

Recombinant plasmid DNA was incubated with φX174 gene A protein
together with φX174 RFI DNA as an internal control. Plasmid DNA which
contains 25 b.p. homologous to the origin was nicked by the φX174
gene A protein. However, plasmid DNAs containing 27 b.p. in either
orientation, 28 b.p. or the complete 30 b.p. origin were nicked by

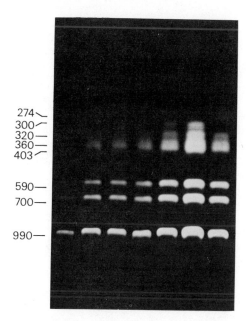

Figure 4: TaqI restriction enzyme digestion of 6 plasmid DNAs and pACYC177 DNA (right lane). Each plasmid DNA contains a single insert in the 274 b.p. TaqI restriction enzyme fragment, which in pACYC177 contains the SmaI site. TaqI restriction endonuclease was from Boehringer. Assay conditions were as described by the manufacturer. Electrophoresis was carried out on a 2% agarose gel. (Lengths are indicated in b.p.).

the gene A protein. Results are shown in Figure 5. So homology to the first 27 b.p. of the φX174 replication origin is sufficient for cleavage by gene A protein.

These results are in agreement with and extend recent results obtained by Brown et al. (15), who showed that a recombinant plasmid DNA containing the first 28 b.p. of the φX174 origin is nicked by the gene A protein and is also successfully replicated in the φX174 in vitro replication system.

Cloning of a fragment corresponding to the first 26 b.p. of the φX174 origin in a plasmid will be necessary to determine exactly the minimum length required for cleavage by gene A protein and is in progress. These plasmids will be tested for their ability to support DNA replication in in vitro and in vivo φX174 DNA replication systems (11, 16).

Table 1: DNA sequences of the φX174 origin and of 6 plasmids containing inserts partially or totally homologous to the φX174 origin and surrounding sequences

φX174	G T G C T C C C C A A C T T G A T A T T A A T A A C A C T A T A G A C C A C C	G C C C C C G A A
pAF25A	C T G T T T C C C A A C T T G A T A T T A A T A A C A C T A T A G - - - -	G G G G A T C G C
pAF25B	C T G T T T C C C A A C T T G A T A T T A A T A A C A C T A T A G - - - -	G G G G A T C G C
pAF27A	T G T T T T C C C A A C T T G A T A T T A A T A A C A C T A T A G A C - - -	G G G G A T C G C
pAF27B	G C C A T C C C C A A C T T G A T A T T A A T A A C A C T A T A G A G - - -	G G G A A A A C A
pAF28	T G T T T T C C C A A C T T G A T A T T A A T A A C A C T A T A G A C C - -	G G G G A T C G C
pAF30	C T G T T T C C C A A C T T G A T A T T A A T A A C A C T A T A G A C C A C C	G G G G A T C G C

Three recombinant plasmid DNAs, pAF25A, pAF25B and pAF30 apparently lost one C-G base pair of the cleaved SmaI site or the first base pair of the synthetic fragment during the enzymatic reactions. One of the analyzed plasmids (pAF28) from the transformation experiment with the 30 b.p. double-stranded fragments contains an insert of 28 b.p.. This is not surprising because of the heterogeneity of the 5'-end of the used oligodeoxyribonucleotide 11-30 of the (-) strand.

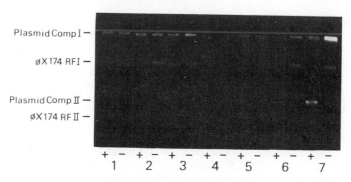

*Figure 5:   Analysis of gene A protein incubations of supercoiled
covalently closed plasmid DNAs containing different parts of the
$\phi X174$ origin. Supercoiled plasmid DNA and $\phi X174$ RFI, 0.1 μg each,
were incubated with (+) and without (-) gene A protein for 30 min
at 30°C in a reaction volume containing: 28 mM Tris-HCl (pH 7.5),
0.6 mM EDTA, 5 mM $MgCl_2$, 125 mM NaCl, 5% glycerol and 0.025 °/oo
nonidet P40. Reactions were terminated by adding 0.25 M EDTA (pH 8.0)
to a concentration of 30 mM. Then, 3 μl of a 5 mg/ml proteinase K
solution was added followed by incubation for 30 min at 37°C. Conversion of component I DNA into component II DNA was analyzed after
electrophoresis on a 1.5% agarose gel. Lane 1: pACYC177; lane 2:
pAF25B; lane 3: pAF25A; lane 4: pAF27B: lane 5: pAF27A; lane 6:
pAF28; lane 7: pAF30.*

ACKNOWLEDGEMENTS

This work was supported in part by The Netherlands Organization
for the Advancement of Pure Research (ZWO) with financial aid from
the Foundation for Chemical Research (SON).

REFERENCES

1. Heidekamp,F., Baas,P.D. and Jansz,H.S. (1982) J. of Virol.
    42, 91.
2. Eisenberg,S., Griffith,J. and Kornberg,A. (1977)
    Proc. Natl. Acad. Sci. USA 74, 3198.
3. Van Mansfeld,A.D.M., Langeveld,S.A., Baas,P.D., Jansz,H.S.,
    Van der Marel,G.A., Veeneman,G.H. and Van Boom,J.H.
    (1980) Nature (London) 288, 561.
4. Baas,P.D., Heidekamp,F., Van Mansfeld,A.D.M., Jansz,H.S.,
    Langeveld,S.A., Van der Marel,G.A., Veeneman,G.H. and
    Van Boom,J.H. (1980) in: "ICN-UCLA Symposia on Molecular
    and Cellular Biology: Mechanistic Studies of DNA Replica-
    tion and Genetic Recombination", Vol. XIX, B. Alberts,
    ed., Academic Press, New York.

5. Baas,P.D., Teertstra,W.R., Van Mansfeld,A.D.M., Jansz,H.S.,
    Van der Marel,G.A., Veeneman,G.H. and Van Boom,J.H.
    (1981) J. Mol. Biol. 152, 615.
6. Heidekamp,F., Baas,P.D., Van Boom,J.H., Veeneman,G.H.,
    Zipursky,S.L. and Jansz,H.S. (1981) Nucl. Acids Res.
    9, 3335.
7. Baas,P.D., Heidekamp,F., Van Mansfeld,A.D.M., Jansz,H.S.,
    Van der Marel,G.A., Veeneman,G.H. and Van Boom,J.H.
    (1981) in: "ICN-UCLA Symposia on Molecular and Cellular
    Biology: The Initiation of DNA Replication", Vol. XXII,
    D.S. Ray, ed., Academic Press, New York.
8. Langeveld,S.A., Van Mansfeld,A.D.M., Baas,P.D., Jansz,H.S.,
    Van Arkel,G.A. and Weisbeek,P.J. (1978) Nature (London)
    271, 417.
9. Sanger,F., Coulson,A.R., Friedman,T., Air,G.M., Barrell,B.G.,
    Brown,N.L., Fiddes,J.C., Hutchinson III,C.A., Slocombe,P.M.
    and Smith,M. (1978) J. Mol. Biol. 125, 255.
10. Van Mansfeld,A.D.M., Baas,P.D. and Jansz,H.S., this volume.
11. Brown,D.R., Reinberg,D., Schmidt-Glenewinkel,T., Zipursky,S.L.
    and Hurwitz,J. (1983) in: "Methods in Enzymology", Vol.
    100, R. Wu, L. Grossman and K. Moldave, eds., Academic
    Press, New York.
12. Kiger Jr.,J.A. and Sinsheimer,R.L. (1969) J. Mol. Biol. 40, 467.
13. Birnboim,H.C. and Doly,A. (1970) Nucl. Acids Res. 7, 1513.
14. Maxam,A.M. and Gilbert,W. (1980) in: "Methods in Enzymology",
    Vol. 65, L. Grossman and K. Moldave, eds., Academic Press,
    New York.
15. Brown,D.R., Schmidt-Grenewinkel,T., Reinberg,D. and Hurwitz,J.
    (1983) J. Biol. Chem. 258, 8402.
16. Van der Ende,A., Teertstra,R. and Weisbeek,P.J. (1982) Nucl.
    Acids Res. 10, 6849.

HOW DOES SV40 T ANTIGEN CONTROL INITIATION OF VIRAL DNA REPLICATION?

E. Fanning, C. Burger[1], B. Huber, U. Markau, S. Sperka,
S. Thompson, E. Vakalopoulou and B. Vogt

Institute for Biochemistry
Karlstrasse 23
D-8000 Munich 2, Germany

[1]Faculty for Biology
University of Constance
D-7750 Konstanz, Germany

INTRODUCTION

Simian virus 40 (SV40) large T antigen is a multifunctional protein involved in the regulation of viral transcription and DNA replication in lytically infected monkey cells and in the establishment and maintenance of cell transformation by SV40 (1). T antigen occurs in several different forms which differ biologically and biochemically. T antigen binds specifically to several sites in the origin region of SV40 DNA, acts as an ATPase and associates stably with a transformation- and cell cycle-related cellular phosphoprotein p53 (1). Yet how these properties enable this remarkable protein to fulfill its many varied functions remains unknown. This communication summarizes the biochemical properties of different forms of T antigen from two groups of mutants defective in initiation of SV40 DNA replication, as compared with those of wild-type T antigen. Based on these results, we suggest a possible mechanism for T antigen function in the initiation of SV40 DNA replication and present the results of initial experiments designed to test this model.

RESULTS AND DISCUSSION

Newly synthesized soluble T antigen occurs as a monomer in SV40-infected and -transformed cells (2). Monomeric T antigen is normally converted in a time-dependent process to a stable highly phosphorylated tetramer which accumulates in productively infected cells. This conversion fails to take place at nonpermissive temperature in cells infected with a tsA mutant which codes for temperature-sensitive T antigen. Similarly, initiation of viral DNA replication in tsA infection ceases within 20 to 30 min after a shift to non-permissive temperature. However, this defect, like the defect in conversion of T antigen to tetramer, can be reversed by a shift-down to permissive temperature (2). These data led us to propose that initiation of viral DNA replication may be dependent on newly synthesized T antigen and perhaps correlated with its aggregation and phosphorylation (Figure 1). Specific binding of T antigen to the origin of replication has been shown to be essential for initiation of replication (1). Thus, if monomeric and tetrameric T antigen have different functions in replication, one might expect the two forms to exihibit differences in their DNA binding properties.

A simple rapid immunoprecipitation assay was developed to study binding of fresh crude extracts of T antigen to viral and cellular DNA (3). The origin binding activity of monomeric soluble lytic T antigen was greater than that of the tetrameric form. However, both forms had nearly equal affinity for origin DNA, as measured by rate dissociation experiments (3). Their affinities for cellular DNA, on the other hand, were strikingly different. Monomeric T antigen dissociated from excess cellular DNA within 2 h and bound to origin DNA, whereas tetrameric T antigen remained bound to cellular DNA and failed to bind to origin DNA, even though it bound to origin in the absence of cellular DNA (Figure 2). Thus, tetrameric T antigen appears to have a greater affinity for cellular DNA than monomeric T antigen. This conclusion was confirmed using chromatographically purified forms of T antigen in a filter binding assay (4). These results suggest that tetrameric T antigen may be unable to locate

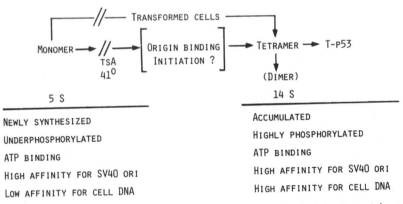

Figure 1: *Monomeric newly synthesized soluble lytic T antigen is converted in a time-dependent process to a tetrameric form. Biochemical properties of each form of T antigen are listed.*

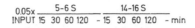

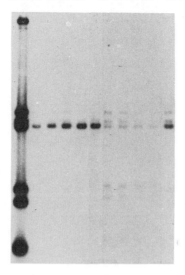

Figure 2: Affinity of monomeric and tetrameric lytic T antigen for viral and cellular DNA. Crude extract from SV40-infected TC7 cells was fractionated by zone velocity sedimentation. End-labeled HindIII fragments of pSV-WT DNA (0.2 µg per assay) were incubated with mono-meric (5-6S) and tetrameric (14-16S) T antigen without (-) and with 10 µg of calf thymus DNA for the time period indicated. Bound DNA was immunoprecipitated and analyzed by electrophoresis in 2% agarose gels and autoradiography. The C fragment contains the viral origin of replication. (reprinted from Fanning et al., (3)).

origin sequences on the few copies of viral DNA present in the nucleus of the cell early in infection, offering a plausible explanation for dependence of SV40 DNA replication on newly synthesized T antigen.

Given that monomeric T antigen efficiently locates and binds to viral origin DNA, what might be the next step in the initiation process? Since aggregation of T antigen appears to be essential for initiation (2), we propose that aggregation, possibly on the DNA, into a tetrameric form may be the next step (Figure 3). The mechanism of initiation might well include transcriptional activation, origi-nally proposed for initiation of phage lambda DNA replication (5), triggered by T antigen binding-related changes in early transcription. In support of this idea, partial deletion of the early SV40 promoter has been shown to cause a dramatic decrease in initiation of DNA replication (6, 7). Phosphorylation, which may be coupled with aggre-gation ((8) and K.H. Scheidtmann, personal communication), could control the amount of newly synthesized T antigen available for spe-cific binding to the origin (Figure 1, 3).

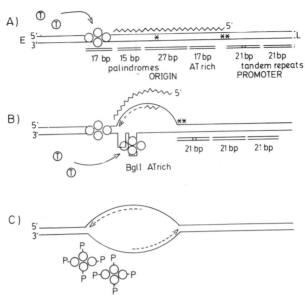

Figure 3: A model for T antigen function in initiation of SV40 DNA replication. (A) Newly synthesized underphosphorylated T antigen binds specifically to the primary binding site near the origin of replication. The 5' end (*) of the early transcript (〰〰) is thus shifted to an upstream site (**) (9, 10). (B) The resulting transiently single-stranded DNA could form a hairpin, which may be stabilized by binding of newly synthesized T antigen to sites in the hairpin. Host enzymes may then initiate leading strand synthesis (- - - -) on one or both template strands (11). (C) Soluble phosphorylated tetrameric T antigen, once dissociated from SV40 DNA, would be unlikely to bind specifically to origin, and would thus no longer function in initiation of viral DNA replication (2, 3).

        In our initial attempts to test these ideas, monomeric T antigen has been treated, in the presence and absence of viral DNA, with the covalent cross-linking agent dimethylsuberimidate (Figure 4). This agent efficiently cross-links T antigen at pH 8.5 through the ε-amino groups of lysine residues (S.T. and E.F., unpublished). Since binding of T antigen to the origin region is not detectable at pH 8.5, binding and cross-linking had to be carried out at pH 7.8, although cross-linking is rather inefficient under these conditions. Monomeric T antigen remained monomeric after cross-linking except when SV40 DNA was added before the reaction: in the presence of viral DNA, traces of dimeric T antigen were formed (Figure 4). Considering that cross-linking was inefficient at pH 7.8, the occurrence of higher oligomers of T antigen after cross-linking suggests that T antigen may, in fact, aggregate at the origin. In support of this interpretation,

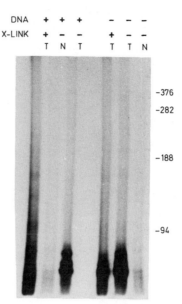

Figure 4:   Chemical cross-linking of monomeric lytic T antigen in the presence of viral DNA. ^{35}S-methionine-labeled monomer was incubated with (+) and without (-) 0.25 µg of HindIII-cleaved pSV-WT DNA and 10 µg of calf thymus DNA at pH 7.8 as described (3). Samples were incubated with (+) and without (-) 5 mM dimethylsuberimidate for 20 min at 23°C, followed by immunoprecipitation with anti-T or control (N) serum, SDS gel electrophoresis and fluorography. Cross-linked phosphorylase was used as a molecular weight marker, as indicated on the right. Some proteolytic degradation of T antigen was apparent in this experiment.

the patterns of DNase footprinting of monomeric, dimeric and tetrameric lytic T antigen bound to SV40 are identical (E.V. and E.F., unpublished).

Further dissection of the initiation process will require nonconditional T antigen mutants deficient in initiation of viral DNA replication, but whose other functions remain as intact as possible, e.g. transcriptional control and cell transformation. We have screened a number of SV40-transformed cells for such mutant T antigens. Polyethylene glycol-mediated fusion of each transformed cell line with uninfected TC7 monkey cells and with cos 1 SV40-transformed cells (12) was used as an assay (B.H. and E.F., unpublished). If the integrated viral DNA in each line contains a functional origin, then fusion with cos 1 cells should result in excision and replication of viral DNA sequences under the control of the replication-competent cos 1 T antigen. All of the lines tested appeared to contain a functional origin by this criterium (Table 1). Assuming that a transformed

Table 1:   Properties of T antigen from SV40-transformed cells

Cell line	Excision after fusion with		Origin DNA binding	Cellular DNA binding (monomer)	ATP affinity label	p53 complex
	cos 1	TC7				
SVT2	+	−	low	high	+	+
VLM	+	−	high	low	+	+
H65/90B	+	−	low	high	+	+
THK-1$_T$	+	+	high	low	+	+
C2	+	−	high	low	+	+
SV80	+	−	low	?	+	+
cos 1	−	−	high	low	+	+

line contains a functional SV40 origin, fusion with TC7 cells leads
to rescue and replication of viral DNA only if the transformed line
expresses a replication-competent T antigen. Five lines expressed
T antigens defective for initiation of viral DNA replication (Table
1). Biochemical analysis of these mutant proteins excluded major
structural defects, failure to oligomerize and inability to associate
with the cellular protein p53 (13). DNA binding properties of each
T antigen were tested in immunoprecipitation assays and compared with
those of cos 1 T antigen (Table 1). The specific binding activity of
T antigen varied widely between lines, but all of those with low
origin binding activity were defective in initiation of SV40 DNA
replication. This defect in binding may be responsible for their loss
of replication activity. Interestingly, several T antigens, such as
VLM, were just as active in origin DNA binding as lytic T antigen
(Table 1, (13)). However, it is conceivable that a more subtle defect
in the DNA binding properties of VLM T antigen, undetectable by the
immunoprecipitation assay, might lead to loss of replication activity.
Alternatively, VLM T antigen might be defective in a subsequent step
in the initiation process, which would make it a useful tool in
identifying that step.

We have thus undertaken a detailed investigation of the DNA
binding properties of VLM T antigen in comparison with those of lytic
T antigen. For example, in an attempt to define the minimal DNA sequence which can be recognized and specifically bound by T antigen,
short sequences identical or similar to segments of the viral origin
have been chemically synthesized and cloned in a plasmid vector. Binding of T antigen to the cloned DNA has been assayed by immunopreci-

pitation, DNase protection and DNase footprinting. A synthetic sequence which functioned as a high affinity binding site after insertion into the vector pAT153 is shown in Figure 5. The monomeric,
dimeric and tetrameric forms of lytic T antigen bound specifically
to the 646 bp end-labeled fragment containing the synthetic sequence
as well as to the 1118 bp SV40 origin fragment. The corresponding
621 bp vector fragment without the synthetic sequence was not bound.
VLM T antigen bound to this minimal binding site and to intact origin

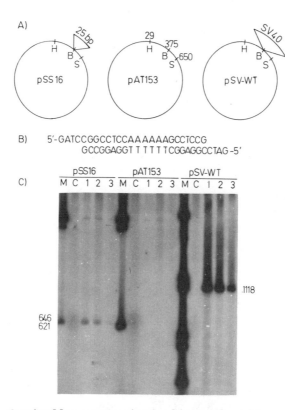

*Figure 5: A chemically synthesized oligonucleotide acts as a binding
site for lytic T antigen. The 25 bp DNA sequence shown in (B) was synthesized according to Dörper and Winnacker (14) and cloned into the
Bam H-I site of the vector pAT153 as shown in (A). (C) Immunoprecipitation DNA binding assays of HindIII-cleaved pSV-WT and HindIII/SalI-
cleaved pSS16 and pAT153 end-labeled DNA with monomeric (1), dimeric
(2) and tetrameric (3) T antigen were performed. As a marker (M), the
DNA fragments used in each assay were run in parallel. C shows a
control without T antigen extract but with DNA, antiserum and S.
aureus. The numbers at the sides refer to the length of the fragment,
in bp, which carries the DNA sequence of interest in this experiment.*

in the same fashion as did lytic T antigen in the immunoprecipitation
and DNase footprint assays (E.V., U.M. and E.F., unpublished). Further
experiments to clarify the nature of the VLM T antigen defect are in
progress.

In conclusion, studies of the different forms of lytic T antigen
and their biochemical properties have enabled us to propose a working
model for T antigen function in SV40 DNA replication. Several features
of this mechanism are being tested by chemical cross-linking, DNase
footprinting and characterization of mutant T antigens unable to
initiate viral DNA replication.

ACKNOWLEDGEMENTS

We thank E. Arnold for outstanding technical assistance, A.
Moosbauer for construction of laboratory equipment, T. Dörper and
E.-L. Winnacker for advice and precursors for chemical oligonucleotide synthesis. This work was supported by the Deutsche Forschungs-
gemeinschaft, Stiftung Volkswagenwerk and Fonds der Chemischen Industrie.

REFERENCES

1.   Rigby,P.W.J. and Lane,D.P. (1983) in: "Advances in Viral
            Oncology", Vol. 3, G. Klein, ed., Raven Press, New York.
2.   Fanning,E., Nowak,B. and Burger,C. (1981) J. Virol. 37, 92.
3.   Fanning,E., Westphal,K.-H., Brauer,D. and Cörlin,D. (1982)
            EMBO J. 1, 1023.
4.   Dorn,A., Brauer,D., Otto,B., Fanning,E. and Knippers,R.
            (1982) Eur. J. Biochem. 128, 53.
5.   Dove,W.F., Inokuchi,H. and Stevens,W.F. (1971) in: "The
            Bacteriophage Lambda", A.D. Hershey, ed., Cold Spring
            Harbor Laboratory.
6.   Bergsma,D.J., Olive,D.M., Hartzell,S.W. and Subramanian,K.N.
            (1982) Proc. Natl. Acad. Sci. USA 79, 381.
7.   Byrne,B.J., Davis,M.S., Yamaguchi,J., Bergsma,D.J. and
            Subramanian,K.N. (1983) Proc. Natl. Acad. Sci. USA
            80, 721.
8.   Baumann,E.A. and Hand,R. (1982) J. Virol. 44, 78.
9.   Ghosh,P.K. and Lebowitz,P. (1981) J. Virol. 40, 224.
10.   Hansen,U., Tenen,D.G., Livingston,D.M. and Sharp,P.A. (1981)
            Cell 27, 603.
11.   Hay,R.T. and DePamphilis,M.L. (1982) Cell 28, 767.
12.   Gluzman,Y. (1981) Cell 23, 175.
13.   Burger,C. and Fanning,E. (1983) Virology 126, 19.
14.   Dörper,T. and Winnacker,E.-L. (1983) Nucl. Acids Res. 11, 2575.

STUDIES ON THE INITIATION OF DNA SYNTHESIS IN PLANT AND ANIMAL CELLS

S. Litvak, J. Graveline, L. Zourgui, P. Carvallo[1],
A. Solari[1], H. Aoyama, M. Castroviejo and
L. Tarrago-Litvak

Institut de Biochimie Cellulaire et Neurochimie du CNRS
1, rue Camille Saint Saëns
33077 Bordeaux cedex, France

[1]Departamento de Bioquimica, Facultad de Medicina
Universidad de Chile
Casilla 6671
Santiago 7, Chile

INTRODUCTION

DNA dependent DNA polymerases (E.C.2.7.7.7) play a key role in
DNA replication by copying with high efficiency and fidelity the
appropriate templates. These enzymes are, however, unable to initiate
DNA synthesis in the absence of a primer able to provide a free 3'OH
starting point (1). Several mechanisms have been described concerning
the initiation of DNA synthesis by DNA polymerases. In Figure 1 we
show schematically the different possibilities a DNA polymerase can
find to start a DNA chain. In normally replicating prokaryotic or
eukaryotic cells, DNA synthesis is initiated from short RNA primers
synthesized by a unique kind of RNA polymerase called DNA primase
(1). Discontinuous DNA synthesis gives raise to the so called Okazaki
fragments which are then joined to give the high molecular weight
nascent DNA. A very different mechanism is observed in the case of
the replication of adenovirus DNA. The adenovirus coded DNA polymerase
initiates DNA synthesis from a dCMP residue covalently linked to a
55 Kd protein (see van der Vliet et al., this volume). Parvovirus
contain a single stranded DNA genome with a high degree of self complementarity; hairpin structures allow the initiation of DNA synthesis in this system by a cell coded DNA polymerase (see Vos et al.,
this volume).

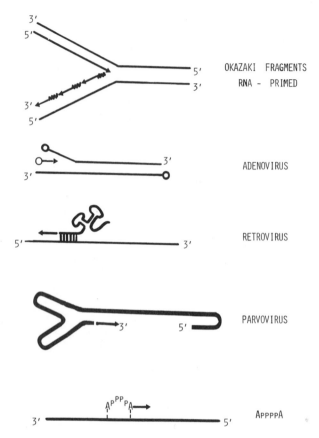

Figure 1:   Initiation of DNA synthesis.

Our laboratory has been interested these last years in the study of different mechanisms of DNA synthesis initiation. In the case of retrovirus it has been shown that RNA directed DNA synthesis is initiated by reverse transcriptase from a specific tRNA partially base paired to the viral genome. A recent review on this subject is reference (2). In this article we will present some evidences concerning the initiation of DNA synthesis by wheat embryos DNA primase and DNA polymerase, as well as the involvement of diadenosine tetraphosphate (Ap_4A) in DNA synthesis in <u>Xenopus laevis</u> oocytes.

RESULTS AND DISCUSSION

<u>The involvement of DNA primase and DNA polymerase γ-like in the
initiation of DNA synthesis in wheat embryos</u>

An RNA polymerase activity able to catalyze the synthesis of
short RNA primers (DNA primase) has been recently described in animal
cells (3-5). RNA primers are then elongated by DNA polymerase α to
give the nascent DNA. Multiple DNA polymerases have been described
in plant cells (6-8), but the study of these enzymes is less advanced
than in animal cells. Here we describe some properties of a DNA primase from wheat embryos. We have also been interested in the study
of the DNA polymerase involved in the initiation of DNA synthesis
from RNA primers, since we have found, in wheat embryos, the surprising result that the two DNA polymerases α-like that we have purified
from this source are not able to initiate DNA synthesis from RNA
primers.
a) <u>Purification of DNA polymerases from wheat embryos</u>. The method has
been described before (7, 8). Four DNA polymerases have been purified
from the 100,000 x g supernatant as seen in the scheme of Figure 2.
DNA polymerase C (α-like) has been separated into two fractions after
linear gradient centrifugation. Some properties of these enzymes are
summarized in Table 1. DNA polymerase B and C_{II} can be classified as
α-like, while DNA polymerase A has many properties of a γ polymerase.
DNA polymerase C_I is a new low molecular weight polymerase from plant
cells we have isolated from the moment we introduced the use of some
protease inhibitors: amino-acetonitrile, leupeptin, aprotinin and
sodium metabisulphite in the first steps of the purification proce-

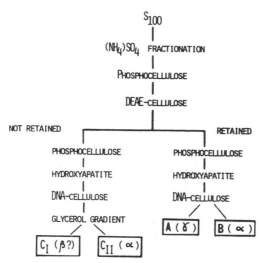

Figure 2:   General scheme of wheat embryo DNA polymerases purifica-
            tion.

Table 1: Some properties of wheat embryo DNA polymerases

| | | DNA POLYMERASES[*] | | |
	A (γ)	B (α)	C_I (β?)	C_{II} (α)
Molecular weight (Kd)	140	110	50	150
Recognition of Poly A-oligo dT	yes	no	no	no
Recognition of Poly dT-oligo rA	yes	no	no	no
% Inhibition { Aphidicolin (50 µg/ml)	0	83	0	92
ddTTP (ddTTp/TTP = 10)	78	0	66	0
Ethidium bromide (10 µM)	65	10	5	14
N-ethyl maleimide (1 mM)	55	61	94	54

[*]DNA polymerase assays: the incubation conditions with several templates have been described before (6-8).

dure. A detailed study concerning the purification and properties of
DNA polymerase C_I will be published elsewhere (M.C. and S.L. in preparation). This enzyme cannot be considered as a β polymerase since
it is extremely sensitive to the SH reagent, N-ethyl maleimide. Beside
these four polymerases, we have described a distinct enzyme isolated
from purified wheat mitochondria (10, 11).

b) Wheat embryo DNA primase is similar to animal cell primases. The
first steps in the purification of this enzyme are common with those
of the purification of DNA polymerases. More than 90% of the DNA polymerase activity is retained in a phosphocellulose column, while the
bulk of DNA primase activity is found in the flow through of this
column, even at very low ionic strength. The enzyme activity is ab-
sorbed in a DEAE-cellulose column and eluted with a 300 mM KCl step.
The active fractions are pooled and concentrated against a buffer
containing 20 mM Tris-HCl pH 7.5, 1 mM DTT and 50% glycerol. It can
be stored at -20°C in 50% glycerol for several months with very little
loss of activity.

Some properties of wheat embryo DNA primase are summarized in
Table 2. It can be seen that single stranded DNA can serve as template
for DNA synthesis in the presence of DNA polymerase and primase more
efficiently if the appropriate rNTPs are present. Wheat primase has
a molecular weight of about 110,000 determined by glycerol gradient
centrifugation and it is completely resistant to alpha-amanitin, in-

dicating that this RNA synthesizing activity is different from the
nuclear RNA polymerases (DNA dependent) I, II and III (12). The size
of the RNA primer synthesized by this enzyme is between 2 and 15
bases, as in the case of the animal DNA primase. By using $(\alpha-^{32}P)$dATP
we have found that the RNA primer is covalently linked to the nascent
DNA (experiment not shown). The DNA polymerase activity found associated with the primase activity is resistant to aphidicolin, but
strongly inhibited by dideoxy-TTP (ddTTP).
c) <u>DNA polymerase A from wheat embryos is able to initiate DNA synthesis from RNA primers</u>. As seen in Table 2 the two DNA polymerases of
the α type from wheat embryos are not able to initiate DNA synthesis
from RNA primers. These results and those of the previous paragraph,
indicating that a DNA polymerase, inhibited by ddTTP, is found associated with the primase activity, prompted us to study the possibility
that, in plant systems, DNA polymerase A (γ-like) would be involved
in the initiation of DNA synthesis. As seen in Table 3 the only DNA
polymerase from wheat embryos able to complement the activity of DNA
primase is polymerase A. <u>Escherichia coli</u> DNA polymerase I is also

Table 2: A. Some properties of wheat DNA primase[*]

Molecular weight (Kd)	110
Size of RNA primer:	2 - 15 mers
Inhibition of DNA synthesis (Primase dependent) by:	DideoxyTTP
Same reaction resistant to:	Aphidicolin Alpha-amanitin
RNA synthesis by DNA primase resistant to	Alpha-amanitin Rifampicin Aphidicolin DideoxyTTP

B. Ribonucleoside triphosphates on DNA synthesis-DNA
primase dependent

Conditions	Endogeneous DNA pol.	E.coli pol I
	(pmoles dAMP incorporated)	
Poly dT alone	1.1	25
Poly dT plus ATP	16.2	203
M$_{13}$ DNA alone	2.0	19
M$_{13}$ DNA plus rNTP	9.0	33

[*]Assays for DNA polymerase and DNA primase were performed as in ref.4.

Table 3: Initiation of DNA primase-dependent DNA synthesis by different DNA polymerases

Addition	pmoles dAMP incorporated
None (endogeneous DNA polymerase)	11
Wheat DNA polymerase A	19
Wheat DNA polymerase B	9
Wheat DNA polymerase C_I	7
Wheat DNA polymerase C_{II}	10
E. coli DNA polymerase I	188
E. coli DNA polymerase I (without primase)	7

very active in elongating the RNA primers synthesized by the plant primase. These results together with the fact that the DNA polymerase associated with the wheat primase behaves like DNA polymerase A concerning the effect of aphidicolin and ddTTP (Table 2) and can recognize very efficiently a template poly rA-oligo dT (not shown) strongly support the idea that DNA polymerase A can play an important role in the step of initiation of DNA synthesis. It is interesting to point out that in animal cells the situation seems to be quite different, since DNA polymerase α is found associated with the primase (13, 14) and is directly involved in initiating DNA synthesis on RNA primers.

d) DNA polymerase A is able to copy natural RNA templates. We have classified DNA polymerase A as belonging to the γ type of polymerases essentially based in the preferential recognition of a template poly rA-oligo dT by this enzyme. The polymerase found in wheat mitochondria shares some properties concerning the behaviour with several inhibitors, but does not recognize a synthetic polyribo template (10). Very recently, it was found that a reverse transcription step was involved in the replication of two DNA virus, hepatitis B virus (HBV) and cauliflower mosaic virus (CaMV) (15, 9). Thus, reverse transcriptase (RNA dependent-DNA polymerase) plays a role, not only in the replication of retroviruses which contain a single stranded RNA genome (16), but also in the expression of HBV and CaMV whose genomes are very different from that of the retrovirus group. Moreover, the DNA polymerase activity partially characterized in the CaMV replication complex has many of the properties of DNA polymerase A studied in our laboratory. It seemed interesting to search for a DNA synthesizing activity of DNA polymerase A not only with synthetic ribo templates,

but also with natural RNAs. As seen in Table 4, DNA polymerase A can
recognize efficiently some natural RNAs like avian myeloblastosis
virus (AMV) RNA, as well as turnip yellow mosaic virus (TYMV) RNA.
This recognition is less important, however, than the activity of
AMV reverse transcriptase under the same conditions. The addition of
oligo dT, able to anneal to the 3' poly A tail of AMV RNA, increases
DNA synthesis with this template. Not shown in this work are our results indicating that polymerase A is able to initiate DNA synthesis
from the endogeneous tRNA primer associated with the AMV genome, as
well as the finding that newly synthesized DNA with oligo dT primer
can copy a region of the AMV template further than the poly A tail.
Although it is not clear at the present time if the reverse transcriptase involved in CaMV replication is coded in the cellular or
viral genome, our results and those of Pfeiffer and Hohn (9) seem to
indicate that a cellular DNA polymerase can mimic the activity of
reverse transcriptase. Whether the activity of this host enzyme can
be stimulated or modified by a virus encoded subunit, remains to be
established.

DNA polymerases and the effect of Ap$_4$A in large oocytes and eggs of
Xenopus laevis

Xenopus laevis oocytes and eggs are giant cells which provide
an interesting system to investigate various aspects of DNA metabolism. Large oocytes (stages V and VI) are highly specialized cells
with some unique properties allowing the study of the intracellular
location of DNA polymerase activities. Even if these cells are very
rich in DNA polymerase activity, they are silent concerning the ability to replicate endogenous or exogenous DNA (18). It is not clear
at the present time whether Xenopus oocytes have the three DNA polymerases found in most animal cells: DNA polymerases α, β and γ, since
it has been reported the presence of these three polymerases in Xenopus oocytes (19), while other group has claimed the absence of DNA
polymerase β in large oocytes (20).

Our first attempts in this system were to search the DNA polymerases present in those cells. We have found that large oocytes contain DNA polymerase α in the nucleus and DNA polymerase γ in the
mitochondria, while the ovarian β polymerase activity previously reported by Joenje and Benbow is most probably located in the follicular cell layer (19).

As mentioned before, large oocytes are silent concerning DNA
synthesis. It seemed interesting to study the possibility of over-
coming this inability to duplicate the genomic material by Ap$_4$A, a
dinucleotide whose level is dramatically increased in tissues under
active proliferation (21). This dinucleotide is able to bind specifically to a subunit of the high molecular form of DNA polymerase α
(22), as well to stimulate DNA synthesis in permeabilized animal
cells (23). In this work we present evidences that Xenopus laevis DNA

Table 4: Recognition of synthetic and natural RNA templates by wheat embryos DNA polymerase A and AMV reverse transcriptase *

Templates	Wheat DNA polymerase A		AMV reverse transcriptase	
	Mg	Mn	Mg	Mn
	(cpm (^{3}H)dAMP incorporated)			
Poly rA-oligo dT	1,802	376	15,980	9,120
Poly rC-oligo dG	250	4,805	N.T.	N.T.
AMV RNA	41,610	7,200	91,950	N.T.
AMV RNA + oligo dT	91,150	18,308	404,570	N.T.
TYMV RNA	54,830	12,115	158,940	N.T.

* Incubation conditions for DNA polymerase A and synthetic polyribonucleotides were as in reference (8), and for reverse transcriptase and AMV RNA as in reference (17). The concentration of Mg was 5 mM and of Mn was 0.5 mM. N.T. = not tested.

polymerase activity is increased in vitro and in vivo by Ap_4A. This
stimulatory effect is probably mediated by the ability of this dinucleotide to act as primer for DNA synthesis creating new origins of
replication.

a) Localisation of DNA polymerase activities in oocytes and eggs.
DNA polymerase activity (using the specific assays for α, β and γ)
was determined in extracts from oocytes, enucleated oocytes, nuclei
and mitochondria. The results are shown in Table 5. Oocytes have a
high level of DNA polymerase α detected exclusively in the nuclei.
When enucleation is made fast enough, no DNA polymerase activity was
found in the enucleated extract. Mitochondria contained all the DNA
polymerase of the γ type in the oocyte. The answer to the β assay was,
in all cases, extremely low. This fact was in contradiction to the
data presented by Fox et al. (25), which found a DNA polymerase β
activity in the enucleated oocytes.

In order to establish if DNA polymerase β was present in oocytes
and unfertilized eggs, we used different procedures to analyze the
content of DNA polymerase in those cells. In Figure 3 it can be seen
the polymerase activity profile after ultracentrifugation in a glycerol gradient of extracts from oocytes and unfertilized eggs. Activated DNA and poly A-oligo dT were used as templates. A low molecular
weight DNA polymerase active with poly A-oligo dT is found in the
oocyte extract, but it is absent in the egg extract. As follicular

Table 5: Localisation of DNA polymerase activity

	Assay α	Assay β (cpm/oocyte)	Assay γ
Oocytes [a]	16,000	550	220
Enucleated oocytes [a]	800	403	250
Nuclei [b]	19,694	237	276
Mitochondria [c]	3,200	362	10,500

[a] Stage VI oocytes and enucleated oocytes were homogenized and centrifuged 10 min at 27,000 x g to separate mitochondria. [b] Nuclei
were prepared inserting a syringe needle into the animal pole of the
oocyte. The oocyte is then squeezed equatorially with forceps until
the translucent germinal vesicle partly emerges from the slit; it is
then allowed to squeeze itself out. [c] Mitochondria were separated
as described by Brun et al. (29) and sonicated in the presence of
0.5% Triton X-100 before the assay. The specific assays for DNA polymerases α, β and γ were used as described by Knopf et al. (24).

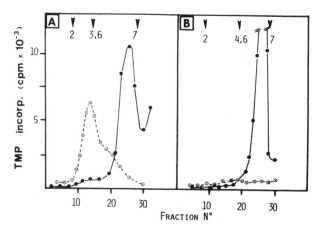

Figure 3: Size of DNA polymerases determined by sedimentation in glycerol gradients. Either oocytes, A, or unfertilized eggs, B, were homogenized with a Potter in the presence of 4 volumes of a buffer containing 20 mM pH 7.5, 1 mM 2-mercaptoethanol, 1 mM phenylmethyl-sulphonyl fluoride, 10 mM sodium metabisulfite, 8% glycerol and centrifuged at 27,000 x g for 20 min. The supernatant was loaded onto a DEAE-cellulose column equilibrated with the same buffer. After washing with 2-3 column volumes, proteins were eluted with 0.6 M KCl. The fractions which contained DNA polymerase activity were pooled and an aliquot (200 µl) was layered onto 10-30% linear glycerol gradients (3.8 ml) made in 50 mM Tris-HCl pH 7.5, 0.1 mM EDTA, 1 mM 2-mercaptoethanol and 300 mM KCl. Cytochrome c from horse heart, ovoalbumin, bovine serum albumin and bovine γ globulin were used as standards. Centrifugation was at 59,000 rpm in a TST 60 rotor for 15-16 hours in a Kontron ultracentrifuge (0-4°C). Fractions (120 µl) were collected from the top of the gradients and 10 µl aliquots of each fraction were assayed for DNA polymerase activity in the presence of activated DNA (●——●) or poly A-oligo dT (⊙--⊙).

cells are lost during hormonal maturation, the absence of DNA poly-merase β in unfertilized eggs can be attributed to the fact that this activity is located in the follicular cells surrounding the oocytes. This idea is supported by the absence of DNA polymerase β in the nuclear fraction of Xenopus oocytes. It is interesting to point out that in all animal cells studied up to now, DNA polymerase β is con-fined to the nuclear compartment. Our data may explain the contra-dictory results of Zimmermann and Weissbach (20) who claim that there is no polymerase β in Xenopus ovaries because no enzymatic activity was obtained in the flow through of a DEAE-cellulose column, and those of Joenje and Benbow (19) who have purified an ovarian DNA po-lymerase β. Our results indicate that DNA polymerase β from ovarian extracts is slightly retained in a DEAE-cellulose column; this enzyme would be follicular in origin and would not be found inside the oocytes as implied by Fox et al. (25). It has been shown by other authors that mature sperm cells contain only DNA polymerases β and γ (26, 27). Thus, in the fertilized eggs DNA polymerase α would be maternal in origin, while DNA polymerase β would be supplied by the male cell.

b) Effect of Ap_4A on DNA synthesis. Large oocytes which are silent concerning DNA synthesis seem to be good system to study the mechanism regulating the onset of DNA synthesis. It has been suggested that Ap_4A can act as a trigger signal for DNA synthesis in animal cells (23). It seemed interesting to see whether this dinucleotide was able to affect the DNA synthetic capacity of Xenopus oocytes. As seen in Table 6, DNA polymerase activity from a crude extract of stage VI oocytes is greatly stimulated by Ap_4A, while this situation was not observed in unfertilized eggs.

Other dinucleotides, like Ap_3A, Gp_3G and Gp_4G, are able to mimic Ap_4A, although to a lesser extent (results not shown). ATP can also stimulate DNA synthesis under the same conditions, but when we followed the fate of labeled ATP and Ap_4A we found that the latter was almost unaffected after incubation, while most of the ATP was transformed to metabolites that are currently being studied (results not shown).

A role of Ap_4A in creating new origins of replication is suggested by the results of Table 7. Only when templates susceptible to anneal Ap_4A were used (poly dT or poly dT-oligo dA), a strong stimulation by Ap_4A was observed. No stimulation was found when poly dA-oligo dT or poly dC-oligo dG were used as templates. With activated DNA, where some clusters of oligo dT should be expected, a strong stimulation was found as seen in Table 6. These results support a role of Ap_4A as primer in DNA synthesis, role which has been suggested previously using both poly dT or DNA as templates (28).

The in vitro effect of Ap_4A summarized in the previous Tables was also found in vivo, since the microinjection of Ap_4A plus the labeled precursor to stage VI oocytes induced a significant stimulation of DNA synthesis (Table 6). A dramatic increase in DNA synthesis was obtained when denatured DNA was micro-injected into oocytes; this effect was amplified when Ap_4A was co-injected. No stimulation by Ap_4A was observed in micro-injected eggs, as in the case of the in vitro assay. Aphidicolin strongly inhibits the stimulation suggesting a role of polymerase α in this stimulation.

Our results indicate that large oocytes of Xenopus laevis contain only two DNA polymerases: polymerase α is found in the nucleus, while the γ enzyme is confined to the mitochondrial organelle. Ovarian DNA polymerase β is located in the follicular cell layer. The activity of DNA polymerase α is strongly stimulated by Ap_4A both in vitro and after micro-injection into large oocytes. Aphidicolin

Table 6:   DNA polymerase activity from oocytes or unfertilized eggs:
           Effect of Ap_4A in vitro (A) and in vivo (B) *

(A) Additions	Oocytes	Unfertilized eggs
	(pmoles)	
Extract (50 µg/ml)	8.0	2.0
Extract (50 µg/ml) + 5 mM Ap_4A	21.0	2.1
Extract (100 µg/ml)	6.0	2.9
Extract (100 µg/ml) + 5 mM Ap_4A	42.0	3.1

(B) Substances injected	Oocytes (cpm/ooc.)	Unfertil. eggs (cpm/egg)
$(\alpha-^{32}P)dTTP$	51	58
$(\alpha-^{32}P)dTTP + Ap_4A$	183	90
$(\alpha-^{32}P)dTTP$ + Denatured DNA	286	1118
$(\alpha-^{32}P)dTTP$ + Denatured DNA + Ap_4A	670	925
$(\alpha-^{32}P)dTTP$ + Denatured DNA + Ap_4A + Aphidicolin	151	223

*
 Either oocytes or unfertilized eggs (previously washed with 2%
cysteine) were homogenized with a Potter in the presence of 2 volumes
of buffer A (20 mM KPi pH 7.5, 1 mM 2-mercaptoethanol, 1 mM phenyl-
methylsulphonyl fluoride and 20% glycerol) and centrifuged at 20,000
x g for 20 min. The supernatant was dialyzed against 50% glycerol in
buffer A. DNA polymerase activity with activated DNA was measured at
37°C for 30 min.

Oocytes were injected in the animal pole aiming for the germinal
vesicle. Unfertilized eggs were irradiated with UV light before
micro-injection. Three batches of 10 oocytes or eggs were injected
with 20-50 µl of a solution containing $(\alpha-^{32}P)dTTP$ (3184 Ci/mmole),
0.8 mM Ap_4A, 0.3 mg/ml of calf thymus denatured DNA, 0.8 µg/ml Aphidicolin, either alone or in a mixture. After 6 hours at 20-22°C the
reaction was stopped by freezing the oocytes at -80°C. DNA was extracted and counted as in (18).

Table 7:   Effect of Ap$_4$A on the utilisation of synthetic
           polynucleotides*

Template	Primer	Radioactive precursor	R
Poly dT	--	dATP	3.0
Poly dT	Oligo dA	dATP	5.0
Poly dA	Oligo dT	dTTP	0.5
Poly dC	Oligo dG	dGTP	0.9

*
The effect of Ap$_4$A on DNA polymerase activity was measured using
the crude extract prepared as described in Table 6. Incubations were
carried out at 37°C for 30 min. Results are given as the ratio of
the activity in the presence of 5 mM Ap$_4$A (in the case of poly dT
and poly dA), or 2.5 mM Ap$_4$A (in the case of poly dC) to the activity
in the absence of the dinucleotide.

inhibits the Ap$_4$A mediated stimulation both with the extract or micro-
injected into oocytes. The effect of Ap$_4$A is probably due to the annealing of Ap$_4$A to dT clusters creating new origins of replication.
We are currently characterizing the protein factor involved in the
dinucleotide effect.

ACKNOWLEDGEMENTS

        This work was supported by grants from CNRS and the University
of Bordeaux II (to S.L.), INSERM (to L.T.-L.) and the University of
Chile (to A.S.). The authors gratefully acknowledge the gift of
aphidicolin by Dr. A.H. Todd (ICI.England) and reverse transcriptase
by Dr. J.W. Beard (Life Science Inc., Florida, USA). One of us (H.A.)
is supported by the Fundaçao de Amparo à Pesquisa do Estado de Sao
Paulo, Brazil.

REFERENCES

1.   Kornberg,A. DNA Replication (and 1982 Supplement), Plenum
         Press, San Francisco (1980).
2.   Litvak,S. and Araya,A. (1982) Trends Biochem. Sci. 7, 361.
3.   Conaway,R.C. and Lehman,I.R. (1982) Proc. Natl. Acad. Sci. USA
         79, 2523.
4.   Yagura,T., Kozo,T. and Seno,T. (1982) J. Biol. Chem. 257, 11121.
5.   Hübscher,U. (1983) EMBO J. 2, 133.

6. Castroviejo,M., Tarrago-Litvak,L. and Litvak,S. (1975)
    Nucleic Acids Res. 2, 2077.
7. Castroviejo,M., Tharaud,D., Tarrago-Litvak,L. and Litvak,S.
    (1979) Biochem. J. 181, 183.
8. Castroviejo,M., Fournier,M., Gatius,M., Gandar,J.C.,
    Labouesse,B. and Litvak,S. (1982) Biochem. Biophys.
    Res. Comm. 107, 294.
9. Pfeiffer,P. and Hohn,T. (1983) Cell 33, 781.
10. Christophe,L., Tarrago-Litvak,L., Castroviejo,M. and Litvak,S.
    (1981) Plant Sience Lett. 21, 181.
11. Ricard,B., Echeverria,M., Christophe,L. and Litvak,S. (1983)
    Plant Mol. Biol. 2, 167.
12. Roeder,R.G. (1976) RNA polymerase, Cold Spring Harbor
    Laboratory.
13. Shioda,M., Nelson,E.M., Bayne,M.L. and Benbow,R.M. (1982)
    Proc. Natl. Acad. Sci. USA 79, 7209.
14. Kaguni,L.S., Rossignol,J.M., Conaway,R.C. and Lehman,I.R.
    (1983) Proc. Natl. Acad. Sci. USA 80, 2221.
15. Summers,J. and Mason,W.S. (1982) Cell 29, 403.
16. Varmus,H. (1982) Science 216, 812.
17. Araya,A., Sarih,L. and Litvak,S. (1979) Nucleic Acids Res.
    6, 3831.
18. Harland,R.M. and Laskey,R.A. Cell 21, 761.
19. Joenje,H. and Benbow,R.M. (1978) J. Biol. Chem. 253, 2640.
20. Zimmermann,W. and Weissbach,A. (1981) Mol. Cell. Biol.
    1, 680.
21. Rapaport,E. and Zamecnik,P.C. (1976) Proc. Natl. Acad. Sci. USA
    73, 3984.
22. Grummt,F., Waltl,G., Jantzen,H.M., Hamprecht,K., Hübscher,U.
    and Kuenzle,C.C. (1979) Proc. Natl. Acad. Sci. USA
    76, 6081.
23. Grummt,F. (1978) Proc. Natl. Acad. Sci. USA 75, 371.
24. Knopf,K.W., Yamada,M. and Weissbach,A. (1976) Biochemistry
    15, 4540.
25. Fox,A.M., Breaux,C.B. and Benbow,R.M. (1980) Develop. Biol.
    80, 79.
26. Philippe,M. and Chevaillier,P. (1980) Biochem. J. 189, 635.
27. Habara,A., Nagano,H. and Mano,Y. (1980) Bioch. Biophys. Acta
    608, 287.
28. Zamecnik,P.C., Rapaport,E. and Baril,E. (1982) Proc. Natl.
    Acad. Sci. USA 79, 1791.
29. Brun,G., Vannier,P., Scovassi,I. and Callen,J.C. (1981)
    Europ. J. Biochem. 118, 407.

IV
DNA Primase

FUNCTION AND PROPERTIES OF RP4 DNA PRIMASE

Erich Lanka and Jens P. Fürste

Max-Planck-Institut für Molekulare Genetik
Abt. Schuster
D-1000 Berlin-Dahlem

Plasmid-specific DNA primase activity was detected by an in vitro assay for the conversion of single-stranded circular DNA of small phages to its duplex form (1). A survey of plasmids representative of most of the known incompatibility groups (2) revealed that several of these plasmids specify this activity (Table 1). Existence of a plasmid-encoded DNA primase explains the discovery that IncI plasmids can partially suppress the effect of temperature-sensitive dnaG mutations in Escherichia coli (3). All the plasmids listed in Table 1 specifying primase activity suppress the dnaG3 mutation as measured by colony-forming ability at the nonpermissive temperature, thus providing a potent in vivo assay for this class of enzymes. It is also known that expression with plasmids derepressed for functions of conjugal DNA transfer and for pilus synthesis leads to higher levels of primase than with the corresponding wild-type plasmids (1, 2, 5, 7). This correlation suggests that the enzyme may play a role in conjugal DNA synthesis. Studies with the IncIα plasmid ColIb have indicated that plasmid primase is required to initiate efficient synthesis of DNA complementary to the transferred strand in the recipient cell, with the protein being supplied by the donor parent and probably transmitted between the mating cells (11). Plasmid primases of three different plasmids (R16, ColIb, RP4) have nonhomologous primase genes (4) and the enzymes are immunologically distinct from each other and from the Escherichia coli dnaG gene product (2), indicating that evolution of these genes has occurred rather independently.

The IncP group plasmids (Table 1) have received a large amount of attention recently because of their exceptionally broad host range. No host specificity could be detected so far for conjugal transfer,

Table 1:   Plasmids specifying DNA primase activity and their ability
           to suppress the Escherichia coli dnaG3 mutation

Plasmid(s)	Inc group	Primase activity	Suppression of dnaG3
R16, R864a	B	+ (2,4,5)	+ (4,5)
R40a	C	+ (6)	+ (6)
ColIb, R64, R144	Iα	+ (1,2,7)	+ (3,8)
R621a	Iγ	+ (2)	+ (4,8)
TP114	Iδ	+ (2)	n.t.
R805a	Iζ	+ (6)	+ (4)
R391	J	− (2)	+ (6,8)
R387	K	+ (2)	+ (6)
R831b, R446b	M	+ (2)	− (6)
RP4, (RP1, R68, R18, RK2)	P	+ (2)	+ (2,9,10)
R1033, R751	P	+	n.t.
RA3	U	+ (6)	+ (6)

n.t. = not tested

replication and maintenance of most of these plasmids among all of
the known Gram-negative bacteria (12). This promiscuous property
makes RP4 an ideal model system to characterize its primase function.

Plasmid RP4 encodes two forms of a DNA primase

The RP4 primase gene (pri) belongs to the Tra1 region of the
plasmid and maps between coordinates 40.3 and 43.5 (Figure 1). The
pri gene (3.2 kb) comprises about 5% of the total RP4 genome. Part
of the Tra1 region is probably transcribed as a polycistronic messenger which is initiated at a promoter site located around coordinate
48.6. Transcription of this region including the primase gene proceeds
counterclockwise with regard to the RP4 map and extends to the
kanamycin resistance gene which is transcribed separately from a
different promoter. The transcript carries information for at least
five proteins of 68, 77, 16.5 kDa, and the two primase gene products
of 80 and 118 kDa (10).

The primase polypeptides have been identified by purification
following the enzymatic activity, immunological techniques and

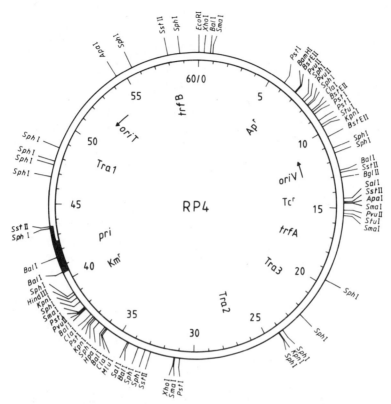

Figure 1: Combined physical and functional map of plasmid RP4. The map is calibrated in kilobases (kb). The restriction enzyme map has been redrawn from Lanka et al. (13) with the addition of the StuI sites. Genetic and phenotypic symbols are taken from Thomas (12) and are correlated to the specific DNA regions relative to the restriction sites. Abbreviations: Ap^r (ampicillin resistance), Tc^r (tetracycline resistance), oriV (origin of vegetative plasmid DNA replication), trf (trans-acting replication function), Tra (regions required for conjugal transfer), oriT (origin of transfer). The location of the DNA primase gene (pri) is marked by a heavy black line. The beginning of the coding regions for the two pri gene products of 118 and 80 kDa are indicated at coordinates 43.5 and 42.5 by the staggered arrangement of the black line.

molecular cloning of the pri gene. Purification of primase activity of cells carrying RP4 revealed that the activity resides in a 118 kDa and possibly an 80 kDa polypeptide (2). Antiserum was prepared against the purified larger polypeptide and reacted with proteins of

crude extracts of RP4 harboring cells using the electrophoretic
blotting technique for proteins as described by Towbin et al. (14).
Two polypeptides reacting with antiserum can be detected on the
pattern produced with proteins from cells carrying RP4 (Figure 2,
lane e). The upper band corresponds to the 118 kDa antigen, whereas
the lower band contains a polypeptide of 80 kDa which crossreacts
with the anti-primase serum. This demonstrated that RP4 specifies
two poypeptides which must share aminoacid sequence homology. Two
polypeptides reacting with anti-primase serum can also be detected
in extracts of cells carrying the Pri[+]-recombinant plasmid pJF107
(Figure 2, lane f). Plasmid pJF107 is a pBR325 derivative containing
the RP4 region between coordinates 39.0 and 43.5 (Figure 1 and 3).
Since the polypeptides produced by BW86(pJF107) migrated with the
same mobility as those of BW86(RP4), we concluded that the intact
RP4 pri gene was inserted into the vector plasmid.

     The existence of five contiguously arranged BssHII sites within
the RP4 portion of pJF107 allowed us to construct partial deletion
derivatives of the plasmid (Figure 3, lower part). Removal of adjacent
BssHII fragments of pJF107 causes truncation of both polypeptides
specified by the deletants Δ2 and Δ67 as demonstrated with the solid
phase immunoassay (Figure 2, lanes g and h). The extent of the deletions corresponds to the decrease in size of the pri polypeptides
encoded by the derivatives Δ2 and Δ67. Deletion of 280 bp (Δ2) and
365 bp (Δ67) results in pairs of polypeptides which are, respectively,
10 kDa and 13 kDa smaller than those of wild type. This indicates
that DNA has been eliminated from the pri gene within the coding
region for both polypeptides. Fortunately, recombinant plasmids
(pMS205 and pMS305, ref. 13) containing the 4.3 kb SphI fragment of
pJF107 which maps between coordinates 39.1 and 43.4 on the RP4 genome,
only specifiy the 80 kDa polypeptide and have a Pri[+]-phenotype. These
two lines of evidence obtained with the BssHII deletion derivatives
of pJF107 and with the plasmids pMS205 and pMS305 demonstrated that
the RP4 pri gene encodes two primase polypeptides which are separate
translation products. The polypeptides of 118 and 80 kDa arise from
a in-phase overlapping gene arrangement containing two translational
start signals.

Suppression of the Eschrichia coli dnaG3 mutant by RP4 DNA primase

     Suppression of dna(ts) mutant defects can be measured by com-
paring plating efficiency of bacteria at permissive and nonpermissive
growth temperatures. RP4 introduced into the Escherichia coli dnaG3
mutant strain BW86 (16) gives rise to significantly higher colony-
forming ability at 40°C as compared to BW86 carrying the vector
plasmid pBR325 (Table 2). Similar results were obtained when the
resident plasmid in BW86 was pJF107, the Pri[+]-recombinant plasmid
described above. Primase levels of the strains BW86 (RP4) and BW86
(pJF107) are comparable, suggesting that the primase is responsible
for the increase in colony formation and DNA synthesis at 40°C

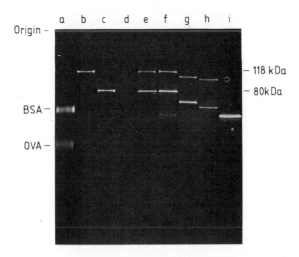

*Figure 2:   Solid phase immunoassay of pri polypeptides specified by
RP4 and recombinant plasmids. Cells of 100 ml cultures (A_{600}=1.0)
were lysed by the Brij-58-lysozyme procedure (15) and the supernatants
after centrifugation were passed over 1 ml Heparin-Sepharose columns.
Proteins were step eluted (10% (w/v) glycerol, 20 mM Tris-HCl/pH 7.6,
0.1 mM EDTA, 1 mM DTT, 0.8 M NaCl) and concentrated by dialysis
against 20% (w/w) polyethylene glycol 20.000 to a residual volume of
100 µl. 10 µl aliquots (35 µg of protein) of each sample were applied
to a 15% (w/v) polyacrylamide-SDS-gel and electrophoresed. Proteins
were then transferred electrophoretically from the gel to a nitrocellulose membrane according to the procedure described by Towbin et al.
(14), reacted with 500-fold diluted anti-RP4 primase serum and subsequently with 50-fold diluted FITC-goat anti-rabbit immunoglobulins
(Nordic Immunochemicals). The figure represents a photograph of the
membrane taken under UV-light at 366 nm. Lane (a) FITC-bovine serum
albumin 0.5 µg, FITC-ovalbumin 0.5 µg; purified RP4 DNA primase polypeptides (b), 118 kDa, 75 ng; (c) 80 kDa, 75 ng; lanes (d)-(i) contain
Heparin-Sepharose fractions of BW86 cells carrying the following plasmids: (d) pBR325; (e) RP4; (f) pJF107; (g) Δ2; (h) Δ67; (i) pMS101.*

(Table 2). Colonies of BW86(RP4) and BW86(pJF107) which formed at
40°C, grew after replating at the non-permissive temperature with an
efficiency of 1, suggesting that mutations must have occurred. Primase
activity of extracts of such mutants was always higher than compared
to the one of the parental strain. Mutants originating from RP4 in
most cases represent deletions or insertions. One example, pMS101,
is a deletion derivative of RP4 beginning within the pri gene and
extending far into the TraI region beyond oriT. This plasmid encodes
a primase polypeptide of 60 kDa as demonstrated by the solid-phase
immunoassay (Figure 2, lane i). We were unable to select such RP4
mutants in a dnaG3 recA-1 strain, indicating that generation of these
mutants might be a recA-dependent process. Such alterations as observed with the RP4 pri mutants representing deletions or insertions

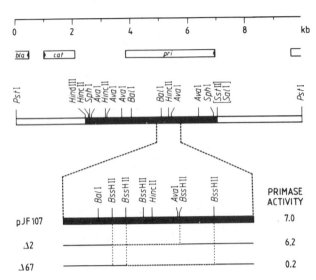

Figure 3: Combined physical and genetic map of plasmid pJF107.
Molecular cloning of the RP4 *pri* gene has been described elsewhere
(10). Restriction endonuclease sites are indicated on the RP4 portion
(▬) of the plasmid. Vector plasmid is pBR325 (▭). The map is
calibrated in kilobases (kb). *Cat* refers to chloramphenicol acetyl
transferase, *bla* to β-lactamase and *pri* to DNA primase. The direction
of transcription is indicated by the arrows. The plasmid region
between coordinates 4.9 and 5.7 is enlarged in the lower part of the
figure. BssHII sites are included. The extent of two deletions
-created by partial digestion of pJF107 with BssHII and religation-
are represented by the dotted lines. The minimal coding region of
the *pri* gene was determined by Bal31 deletion analysis. Primase activity is given in mU/mg of crude cell extract (2).

affecting the *pri* gene have not been found with mutants isolated of
BW86 (pJF107). An example of such a mutant is BW86(pJF107-1). The only
change to be seen was a 20-fold increase in primase activity (Table 2).
DNA, as far as cleavage pattern was concerned or the *pri* polypeptide
sizes remained unchanged.

The primases of pMS101 and pJF107-1 complement the *dnaG3* mutation of *Escherichia coli* indicating that these mutant enzymes can
replace the thermo-inactivated host primase acting in discontinous
DNA replication of plasmid and chromosomal DNA. Therefore, efficient
suppression of *dnaG3* seems to require an altered primase and/or overproduction of the plasmid-encoded enzymes. Reduction in polypeptide
molecular weight could facilitate interaction with host components
of the replication apparatus, i.e. the primosome, or its components;
overproduction, on the other hand might compensate for a lower efficiency of the plasmid priming system.

Table 2:   Properties of <u>Escherichia coli</u> <u>dnaG3</u> strain BW86
           harboring Pri$^+$-plasmids

Plasmid	Primase activity	DNA synthesis		Colony formation 40°/30°C
	mU/mg CCE	10^{-3}cpm(ml.A$_{450}$unit)$^{-1}$		
		30°C	40°C	
RP4	6	307	96	3×10^{-5}
pMS101	19	not tested		1
pJF107	7	308	167	6×10^{-4}
pJF107-1	135	118	246	1
pBR325	0	296	8	10^{-7}

Primase activity in crude cell extracts (CCE) was assayed as described
previously. 1 U = 1 µmol dTMP incorporated into acid-insoluble material in 30 min at 30°C (2). DNA synthesis was measured according to
Wilkins et al. (7). For quantitative analysis of colony formation
overnight cultures were plated at suitable dilutions on prewarmed
plates which were incubated at either 30° or 40°C. Colony formation
is the action of the yields of colonies formed at these temperatures.

Purification of RP4 DNA primase polypeptides

     To characterize the primase function of the two <u>pri</u> gene pro-
ducts of 118 and 80 kDa, the proteins were purified using an in vitro
assay (1, 2). The enzymes can initiate DNA synthesis de novo on
single-stranded DNA of small phages like φX174, G4 or fd. <u>Escherichia
coli</u> proteins precipitated by 43% saturated ammonium sulfate from
crude cell extracts contained a sufficient level of replication proteins to mediate (-)strand synthesis on these templates. Conversion
of φX174 ssDNA in the presence of plasmid-encoded primase was
rifampicin-resistant and independent of the products of the <u>dnaB</u>,
<u>dnaC</u> and <u>dnaG</u> gene of <u>Escherichia coli</u>, indicating that the primase
functions independently of the primosome. In vivo, however, only the
<u>dnaG</u> but not <u>dnaB</u> or <u>dnaC</u> mutations can be suppressed by plasmid-
specified primase.

     Insertion of RP4 DNA fragments containing the <u>pri</u> gene into
the <u>tac</u>-vector-plasmid pKK223-3, the transcription of which is under
inducible control of the <u>tac</u>-promoter (17), greatly facilitated the

purification of the enzymes in large amounts and high yields. The enzymes were purified to near homogeneity (Figure 4, lanes b and d) according to procedures described previously (1, 2). Starting with the supernatant of a crude cell extract prepared by the Brij-lysozyme procedure (15), heparin-Sepharose chromatography was remarkably useful in a first purification step (2, 7). By this means, a 20-50-fold purification of the proteins was achieved (Figure 4, lanes a and c). The primases bind to ssDNA-agarose and elute around concentrations of 0.3 M NaCl.

The 118 kDa polypeptide can be separated of a mixture of both polypeptides by glycerol gradient centrifugation or by CM-Sepharose

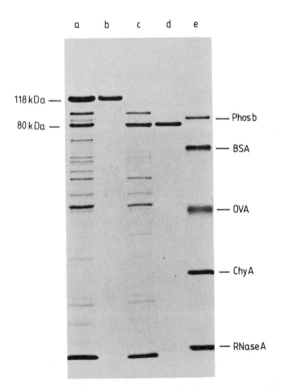

Figure 4: Gel electrophoresis under denaturing conditions of RP4 DNA primase polypeptides. Samples were loaded on a 15% (w/v) poly-acrylamide-SDS-slab gel according to Laemmli (18). Heparin-Sepharose fractions of primase overproducing strains: lane (a) 118-80 kDa, 10 µg, and (c) 80 kDa, 7 µg; purified RP4 DNA primase polypeptides (b) 118 kDa, 1 µg, and (d) 80 kDa, 1 µg; lane (e) Size markers: phosphorylase b (Phosb, 92.5 kDa), bovine serum albumin (BSA, 68 kDa), ovalbumin (OVA, 43 kDa), chymotrypsinogen A (ChyA, 25.7 kDa), and RNase A (13.6 kDa).

chromatography to yield fractions of the 118 kDa protein essentially free of the 80 kDa polypeptide. The 80 kDa primase has been purified from cells carrying Pri[+]-recombinant plasmids which only produce this enzyme. This has been achieved by removing the translational initiation signal for the larger polypeptide. Since the two proteins in a mixture have different sedimentation velocities, it seems unlikely that they are arranged in a complex. The two enzymes sediment in a glycerol gradient with 5.2S (118 kDa) and 4.2S (80 kDa) suggesting that they function as monomers.

The two pri polypeptides have identical specific activities

To characterize the priming activity of the two pri gene products of RP4, specific activities of the purified enzymes were compared by measuring complementary strand synthesis on φX174 DNA in an extract of the Escherichia coli dnaG mutant BT308 (19). The two polypeptides alone and a 1:1 (w/w) mixture were assayed (Figure 5). No significant differences were observed, indicating that the specific activities of the two enzymes are identical. This result demonstrates also that a large portion of the N-terminus of the 118 kDa primase is not essential for the priming function in vitro and therefore might serve another as yet unknown function.

Plasmid-encoded primases can function on a variety of ssDNA templates

Purified DNA primases of RP4(IncP) and pLG214(IncIα, ref. 7) were tested for their ability to initiate DNA synthesis on several ssDNA templates in an extract of BT308. DNA of the two polyhedral phages φX174 and G4 and the three filamentous phages fd (F-pili specific), Ifl and Pr64FS (both I-pili specific, ref. 20, 21) were used as templates. The sizes of the DNA of Ifl and PR64FS are similar to fd DNA (6,408 nucleotides) as judged by agarose gel electrophoresis and electron microscopy. Complementary strand synthesis takes place on all five templates regardless whether RP4 primases or an IncIα-type primase are used (Table 3). However, in order to obtain similar levels of DNA synthesis 100 ng of the RP4 enzymes had to be applied for φX174 DNA instead of 10 ng for the other DNAs indicating that there might be differences regarding the efficiency of template utilization. Such an effect has not been observed with the I-type enzyme which on all the DNAs has a 10 to 100-fold higher specific activity than the RP4 primases. These data suggest that the plasmid primases can initiate complementary strand synthesis on all natural single-stranded DNAs.

Oligoribonucleotide synthesis by plasmid-encoded primases starts with 5'-cytidine monophosphate or cytidine

Priming of φX174(-) strand synthesis by plasmid primases in crude ammoniumsulfate fractions is absolutely dependent on the presence of ribonucleotide triphosphates. Maximum DNA synthesis was

E. LANKA AND J. P. FURSTE

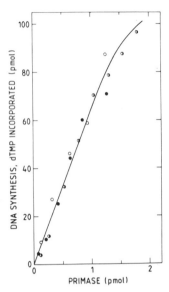

Figure 5: φX174 DNA synthesis in an extract of Escherichia coli
BT308 in the presence of RP4 DNA primase polypeptides. Assay mixtures
(50 µl) contained 20 mM Tris-HCl (pH 7.5), 25 mM KCl, 5 mM MgCl$_2$,
0.2 mM EDTA, 1 mM dithiothreitol, 2.5 mM spermidine-3 HCl, 25 µg of
rifampicin per ml, 2 mM ATP, 125 µM each of CTP, GTP and UTP, 12.5
µM each of dATP, dCTP, dGTP and (methyl-^{3}H)dTTP (1200 cpm/pmol),
0.48 nmol of φX174 DNA, 52 µg of protein of a cell extract of
Escherichia coli BT308 precipitated with 43% saturated ammoniumsulfate, and purified RP4 DNA primase polypeptides, (●) 118 kDa, (o) 80
kDa, and (◉) 1:1 mixture of the 118 kDa and 80 kDa polypeptide.
(^{3}H)dTMP incorporation into acid-insoluble material was measured
after incubation for 30 min at 30°C.

observed in the presence of all four rNTPs. The requirement for ATP
and CTP appeared to be crucial because any combination lacking ATP
or CTP was ineffective (1). This dependence of the replication reaction on rNTPs reflects need for RNA primer synthesis.

To elucidate the RNA primer structures, reactions catalyzed by
plasmid-specified primase on single-stranded DNA templates in the
presence of ATP and (α-^{32}P)CTP were analyzed by polyacrylamide gel
electrophoresis. Small but significant amounts of product could be
seen only after prolonged exposure of the film to the gel. This material migrated to a position corresponding to a dinucleotide with
a 5'-phosphate at the end and was sensitive to RNase A digestion,
suggesting the formation of p̌CpA but not of pÅp̌C. Therefore, the
experiment has been repeated after replacing cytidine 5'-triphosphate
by its 5'-mono-phosphate, with (α-^{32}P)ATP as tracer, and φX174 DNA

Table 3: DNA synthesis mediated by plasmid DNA primases on a
          variety of ssDNA templates

ssDNA	P-80kDa	DNA primase P-118 kDa	I-87 kDa
	dTMP(pmol)		
φX174	32.3*	24.8*	23.3
G4	58.7	51.4	51.6
fd	28.9	23.9	25.2
If1	31.4	29.3	35.3
PR64FS	39.8	31.2	27.2

Assay conditions were as described in the legend to Figure 5. The
amount of template DNA per assay was 0.48 nmol. 10 ng (*100 ng) of
purified RP4 DNA primase polypeptides or 1 ng of Iα DNA primase
(BW86(pLG214), ref. 7) were added per assay.

as template. This reaction resulted in oligoribonucleotides of the
structure pC(p̃A)$_n$, n $\leq$ 4 (Figure 6, lane b). Since the products
synthesized with CTP or CMP (Figure 6, lanes a, b) have the same
mobility, but the reaction being most efficient with CMP, it is concluded that the RNA primers begin with CMP at their 5'-terminal ends.
The products synthesized in the presence of cytidine are Cp̃A and
C(p̃A)$_2$, indicating that cytidine, instead of CMP can also be used
as a substrate by the plasmid primases (Figure 6, lane d). The structure of these oligonucleotides was confirmed by removal of the 5'-
terminal phosphates with alkaline phosphatase of pC(p̃A)$_n$ which resulted in products migrating to same positions as Cp̃A and C(p̃A)$_2$.
The nucleosides dC, A, G or U cannot be utilized for initiation of
the oligoribonucleotide synthesis by plasmid primase (Figure 6,
lanes f-i). AMP is predominantly found in the second position of
the products and to a much lesser extent GMP as indicated by experiments containing (α-^{32}P)GTP or (α-^{32}P)UTP in the reaction mixture
instead of (α-^{32}P)ATP. Deoxynucleosidetriphosphates cannot function
as substrates for plasmid primases under the conditions used.

     φX174 (-) strand synthesis measured in a reaction with a DNA
polymerase III fraction (1), plasmid-specified primase, ATP, CTP
and GTP reaches 75% of the value when all rNTPs are present. In this
reaction CTP can be replaced by CDP, CMP or cytidine without reducing
the amounts of DNA synthesized. However, when the cytidine compund

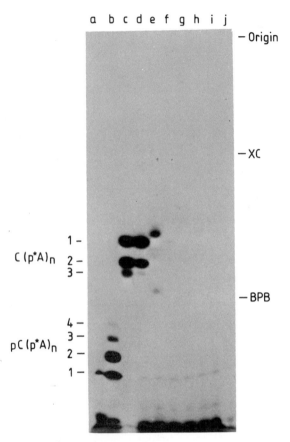

Figure 6: Oligoribonucleotides synthesized on ϕX174 DNA by purified
RP4 DNA primase-80 kDa. Reaction mixtures (20 μl) contained 40 mM
Tris-acetic acid (pH 7.0), 100 mM NaCl, 1 mM $MnCl_2$, 1 mM dithiothreitol, 100 μM (α-^{32}P)ATP (1000 cpm/pmol), 7.5 μg/ml ϕX174 DNA, and
50 μg/ml purified RP4 DNA primase-80 kDa. Incubation was for 10 min
at 30°C. The reaction was stopped by heating for 2 min at 100°C.
Samples were analyzed by electrophoresis at 30 V/cm on a 20% (w/v)
polyacrylamide-7M urea slab gel (22). The figure shows an autoradiograph of the gel. The following compounds were present in the reaction
mixture at a concentration of 100 μM: Lane (a) CTP, (b) CMP, (c) CMP;
products were treated with alkaline phosphatase prior to electrophoresis, (d) cytidine, (e) cytidine, products were digested with RNase A
prior to electrophoresis, (f) 2'-deocytidine, (g) adenosine, (h)
guanosine, (i) uridine, and (j) CMP, RP4 DNA primase-80 kDa ommitted.
Xylene cyanol (XC) and bromphenol blue (BPB) migrated to positions as
indicated. The sizes of the products are marked on the left of the
figure.

is omitted, no synthesis takes place indicating the necessity of
CMP or cytidine for functional primers. Degradation of di- and triphosphates to monophosphates or nucleosides most probably is responsible for the synthesis measured in the presence of CTP or CDP,
because it seems rather unlikely that CMP or cytidine could be phosphorylated by the DNA polymerase III fraction used.

Oligoribonucleotide synthesis by plasmid primase on short oligo-
deoxynucleotide templates of defined sequence indicated that the
enzymes can recognize dTdG sequences in $(dTDG)_n$, n = 1-9, and that
the products are complementary to the template DNA. The product of
such reactions is $pC\overset{*}{p}A$ if CMP and $(\alpha-32P)ATP$ are the substrates, but
if CTP in addition is present in the reaction mixture oligoribonucleotides of the composition $(pC\overset{*}{p}A)_n$, $n \leq 5$ are synthesized by the enzyme, suggesting that CMP can be incorporated into internal positions
of a chain. When dT_8dG was used as template, products were $pC(\overset{*}{p}A)_n$,
$n \leq 8$, which shows that the first 3'-nucleotide in a linear template,
in this case dG, can be recognized and transcribed. Oligoribonucleotides cannot be detected in considerable amounts in reactions with
templates lacking dTdG sequences. dT_ndG sequences are also recognized
in natural ssDNA (Figure 6 and 7). The frequency of existence of
dT_ndG sequences in the various templates (29 dT_3dG in $\phi X174$ DNA, 40
in fd, and 7 in G4, respectively) seems to correlate with the amount
of the corresponding products seen on the gel. The product $pC(\overset{*}{p}A)_3$
with G4 DNA is made in much smaller amounts than with fd or $\phi X174$
DNA (Figure 7, lane b).

The ability of the plasmid-encoded DNA primases to synthesize
oligoribonucleotides with the sequence (p)CpA... without the assistance of additional protein components by recognition of 3'-dGdT...
sequences at the end or within single-stranded linear templates and
by incorporation of CMP or cytidine as the substrate for the initial
nucleotide of the primers makes these enzymes clearly distinct from
the Escherichia coli primase - the dnaG protein (for review see ref.
23). Despite of these enzymatic differences, the plasmid DNA primase
can apparently replace the dnaG protein during discontinous replication of the chromosome in vivo as suggested by the results of suppression studies (vide supra). The major differences of the plasmid-
specified enzymes as compared to other primases and priming systems
like the phage T7 DNA primase, the phage T4-priming system, the dnaG
requiring systems, or the priming of fd/M13 complementary strand
synthesis by RNA polymerase is the exceptional requirement for CMP
or cytidine instead of a purine triphosphate (ATP) In addition, the
property that plasmid primases can initiate primer synthesis from
a 3'-dGdT... sequence at the very end of a polydeoxynucleotide chain,
may prove to be significant during the proposed role of the enzyme
in conjugal DNA replication.

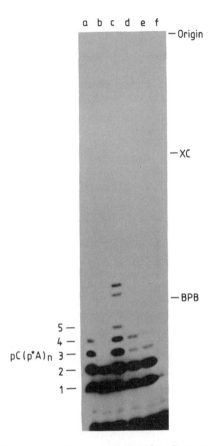

Figure 7: Oligoribonucleotides synthesized on various natural single-stranded phage DNAs by purified RP4 DNA primase - 80 kDa. Reaction conditions were as described in the legend to Figure 6. The reaction mixture contained 1 mM CMP. Phage DNAs were present at a concentration of 7.5 μg/ml: Lane (a) φX174, (b) G4, (c) fd, (d) If1, (e) PR64FS, and (f) DNA omitted from the reaction mixture.

Primers synthesized by plasmid-encoded primases can be elongated by several DNA polymerases

The primase appears to cooperate specifically with DNA polymerase III holoenzyme, as indicated by its ability to stimulate DNA synthesis in extracts deficient in DNA polymerase I and II and by its inability to restore DNA synthesis in a dnaZ mutant extract. This reaction, even in the presence of DNA polymerase I and II, was strictly dependent on ssb protein further indicating the requirement for the DNA polymerase III elongation system. The products synthesized under these conditions on φX174 DNA corresponds to replicative form II (1, 2). Plasmid primase and DNA polymerase I (large fragment) as purified proteins also mediate conversion of single-stranded DNA into duplex form when CMP or cytidine is present in the reaction mixture (Figure 8, lanes b-c and f-h), indicating that the primers must be long enough (about 5 nucleotides, ref. 24) to be efficiently elongated. DNA polymerase of phage T7 similarly under these conditions catalyzes the formation of RFII of φX174 DNA. Universality of the primase for a variety of elongation systems might be one of the properties which account for the broad host range character of RP4.

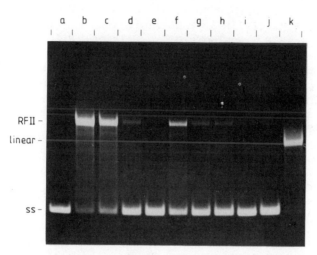

Figure 8:   fd DNA synthesis in the presence of purified RP4 DNA primase - 118 kDa and DNA polymerase I. The complete reaction mixture (50 μl) contained 50 mM potassium phosphate (pH 7.4), 10 mM MgCl₂, 1 mM dithiothreitol, 25 μg/ml bovine serum albumin, 50 μM dNTPs, 25 μM ATP and GTP, 2.5 μM CMP, 10 μg/ml fd DNA, 10 ng of purified RP4 DNA primase - 118 kDa and 1 U of DNA polymerase I (large fragment, Bethesda Research Lab.). Incubation was for 60 min at 30°C. The reaction was stopped by addition of SDS and EDTA to concentrations of 1% and 10 mM, respectively. Samples were analyzed by electrophoresis at 2 V/cm on a 1% (w/v) agarose gel in 40 mM Tris-acetic acid (pH 7.9). The figure shows the photograph of the gel stained with ethidium bromide. Lane (a) 0.5 μg of fd DNA, (b) complete reaction, (c) 1 μM CMP, (d) 0.1 μM CMP, (e) CMP omitted, (f) - (h) CMP replaced by cytidine (f), 50 μM, (g) 10 μM, (h) 5 μM, (i) complete reaction primase omitted, (j) complete reaction DNA polymerase I omitted, and (k) complete reaction, products digested with HpaI prior to electrophoresis.

ACKNOWLEDGEMENTS

We are grateful to M. Schlicht for excellent technical assistance, to B.M. Wilkins for the gifts of bacterial strains and phages, to J. Brosius for providing plasmid pKK223-3, and to H. Schuster for generous support and helpful criticisms of this manuscript.

REFERENCES

1. Lanka,E., Scherzinger,E., Günther,E. and Schuster,H. (1979)
        Proc. Natl. Acad. Sci. USA 76, 3632.
2. Lanka,E. and Barth,P.T. (1981) J. Bacteriol 148, 769.
3. Wilkins,B.M. (1975) J. Bacteriol. 122, 899.
4. Dalrymple,B.P., Boulnois,G.J., Wilkins,B.M., Orr,E. and
        Williams,P.H. (1982) J. Bacteriol. 151, 1.
5. Dalrymple,B.P. and Williams,P.H. (1982) J. Bacteriol. 152, 901.
6. Dalrymple,B.P. (1982) Ph. D. thesis: University of Leicester,
        England.
7. Wilkins,B.M., Boulnois,G.J. and Lanka,E. (1980) Nature
        290, 217.
8. Sasakawa,C. and Yoshikawa,M. (1978) J. Bacteriol. 133, 485.
9. Ludwig,R.A. and Johansen,E. (1980) Plasmid 3, 359.
10. Lanka,E., Lurz,R., Kröger,M. and Fürste,J.P.
        Mol. Gen. Genet, in press.
11. Chatfield,L.K., Orr,E., Boulnois,G.J. and Wilkins,B.M.
        (1982) J. Bacteriol. 152, 1188.
12. Thomas,C.M. (1981) Plasmid 5, 10.
13. Lanka,E., Lurz,R. and Fürste,J.P. Plasmid, in press.
14. Towbin,H., Staehelin,T. and Gordon,J. (1979)
        Proc. Natl. Acad. Sci. USA 76, 4350.
15. Godson,G.N. and Sinsheimer,R.L. (1967) Biochim. Biophys. Acta
        149, 476.
16. Boulnois,G.J. and Wilkins,B.M. (1979) Mol. Gen. Genet.
        175, 275.
17. de Boer,H.A., Comstock,L.J. and Vasser,M. (1983)
        Proc. Natl. Acad. Sci. USA 80, 21.
18. Laemmli,U.K. (1970) Nature 227, 680.
19. Wechsler,J.A. and Gross,J.D. (1971) Mol. Gen. Genet. 113, 273.
20. Meynell,G.G. and Lawn,A.M. (1968) Nature 217, 1186.
21. Coetzee,J.N., Sirgel,F.A. and Lecatsas,G. (1980) J. Gen.
        Microbiol. 117, 547.
22. Maxam,A.M. and Gilbert,W. (1977) Proc. Natl. Acad. Sci. USA
        74, 560.
23. Nossal,N.G. (1983) Ann. Rev. Biochem. 53, 581.
24. Oertel,W. and Schaller,H. (1973) Eur. J. Biochem. 35, 106.

DE NOVO DNA SYNTHESIS BY YEAST DNA POLYMERASE I ASSOCIATED WITH

PRIMASE ACTIVITY

P.Plevani, G.Magni, M.Foiani, L.M.S.Chang[1]
and G.Badaracco

Sezione di Genetica e Microbiologia
Dipartimento di Biologia
Università di Milano
Via Celoria 26
I-20133 Milan, Italy

[1]Department of Biochemistry
U.S.U.H.S.
Bethesda, Maryland 20814, U.S.A.

INTRODUCTION

DNA synthesis at the replication fork is carried out continuously on the leading-strand and discontinuously on the lagging-strand (1). The nascent DNA chains (Okazaki fragments) on the lagging-strand are synthesized by elongation of RNA initiators. Primase, the product of dnaG gene, synthesizes short ribo- or mixed ribodeoxyribo-oligonucleotide which are elongated by DNA polymerase III holoenzyme during the replication of the Escherichia coli chromosome and priming activities are also involved in the replication of several bacterio-phage genomes (1). A similar activity has been implicated in the initiation of Okazaki fragments during the semidiscontinuous repli-cation of polyoma DNA in intact nuclei (2), and some very recent reports have suggested that DNA primase activity may be associated with DNA polymerase α from eukaryotic organisms (3-6). The budding yeast Saccharomyces cervisiae contains two non-mitochondrial DNA polymerases, named DNA polymerase I and II (7-9). DNA polymerase I is the major enzyme and has biochemical properties similar to those of eukaryotic DNA polymerase α. DNA polymerase II resembles pro-karyotic DNA polymerases because of the presence of an associated proofreading 3'-exonuclease. The two yeast enzymes can be unambi-guously distinguished on the basis of the lack of immunological

281

cross-reactivity (7, 9). We have recently devised a new prification
protocol (9), which allows the isolation of an high molecular weight
form of yeast DNA polymerase I, by preventing uncontrolled proteolysis
during purification. We noted that an activity catalyzing a ribonucleotide-dependent DNA synthesis on synthetic and natural single-
stranded DNA template was associated with the high molecular weight
form of yeast DNA polymerase I. This unusual finding has been investigated and some preliminary results are discussed in the present
report.

RESULTS AND DISCUSSION

Earlier studies by Chang (7) identified a yeast DNA polymerase
I that was estimated to be approximately $Mr = 150,000$, but two major
bands of $Mr = \sim 70,000$ were identified by polyacrylamide gel electrophoresis in the presence of sodium dodecyl sulfate. This enzymatic
form can use oligoriboadenylate as an effective initiator for poly
(dT) replication (7) and can efficiently elongate RNA molecules
synthesized by yeast RNA polymerase on single-stranded DNA templates
in a yeast RNA polymerase-DNA polymerase coupled reaction, while DNA
polymerase II is uneffective with such substrates (10, 11). The development of new immunological and biochemical methods (12, 13) capable
of identifying a given antigen in a complex mixture of proteins allowed us to detect proteolysis of DNA polymerase I during purification and were used as an aid in developing a new purification protocol that yielded a DNA polymerase I preparation containing two major
polypeptides of $Mr = 140,000$ and $Mr = 110,000$ (9). In analyzing the
primer-template specificities of the high molecular weight DNA polymerase I preparations, we found that incubation of single-stranded
M13 DNA or poly(dT) with the high molecular weight form of the enzyme
resulted in incorporation of deoxynucleotides into acid-precipitable
products, provided that the corresponding rNTPs were present in the
reaction mixture. The timecourse of in vitro DNA synthesis catalyzed
by DNA polymerase I with M13 DNA and poly(dT) as template in the presence or absence of the 4 rNTPs and ATP respectively, is shown in
Figure 1.

With M13 DNA as template, maximum DNA synthesis was observed
when all the 4 rNTPs were present, although ATP alone supported DNA
synthesis to 70% of the maximum rate. GTP, CTP or UTP cannot substitute for the ATP requirement. With poly(dT) as substrate, ATP (1mM)
was absolutely required.

The replicative form of M13 DNA, as well as plasmid DNA, were
inactive as template of the reaction. Analogously to what found with
Drosophila melanogaster DNA polymerase α (3), very low concentrations
of yeast DNA polymerase I can stimulate an ATP-dependent poly(dT)
replication catalyzed by Escherichia coli DNA polymerase I. The
previous data suggest that yeast DNA polymerase I can catalyze the

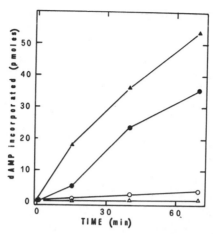

Figure 1:   Time-course of M13 DNA and poly(dT) replication.
Phage M13 single-stranded DNA was extracted from CsCl purified phage
with phenol. Poly(dT) $\frac{}{1000}$ *was prepared with calf thymus terminal*
transferase. The complete reactions (0.1 ml) were 50 mM TRIS-HCl at
pH 8.0, 2 mM 2-mercaptoethanol, 8 mM MgCl, 0.1 mg/ml bovine serum
albumin, 0.020 mg/ml of single-stranded M13 DNA or poly(dT), 0.010
ml of enzyme in buffer with 50% glycerol. With M13 DNA as template
the mixtures usually contained each of the four dNTPs at 0.1 mM with
(^{3}H)-dATP at 400 cpm/pmol, 1 mM ATP and 0.1 mM each GTP, CTP and UTP.
With Poly(dT) as template the mixtures contained (^{3}H)-dATP (0.1 mM;
300 cpm/pmol) and ATP (1 mM). Reaction mixtures were incubated at
35ºC, aliquots were taken at the indicated time and acid precipitable
products were processed on glass-fiber disks as previously described
(10). (●———●) M13 DNA replication, complete reaction; (O———O) M13
DNA replication, minus the four rNTPs; (△———△) poly(dT) replication,
complete reaction; (▲———▲) poly(dT) replication, minus ATP.

synthesis of RNA initiators that can be elongated by the yeast enzyme
itself or by primer-dependent DNA polymerase from other sources.

When yeast DNA polymerase I was incubated with single-stranded
M13 DNA, the rNTPs and the 4 dNTPs in the presence of (a^{32}P)-dATP
and then subjected to electrophoresis in 1% agarose gel and autoradiography, most of the radioactivity comigrated with single-stranded M13 DNA markers, even if a smear of labeled products migrating
as replicative forms were also detected, expecially at longer incubation times. However, the products of the reaction were not covalently linked to the template DNA, because they could be separated
from the template by two-dimensional agarose gel electrophoresis in
which the products were separated in the first dimension at neutral
pH and in the second dimension at alkaline pH, suggesting that

de novo initiation of DNA chains is catalyzed by the DNA polymerase I associated primase activity (Figure 2).

The covalent linkage of RNA initiators with DNA products in the in vitro replication of M13 DNA was demonstrated by ^{32}P-transfer experiments.

Table 1 shows that all four ribonucleotides were present at the RNA-DNA junctions adjacent to dAMP, but their relative frequencies were not random, being the frequency of CMP<<GMP<UMP<AMP.

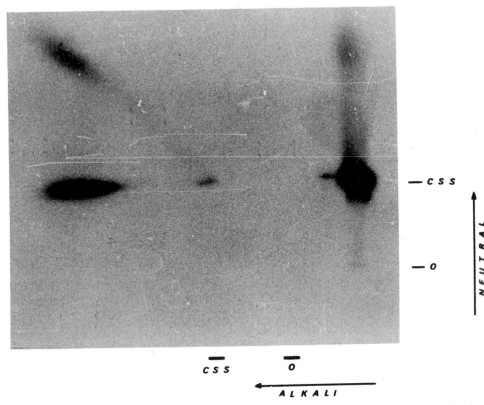

Figure 2: Two-dimensional agarose gel electrophoresis of the M13 DNA replication products. A 0.020 ml M13 DNA replication reaction was carried out as described in legend to Figure 1 except that (α^{32}P)-dATP (3000 Ci/mmol) at 0.1 mCi/ml was present as the labeled nucleotide. After 30 min of incubation the reaction was made 1% in sodium dodecyl sulfate and 20 mM in EDTA and the DNA products were isolated through a Sephadex-G75 column. The DNA in the excluded volume was collected and concentrated. Two-dimensional agarose gel electrophoresis and autoradiography was essentially carried out as described by Gaurlie and Pigiet (15).

Table 1: Nearest-neighbour analysis with ^{32}P transfer from
          $(\alpha-^{32}P)$dATP to 2' (3')NMP

Ribonucleotide	c.p.m.	%
Ap	2578	52
Up	1140	23
Gp	1020	21
Cp	186	4
total	4924	100

The products were synthesized in a M13 DNA replication reaction as
described in the legend to Figure 1, except $(\alpha^{32}P)$-dATP was used
as the labeled nucleotide. The products were isolated as in the
legend to Figure 2 and the isotope transfer analysis was performed
as previously described (11).

Because all four rNMPs are present at the RNA-DNA junctions the
length of RNA initiators in the M13 DNA or poly(dT) replication
reaction was also analyzed. The RNA molecules were labeled with
$(\alpha^{32}P)$-ATP or $(\gamma^{32}P)$-ATP and their length was determined by polyacrylamide gel electrophoresis under denaturing conditions, as described by Conaway and Lehman (14). Both with poly(dT) and M13 DNA,
oligonucleotides ranging from 15 to 30 residues were synthesized.
After DNase I digestion the length of the initiators was reduced,
and oligonucleotides of 8-10 residues were generated. Such short
RNA molecules are the only ones that can be detected when RNA synthesis is coupled to DNA synthesis by addition of the four dNTPs
and the products of DNA synthesis were digested with DNase I before
electrophoresis. These results suggest that yeast DNA polymerase I
can catalyze the synthesis of mixed ribo-deoxyribooligonucleotides.

We believe that the ribonucleotide-dependent poly(dT) or M13
DNA replication reaction provides a possible assay for the isolation
of a DNA polymerase holenzyme containing an associated primase activity. However, the isolation of larger amounts of purified protein
and a detailed biochemical and immunological analysis are required
to identify the putative structure of a yeast DNA polymerase holoenzyme and to discriminate whether the primase and DNA polymerase
activity reside on the same, or different polypeptides.

ACKNOWLEDGEMENTS

     This research was supported by Consiglio Nazionale delle
Ricerche, Rome, Italy (Progetto Finalizzato Ingegneria Genetica e
Gruppo Nazionale Biologia Molecolare ed Evoluzionistica) and by
Grant CA 23365 from the National Cancer Institute, United States
Department of Health and Human Services to L.M.S.C..

REFERENCES

1.   Kornberg,A. (1980) "DNA replication", W.H. Freeman and Co.,
          San Francisco.
2.   Reichard,P. and Eliasson,R. (1978) Cold Spring Harbor Symp.
          Quant. Biol. 43, 271.
3.   Conaway,R.D. and Lehman,I.R. (1982) Proc. Natl. Acad. Sci.
          USA 79, 2523.
4.   Yagura,T., Kozu,T. and Seno,T. (1982) J. Biol. Chem.
          257, 11121.
5.   Shioda,M., Nelson,E.M., Bayne,M.L. and Benbow,R.M. (1982)
          Proc. Natl. Acad. Sci. USA 79, 7209.
6.   Hübscher,U. (1983) EMBO J. 2, 133.
7.   Chang,L.M.S. (1977) J. Biol. Chem. 252, 1873.
8.   Wintersberger,W. (1978) Eur. J. Biochem. 84, 167.
9.   Badaracco,G., Capucci,L., Plevani,P. and Chang,L.M.S.
          J. Biol. Chem., in press.
10.   Plevani,P. and Chang,L.M.S. (1977) Proc. Natl. Acad. Sci.
          USA 74, 1937.
11.   Plevani,P. and Chang,L.M.S. (1978) Biochemistry 17, 2530.
12.   Towbin,H., Staehlin,T. and Gordon,J. (1979) Proc. Natl. Acad.
          Sci. USA 76, 4350.
13.   Spanos,A., Sedgwick,S.G., Yarranton,G.T., Hübscher,U. and
          Banks,G.R. (1981) Nucleic Acids Res. 9, 1825.
14.   Conaway,R.C. and Lehman,I.R. (1982) Proc. Natl. Acad. Sci.
          USA 79, 4585.
15.   Gourlie,B.B. and Pigiet,V.P. (1983) J. Virol. 45, 585.

PARADOXES OF IN SITU POLYACRYLAMIDE GEL ASSAYS FOR

DNA POLYMERASE PRIMING

Geoffrey R. Banks

National Institute for Medical Research
The Ridgeway
Mill Hill, London, U.K.

The DNA polymerase-α from Drosophila melanogaster embryos is
now one of the most extensively characterised eukaryotic multisubunit
polymerases because it can be isolated in high yields in an intact,
yet highly purified form. It has proved to be amenable to detailed
biochemical studies of its subunit structure and subunit contributions to the overall DNA synthesis reaction, primer and template
requirements, processivity, cellular functions and immunological
properties (1-10). The holoenzyme utilises not only gapped duplex
DNA as a primer-template system, but also pre-DNA and pre-RNA primed
single-stranded circular phage DNAs (6, 10). Furthermore, it will
synthesise DNA on an unpreprimed single DNA strand in the presence
of both rNTPs and dNTPs because it also catalyses the synthesis of
short RNA and RNA-DNA oligonucleotides which prime subsequent DNA
chain elongation (8, 9). Whether this composite reaction is related
to Okazaki fragment initiation and synthesis in vivo remains to be
determined.

The holoenzyme comprises a complex of at least three polypeptide subunits - α (Mr 182,000), β (Mr 60,000), γ (Mr 46,000) and
possibly a fourth (Mr 73,000), all of which appear to be antigenically unrelated to each other (2-4). Using the technique of glycerol
gradient sedimentation in the presence of urea to dissociate and
separate the holoenzyme subunits, Villani et al. (3) concluded that
the α subunit catalysed DNA chain elongation. The resulting isolated
α subunit was, however, substantially less active and was less able
to replicate extensive single-stranded DNA template regions, perhaps
because its processivity decreased, than when it was associated with
the other subunits (3, 10). This conclusion was substantiated by an
in situ polyacrylamide gel assay of polymerase activity after elec-

287

trophoresis of the holoenzyme in the presence of SDS (11). DNA chain elongation activity was associated solely with the α subunit. In collaboration with Professor I.R. Lehman's group, I have attempted to determine which subunit catalyses RNA priming using a modification of the in situ gel assay technique (12). Briefly, this involves the electrophoresis of DNA polymerase enzymes through an SDS-polyacrylamide gel containing gapped calf thymus DNA to separate cleanly any subunits, followed by detergent removal and polypeptide renaturation. The gel is then incubated in a polymerase reaction mixture containing $(\alpha-^{32}P)dTTP$ of high specific activity and the other dNTPs. $(\alpha-^{32}P)dTMP$ covalently incorporated in the DNA within the gel by an active DNA polymerase polypeptide is detected by autoradiography of the dehydrated gel after removal of unincorporated radiolabel. Figure 1 shows an autoradiogram after assay of the <u>Drosophila melanogaster</u> DNA polymerase-α holoenzyme by this technique.

The RNA priming capacity of the holoenzyme can be assayed in the test tube by its ability to catalyse extensive DNA synthesis on single-stranded M13 DNA in the presence of ATP or GTP, or to synthe-

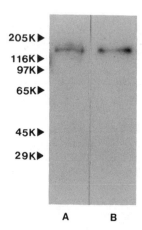

A   B

Figure 1: In situ gel assay of <u>Drosophila melanogaster</u> DNA polymerase-α activity on a gapped duplex DNA template. Technical details have been described by Spanos et al. (12). The following modifications were introduced - 0.5 units of DNA polymerase-α holoenzyme were electrophoresed through a 7.5% SDS-polyacrylamide gel (50 x 40 x 0.75mm) containing gapped calf thymus DNA (30 μg/ml). The gel was washed at 5ºC in five 50 ml aliquots of 50 mM Tris-HCl pH 7.5, 4 mM MSH for 30 min each. The required gel lanes were excised and sealed in a plastic bag containing (0.5 ml/gel lane) 50 mM Tris-HCl pH 8.5, 4 mM DTT, 10 mM MgCl₂, 40 μM each of dATP, dGTP and dCTP, 30 μCi $(\alpha-^{32}P)dTTP$ (3000 Ci/mmole) and 10% glycerol (Lane A). The mixture for lane B also contained 500 μM ATP. Incubation was at 37ºC for 3 hr. Arrows show the migration positions of protein Mr markers run on the same gel and stained with Coomassie blue dye.

sise polydA on polydT in the presence of ATP, in addition to the
necessary dNTPs. Alternatively, Escherichia coli DNA polymerase I
can be added to DNA chain elongate the presumptive primers synthesised by the Drosophila melanogaster holoenzyme (Conaway and Lehman,
1982a, 1982b). Such synthesis is virtually abolished if the rNTPs
are omitted. Extensive analysis has established that the reaction
products are oligoribonucleotides, up to fifteen nucleotides in
length, covalently joined to the 5'-terminus of a DNA chain. Mixed
ribo- and deoxyribo-oligonucleotides can also be generated (8, 9).
The gel assay technique was, therefore, modified in order to mimic
the specificity of these reactions.

Aliquots of the Drosophila melanogaster holoenzyme were electrophoresed through an SDS-polyacrylamide gel containing M13 DNA.
One enzyme-containing gel lane was excised and assayed in the presence of $(\alpha\text{-}^{32}P)$dTTP, dATP, dGTP and dCTP. Another, identical gel
lane was assayed in the presence of ATP in addition to the dNTPs.
Although DNA synthesis by the α subunit was detectable in the
absence of ATP, it was markedly stimulated when ATP was present
(Fig. 2). Control experiments confirmed that no ATP-induced stimulation of DNA chain elongation occurred when gapped calf thymus
DNA was the primer-template in the gel (Figure 1). In a second series
of experiments, holoenzyme was electrophoresed through a gel containing polydT. One enzyme-containing gel lane was incubated in the
presence of $(\alpha\text{-}^{32}P)$dATP and another in $(\alpha\text{-}^{32}P)$dATP and ATP. PolydA
synthesis by the α subunit was readily detected when ATP was present,

A   B

Figure 2: In situ gel assay of Drosophila melanogaster DNA polymerase-α activity on M13 single-stranded DNA. Technical details as
in the legend to Figure 1, except that 0.7 units of the enzyme was
electrophoresed through a gel containing 30 µg/ml M13 DNA. Lane A
-the assay mixture contained dNTPs only; lane B, it contained 500 µM
ATP in addition. Escherichia coli DNA polymerase I run on the same
gel as an activity marker verified that the Drosophila melanogaster
activity bands comigrated with the α subunit.

but not when it was absent. Furthermore, when GTP replaced ATP in the presence of $(\alpha-^{32}P)dATP$, or when $(\alpha-^{32}P)dTTP$ replaced $(\alpha-^{32}P)dATP$ in the presence of ATP, no synthesis was again detectable (results not shown). Complementarity of both ribo- and deoxyribonucleoside triphosphates with the polydT template is thus strictly observed for detectable synthesis and so mimic the specificity of the holoenzyme reactions (8, 9). Attempts to demonstrate convincingly the actual incorporation of rNTPs into the presumptive RNA primer of the RNA-DNA products by the α subunit have been uniformly unsuccessful. In order to maintain the specific activities of the $(\alpha-^{32}P)rNTPs$ (ATP or GTP with M13 and ATP with polydT templates) at a sufficiently high level, their low concentrations (approx. 10 nM) would result in extremely inefficient priming (the holoenzyme's Km for ATP is in the order of 300 μM). Nevertheless, the specificity of the above results suggest that the α subunit catalyses both RNA primer synthesis and DNA chain elongation.

Kaguni et al. (13) have again employed glycerol gradient sedimentation in the presence of urea to dissociate and separate the holoenzyme subunits. As described in Professor Lehman's contribution to this workshop, the appropriate assays of the resulting gradient fractions clearly reveal that the α subunit catalyses DNA chain elongation only and the β/γ subunit RNA priming only. Such a division of labour would be undetected by the in situ gel assay technique because an RNA primer synthesised by β/γ would remain unlabelled when DNA synthesis was assayed and, if RNA labelling had been possible, the primers would probably be lost unless stabilised by subsequent DNA chain elongation. To circumvent this limitation, experiments similar to those illustrated in Figures 2 and 3 were run, but in which the gel assay mixture contained either Escherichia coli DNA polymerase III core enzyme or DNA polymerase I. Both enzymes should stabilise β/γ synthesised RNA primers by DNA synthesis. The results suggested that the large core polymerase did not penetrate the gel, whilst with polymerase I present, a high background radioactivity precluded an identification of specific ATP dependent activity. This background activity is, in fact, a result of DNA synthesis by polymerase I on M13 DNA and polydT templates. When this polymerase was electrophoresed and assayed in gels containing either M13 DNA or polydT, DNA synthesis by the 109,000 dalton polypeptide is clearly evident, both in the presence and absence of ATP in the assay mixture (Figure 4). Synthesis on gapped duplex and M13 DNA templates is significantly inhibited by ATP and that on polydT slightly stimulated. The level of synthesis was also time-dependent in that it increased to a maximum after 3-5 hours of incubation in the gel assay mixture and then decreased, presumably a result of polymerase I associated exonuclease degradation.

What mechanisms might accommodate the different results from the two methods used to determine the identity of the Drosophila melanogaster DNA polymerase-α subunit responsible for RNA priming?

A   B

Figure 3: In situ gel assay of Drosophila melanogaster DNA poly-
merase-α activity on polydT. Technical details as in the legend to
Figure 1, except that 0.7 units of enzyme was electrophoresed through
a gel containing 36 μg/ml polydT. Lane A -the assay mixture contained
30 mCi (α-^{32}P)dATP (3000 Ci/mmole); lane B -it contained 500 μM ATP
in addition.

In the gel experiments, it is unlikely that short complementary
oligonucleotides contaminating both templates used could be responsible for the observed synthesis because it is stimulated by (for
M13 DNA templates) or dependent on (for polydT templates) the presence
of rNTPs, the M13 DNA used in both sets of experiments was prepared

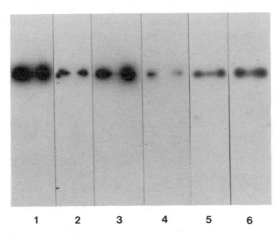

1   2   3   4   5   6

Figure 4:   In situ gel assays of Escherichia coli DNA polymerase I.
0.3 units of the enzyme were electrophoresed in each case. Technical
details are in the legends to Figure 1, 2 and 3, respectively, for
gels containing: lanes 1 and 2, gapped calf thymus DNA; lanes 3 and
4, M13 DNA; lanes 5 and 6, polydT. Assay mixtures for lanes 2, 4 and
6 contained ATP, those for 1, 3 and 5 did not.

in an identical manner and short oligonucleotides are lost from the
polyacrylamide gels before their assay (A. Spanos, personal communication). ATP stimulation/dependence might also arise by oligonucleotide contamination of the ATP, but this was obtained from the same
commercial source (PL Biochemicals Inc.) for both sets of experiments.
The major difference between the two methods may be in the environments
of the assays. The gel assay involves essentially immobilised DNA
templates and enzyme polypeptides, the latter being at a 60-fold
higher concentration than in a typical test tube assay. In addition,
we do not know to what extent the native polypeptide conformation is
achieved after renaturation in the gels. It may not be surprising,
therefore, if the properties of the α subunit in solution differ
somewhat from those in the gel so that some RNA synthesis can occur
in the latter, but not in the former enviroment. It is also possible
that incipient RNA primers are synthesised by the α subunit in solution, but immediately dissociate from the template before stabilisation is accomplished by DNA chain elongation. Such stabilisation
could be a function of β/γ in the holoenzyme in solution, whereas in
their absence in the gel, stabilisation might result from an immobile
gel environment. Alternatively, even if the α subunit synthesises
unstable RNA primers in solution, it is the β/γ subunit which synthesises functional, stable primers for subsequent DNA chain elongation.

The results shown in Figure 4 might imply that Escherichia coli
DNA polymerase I initiates de novo DNA synthesis, even though in solution it apparently does not occur (14). Similar considerations to
those outlined above for the Drosophila melanogaster α subunit may
again pertain although, in contrast, the polymerase I catalysed reactions exhibit no interpretable rNTP dependence. It has been established that when Mn^{++} substitutes for Mg^{++} in the assay mixture,
rNTPs are incorporated into the reaction products (15, 16). Low concentrations of rNTPs contaminating the dNTPs might be incorporated
if the appropriate Kms were low enough. It should be emphasised,
however, that even under these conditions, there is no evidence for
de novo DNA synthesis.

In summary, it is clear from the available evidence, that under
conditions perhaps most closely resembling those found in the intact
cell, that both the α and β/γ subunits of the Drosophila melanogaster
holoenzyme are required for de novo DNA synthesis directed by single-
stranded DNA templates.

ACKNOWLEDGEMENTS

I am indebted to Ad Spanos, U. Hübscher, I.R. Lehman, L.S.
Kaguni, R.C. Conaway and J-M. Rossignol for materials and advice,
and S.G. Sedgwick for critical reading of the manuscript.

REFERENCES

1. Brakel,C.L. and Blumenthal,A.B. (1977) Biochemistry, 16, 3137.
2. Banks,G.R., Boezi,J.A. and Lehman,I.R. (1979) J. Biol. Chem.
    254, 9886.
3. Villani,G., Sauer,B. and Lehman,I.R. (1980) J. Biol. Chem.
    255, 9479.
4. Kaguni,L.S., Rossignol,J-M., Conaway,R.C. and Lehman,I.R.
    (1983) Proc. Natl. Acad. Sci. USA 80, 2221.
5. Sauer,B. and Lehman,I.R. (1982) J. Biol. Chem. 257, 12394.
6. Kaguni,L.S. and Clayton,D.A. (1982) Proc. Natl. Acad. Sci.
    USA 79, 983.
7. Sugino,A. and Nakayama,K. (1980) Proc. Natl. Acad. Sci. USA
    77, 7049.
8. Conaway,R.C. and Lehman,I.R. (1982) Proc. Natl. Acad. Sci.
    USA 79, 2523.
9. Conaway,R.C. and Lehman,I.R. (1982) Proc. Natl. Acad. Sci.
    USA 79, 4585.
10. Villani,G., Fay,P.J., Bambara,R.A. and Lehman,I.R. (1981)
    J. Biol. Chem. 256, 8202.
11. Hübscher,U., Spanos,A., Albert,W., Grummt,F. and Banks,G.R.
    (1981) Proc. Natl. Acad. Sci. USA 78, 6771.
12. Spanos,A., Sedgwick,S., Yarranton,G.T., Hübscher,U. and
    Banks,G.R. (1981) Nucleic Acids Res. 8, 1825.
13. Kaguni,L.S., Rossignol,J-M., Conaway,R.C., Banks,G.R. and
    Lehman,I.R. (1983) J. Biol. Chem. 258, 9037.
14. Kornberg,A. (1980) in:"DNA Replication", W.H. Freeman and Co.,
    San Francisco.
15. Berg,P., Fancher,H. and Chamberlin,M. (1963) in:"Informational
    Macromolecules" (Vogel,J.H., Bryson,B. and Lampen,J.O.,
    eds.), p. 467, Academic Press, New York.
16. van de Sande,J.H., Loewen,P.C. and Khorana,H.G. (1972)
    J. Biol. Chem. 247, 6140.

ASSOCIATION BETWEEN PRIMASE AND DNA POLYMERASE α IN MURINE CELLS

Michel Philippe, Rose Sheinin[1] and
Anne-Marie De Recondo

Unité de Biologie et Génétique Moléculaires
Institut de Recherches Scientifiques sur le Cancer
B.P. NO 8
94802 Villejuif Cédex, France

[1]Department of Microbiology and Parasitology
University of Toronto
Toronto, Ontario M5S 1A1, Canada

INTRODUCTION

In prokaryotes as well as in eukaryotes, extensive analysis of purified replicating DNA has shown that the Okazaki fragments are initiated by RNA primers (1). They are oligoribonucleotides, approximately 10 nucleotides in length, which are covalently attached to 5'-termini of newly synthesized DNA. They are not synthesized by any RNA polymerase involved in transcription but by another enzyme called primase. Such a primase has been described in eukaryotic cells by several authors (2-10). The primase purified from mouse hybridoma cells (10) was found to be very easily separable from DNA polymerase α, and the most highly purified fraction contained two major protein components of 56,000 and 46,000 daltons. In most cases however, the primase appeared to remain closely associated with DNA polymerase α (2-9) throughout all the steps in purification of the holoenzyme (3-9).

Hübscher (9) has reported that the 125,000 dalton subunit of the calf thymus DNA polymerase α contains both DNA polymerase and DNA primase. On the contrary Kaguni et al. (11) have shown that primase activity resides in the 60,000 dalton and (or) in the 50,000 dalton subunit of Drosophila melanogaster DNA polymerase α. We summarize here the results obtained in a study of the association bet-

ween the primase and DNA polymerase α using two different biological
systems. We first studied regenerating rat liver and then a temperature-sensitive mutant (ts2) of the mouse Balb/C 3T3 cell line which
is affected in DNA synthesis.

REGENERATING RAT LIVER

    DNA polymerase α from regenerating rat liver was purified to
near homogeneity by Méchali et al. (12). Five polypeptides were re-
producibly resolved on denaturing gels, corresponding to molecular
weights of 156,000, 64,000, 61,000, 58,000 and 54,000 daltons. The
catalytic activity was shown to be associated with the 156,000 dalton
polypeptide. As shown in Table 1, we were able to detect a primase
activity in aliquots from the different steps of the purification
procedure. Therefore the primase activity appeared to be tightly
associated with DNA polymerase α in regenerating rat liver. The primase was characterized as an activity that catalyzed ribonucleoside
triphosphate-dependent synthesis, and was unaffected by α-amanitin,
an inhibitor of RNA polymerases II and III. Analysis of the RNA primers by gel electrophoresis showed that fragments of about 10 nucleotides in length were synthesized by the primase (13). To investigate
the relationship between the primase and DNA polymerase α we have
chosen two experimental approaches. First we followed the time course
of primase and DNA polymerase α induction as a function of regeneration time after partial hepatectomy. As shown in Table 2, primase
and DNA polymerase (α and β) activities were measured after centrifugation of the cytoplasmic supernatant through a 5-20% sucrose gradient. As expected, the level of DNA polymerase β did not change
significantly during liver regeneration. In contrast, the DNA polymerase α and primase levels increased. The maxima of activities were
reached 43 hours after partial hepatectomy. The curve showing the
appearance of both activities is roughly the same. The sensitivity
of the assay was increased by the addition of exogenous Escherichia
coli DNA polymerase I to the reaction mixture as described by Conaway
and Lehman (3). This was especially true at the early times, when
the DNA polymerase α level was still low (13). These results do not
allow any conclusion to be made about the possible association between the two activities, but they strongly suggest that primase is
involved, as is DNA polymerase α, in the DNA replication process
in vivo.

    To further investigate the relationship between primase and
DNA polymerase α, we have measured their activities through several
purification steps. As shown in Figure 1, the cytoplasmic extract
was fractionated by ammonium sulfate precipitation using the back
wash technique as previously described by Méchali et al. (12). The
distribution of primase and DNA polymerase α was not exactly the
same among the fractions. The maximum value of DNA polymerase α
activity was found in the fractions which correspond to 15-17% (w/v)

Table 1:   DNA polymerase α purification and primase activity[*]

| | DNA polymerase activity | | | |
Fraction	Total protein (mg)	Total activity (units)	Specific activity (units/mg)	Primase activity
I - Cytoplasmic extract	5,400	81,000	15	+
II - Ammonium sulfate	327	63,000	192	+
III - DEAE-cellulose	75	44,500	594	+
IV - Phosphocellulose	4.0	24,000	6,000	+
V - Hydroxylapatite	0.40	16,000	40,100	+
VI - DNA-cellulose	0.059	2,673	45,300	+

[*]Primase activity was measured in aliquots from each of the different steps used in the purification of DNA polymerase α as described by Méchali et at. (12). Primase was detected according to Conaway and Lehman (3) as an activity that catalyzed the ribonucleoside triphosphate-dependent synthesis of DNA with single-stranded M13 DNA or poly dT as template. DNA synthesis was absolutely dependent on the presence of ribonucleotides and was unaffected by α amanitin, an inhibitor of RNA polymerases II and III (13). Extensive analysis of the primers by gel electrophoresis using (^{32}P)-labelled ATP in the absence of deoxyribonucleotides, showed that fragments of about 10 nucleotides in length were synthesized by the primase. The same value was found for primers synthesized in the presence of deoxyribonucleotides, after digestion of the RNA-DNA products by DNase I (13).

$(NH_4)_2 SO_4$ in the elution buffer. In contrast the fraction containing the highest level of primase activity corresponded to 13% $(NH_4)_2 SO_4$. But each fraction contained both activities.

Moreover, by centrifugation analysis on sucrose gradients, there was no detectable difference in the sedimentation coefficient of the DNA polymerase obtained from the different fractions (not shown). This result suggests that the low level of primase compared to DNA polymerase activity in certain fractions is not due to significant proteolysis of the enzyme. We then attempted to separate primase from DNA polymerase α by chromatography of the fraction containing the highest level of primase activity on a Heparin-Sepharose column. As shown in Figure 1, both activities were eluted at exactly

M. PHILIPPE ET AL.

Table 2:   Variations of primase and DNA polymerase (α and β) activities as a function of regeneration time *

Regeneration time (hours)	Incorporated dXTP							
	Picomoles per mg of liver DNA polymerase				Picomoles per mg of proteins DNA polymerase			
	α	β	α+β	primase	α	β	α+β	primase
0	0	8006	8006	0	0	231	231	0
18	3944	14165	18109	10010	77	276	353	196
24	11600	9138	20738	10509	343	270	613	311
43	51688	10958	62646	24550	14711	312	1783	699
72	18534	16245	34779	7061	431	378	809	164

*The results correspond to the measurements of primase and DNA polymerase α and β activities after centrifugation of the cytoplasmic fraction through a 5-20% sucrose gradient containing 250 mM KCl. DNA polymerase I from Escherichia coli was added in the primase assay according to Conaway and Lehman (3) to ensure the detection of the full primase activity even in the absence of DNA polymerase α.

the same position (220 mM KCl), although a small percentage of both
activities was found in the flow through. The peak eluted from the
Heparin Sepharose column was divided into two aliquots: one aliquot
was chromatographed on Hydroxylapatite, the other on a single-stranded
DNA cellulose column (13). In these two experiments (not shown), primase and DNA polymerase α still eluted together and could not be
separated from one another.

We conclude that primase and DNA polymerase α, which are absent
in adult rat liver, are induced with similar kinetics after partial
hepatectomy. We were unable to separate primase from DNA polymerase
α by using either sucrose gradient centrifugation or chromatography
(Heparin-Sepharose, Hydroxylapatite, DNA cellulose). This suggests
that both activities are tightly associated with one another in the
DNA polymerase α holoenzyme.

BALB/C-3T3 ts2 CELLS

The ts2 mutant cells of Balb/C-3T3 fibroblasts isolated and
first characterized by Slater and Ozer (14) have now been analyzed
further by Sheinin and Lewis (15) and Sheinin et al. (16). The principal characteristics of the ts2 mutants are summarized in Figure 2.
When shifted to the non-permissive temperature, the cells complete
ongoing normal DNA replication and subsequent cell division but they
are unable to re-enter S phase. Arrested early in S phase with
hydroxyurea, at the permissive temperature, they are able to synthesize DNA at the non-permissive temperature after the release of the
block, and to perform cell division prior to cell cycle arrest.
Arrested in mid G1 phase by isoleucine starvation, ts2 cells synthesize DNA and divide only when they are given fresh medium containing
isoleucine and incubated at the permissive temperature. Incubation
at the non-permissive temperature prior to this restriction point
(isoleucine starvation) does not affect DNA replication. Thus, it
can be postulated that the ts2 locus encodes information for a protein which is expressed late in G1 and promotes the traverse of G1
to S.

The lower part of Figure 2 shows the uncoupling of synthesis
of histones and non-histone chromatin proteins (NHCP'S) from the
temperature-inactivated DNA synthesis. Moreover, the ts2 cells accumulate proteins and RNA at the non-permissive temperature whereas
the cellular DNA content remains relatively constant. This result
clearly indicates that the defect in ts2 cells is more specifically
related to DNA synthesis than to RNA or protein synthesis.

We have studied primase and DNA polymerase α activities in
ts2 mutants at the permissive temperature (34°C), and 24 hours after
the shift to the non-permissive temperature (38.5°C). Postmicrosomal
fractions of the cells were prepared, dialyzed, and chromatographed

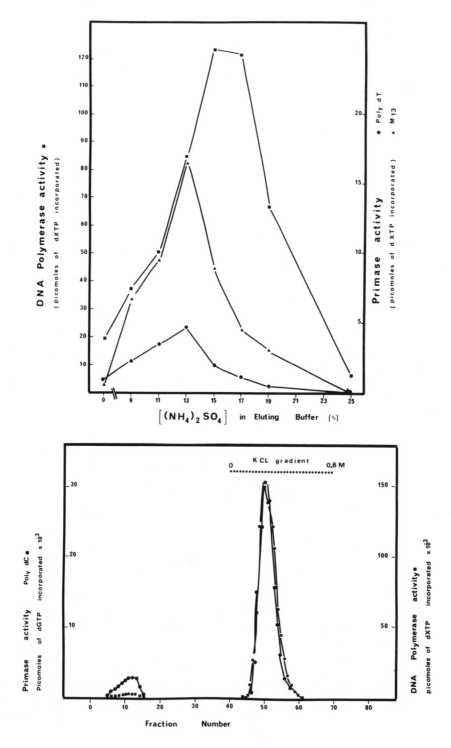

on a Heparin-Sepharose column (18). The results are shown in Figure 3. When ts2 cells were incubated at the permissive temperature, the primase and DNA polymerase α activities were eluted simultaneously from the Heparin-Sepharose column, as was also observed for regenerating rat liver. A higher level of primase activity was detected when poly dC was used as a template instead of poly dT; this was also the case for regenerating rat liver. A second peak of polymerase activity was eluted at higher salt concentrations and was characterized as DNA polymerase β, in accord with the results of Brennessel et al. (17). The elution profile of Heparin-Sepharose chromatography of cytoplasmic extracts from ts2 cells grown at the non-permissive temperature is shown in Figure 3. DNA polymerase β activity was unchanged at the non-permissive temperature, but DNA polymerase α activity was significantly reduced. However, the elution profiles of both polymerase activities were the same at both the permissive and non-permissive temperatures. At the non-permissive temperature however, the primase was not retained on the column and was found in the flow through. These results clearly indicate that the primase and DNA polymerase α activities are not on the same subunit. Extracts from wild type cells, (Balb/C-3T3), grown at $34^{O}C$ or $38.5^{O}C$, were chromatographed on Heparin-Sepharose as a control (not shown). In both cases the primase and DNA polymerase α activities were eluted simultaneously from the column, as was also the case when the ts2 cells were incubated at the permissive temperature. Thus the primase is tightly associated with polymerase α in ts2 cells synthesizing DNA at $34^{O}C$, but becomes separable from the polymerase at the non-permissive temperature, when the cells are unable to enter S phase. The dissociation of the complex in ts2 cells at the non-permissive temperature remains to be explained. Several hypotheses can be made. First, one can imagine that the high temperature induces a partial denaturation of primase which alters its ability to bind to the holoenzyme complex, and therefore the primase is not retained on Heparin-Sepharose. Another simple hypothesis is that the association

Figure 1: Upper part: Ammonium sulfate fractionation of primase and DNA polymerase α activities from a crude extract. The fractionation was performed by a salting-out process as previously described by Méchali et al. (12). Primase activity and DNA polymerase α were expressed with respect to the protein concentration.
Lower part: Heparin Sepharose chromatography of the fraction containing the highest level of primase which corresponds to 13% (w/v) $(NH_4)_2SO_4$ in the elution buffer. Chromatography was performed on a Heparin-Sepharose CL-6B (Pharmacia Fine Chemicals) using a 0-0.5 M KCl gradient. Both activities were eluted from the column at 0.22 M KCl.

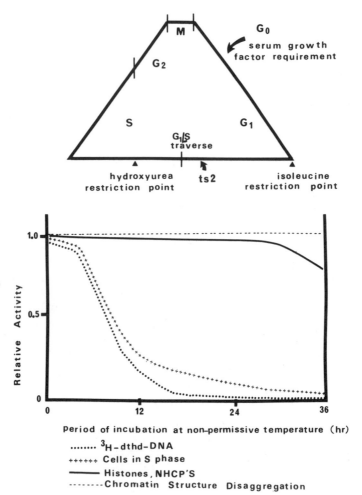

*Figure 2: Principal characteristics of ts2 mutant cells. Top: cell
cycle; bottom: characteristics of ts2 cells. For explanation see text.*

between the primase and polymerase α subunits requires another polypeptide which is temperature sensitive.

        In summary, we conclude that the primase and DNA polymerase α
activities present in mammalian cells that are synthesizing DNA can
not be dissociated by all purification procedures used. Therefore
they could be an integral part of the native DNA polymerase α holoenzyme. The results obtained with Balb/C-3T3 ts2 cells (ts in a
function required for G1/S traverse) demonstrate that the two activities reside on distinct polypeptides. Balb/C-3T3 ts2 cells seem
to be a good system for studying interactions between the primase

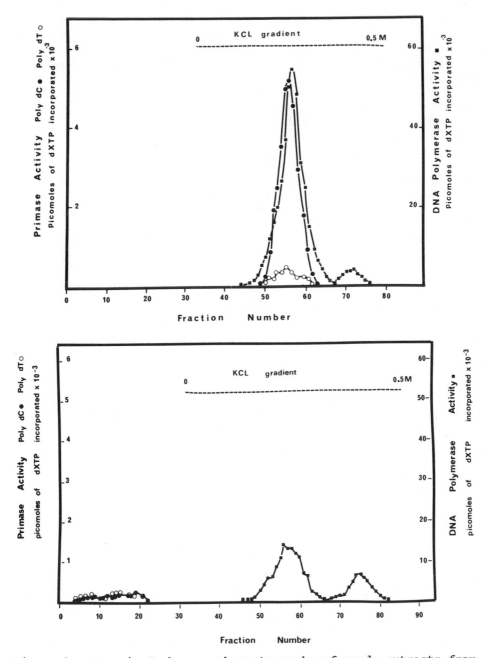

Figure 3: *Heparin-Sepharose chromatography of crude extracts from* *ts2* *mutant cells. Mutant* *ts2* *cells (2x10^8), grown at 34°C or 38.5°C,* *were homogenized; after pelleting the nuclei and the mitochondria;* *a post microsomal fraction was obtained by centrifugation for 3 hours*

at 105,000 g. After extensive dialysis against buffer A (50 mM Tris-HCl pH 7.5, 5 mM 2-mercaptoethanol, 0.1 mM EDTA, 30% glycerol), the sample was loaded on a Heparin-Sepharose CL6B column (Pharmacia Fine Chemicals). After washing with buffer A, the elution was done using a 0-0.5 M KCl gradient. DNA polymerase activity was measured using activated DNA. Primase activity was measured using poly dC or poly dT. Upper part shows the profile of DNA polymerase α and primase activities obtained with ts2 cells grown at 34°C. Lower part shows the profile of the activities obtained with ts2 cells grown at 38.5°C.

and polymerase α in the replication complex. One may also be able with such a system to define precisely the time at which the replisome is assembled during the cell cycle, since, at the non-permissive temperature, the primase activity becomes separable from the polymerase activity.

ACKNOWLEDGEMENTS

We are very grateful to Janine Abadiedebat, Margaret Dubsky, José Sigoiin and Jeanne Tillit for expert technical assistance. We would like to thank I.R. Lehman (Stanford University) for providing manuscripts of his work prior to publication and to James Clark for reading the manuscript.

This work was supported by grants from INSERM (PRC 124022), from the Association pour la Recherche sur le Cancer, from the Medical Research Council of Canada and from the National Cancer Institute of Canada. During her sabbatical year in Villejuif, Rose Sheinin was supported by founds from the Ligue Nationale de Recherche sur le Cancer and from Giofiah Macy Jr. Fundation.

REFERENCES

1.   Kornberg, A. (1982) DNA replication, W.H. Freeman and co., San Francisco.
2.   Yagura,T., Kozu,T. and Seno,T. (1982) J. Biochem. 91, 607.
3.   Conaway,R.C. and Lehman,I.R. (1982) Proc. Natl. Acad. Sci. USA 79, 2523.
4.   Méchali,M. and Harland,R.M. (1982) Cell 30, 93.
5.   Tseng,B.Y. and Ahlem,C.N. (1982) J. Biol. Chem. 257, 7280.
6.   Yagura,T., Kozu,T. and Seno,T. (1982) J. Biol. Chem. 257, 11121.
7.   Riedel,H.D., König,H., Stahl,H. and Knippers,R. (1982) Nucleic Acids Res. 10, 5621.
8.   Shioda,M., Nelson,E.M., Bayne,M.L. and Benbow,R.M. (1982) Proc. Natl. Acad. Sci. USA 79, 7209.

9.   Hübscher,U. (1983) EMBO J. 2, 133.
10. Tseng,B.Y. and Ahlem,C.N. (1983) J. Biol. Chem. 258, 9845.
11. Kaguni,L.S., Rossignol,J.M., Conaway,R.C., Banks,G.R. and
         Lehman,I.R. (1983) J. Biol. Chem. 258, 9037.
12. Méchali,M., Abadiedebat,J. and de Recondo,A-M. (1980)
         J. Biol. Chem. 255, 2114.
13. Philippe,M., Abadiedebat,J., Tillit,J. and de Recondo,A-M.,
         in preparation.
14. Slater,M.L. and Ozer,H.L. (1976) Cell 7, 289.
15. Sheinin,R. and Lewis,P.N. (1980) Somat. Cell Genet. 6, 227.
16. Sheinin,R., Dardick,I. and Doane,F.W., submitted.
17. Brennessel,B.A., Buhrer,D.P. and Gottlieb,A.A. (1978)
         Anal. Biochem. 87, 411.
18. Philippe,M., Sheinin,R. and de Recondo,A-M., in preparation.

PROPERTIES OF THE PRIMASE ACTIVITY OF THE 9 S DNA POLYMERASE α

FROM CALF THYMUS

Gerhard Krauss and Frank Grosse

Abteilung Biophysikalische Chemie, Zentrum Biochemie
Medizinische Hochschule Hannover
Konstanty-Gutschow-Strasse 8
D-3000 Hannover 61 F.R.G.

INTRODUCTION

Enzymes called primases are required for the initiation of DNA synthesis on the lagging strand of the replication fork. Primases are capable of synthesizing RNA primers from ribonucleoside triphosphates with single-stranded DNA as a template. In Escherichia coli, the primase has been identified and characterized (1). It is not part of the DNA polymerase III holoenzyme, the replicative DNA polymerase. In eukaryotes, a primase activity could be demonstrated to be part of a high molecular weight form of DNA polymerase α in several systems (2-7). We have recently described a 9 S DNA polymerase α from calf thymus that carries a catalytically active subunit of 148 kd and further subunits of 59, 55 and 48 kd (8). We now report the properties of a primase activity associated with the 9 S enzyme.

RESULTS

Template utilization

We have investigated different synthetic homopolymers and single-stranded M13mp7 DNA for their ability to serve as template for the primase activity of the 9 S enzyme. In the absence of added primer, a ribonucleotide triphosphate dependent DNA synthesis is observed with poly(dC), poly(dC,T) and single-stranded M13 DNA. Poly(dT) is a poor template. Poly(dA), poly(dA,C) and poly(dI) cannot serve as a substrate for the priming reaction (Table 1).

Table 1.   Template utilization and inhibition studies of the
            primase activity of the 9 S enzyme*

Template	Conditions	dNMP incorp. (pmol)	AMP incorp. (pmol)
M13mp7 (+)-DNA	ATP, UTP, GTP, CTP	40	0.08
	ATP, GTP	38	
	ATP, GTP 100µg/ml α-Amanitin	39	
	ATP, GTP 50µg/ml Actinomycin	8	
	ATP	22	
	GTP	19	
	UTP, CTP	2	
	no NTPs	2	
	100µM (rA)$_6$	80	
	100µM (rA)$_6$ 50µg/ml Actinomycin	75	
poly d(C,T)	ATP, GTP	250	0.4
	ATP, GTP 150 mM KCl	50	0.08
	ATP, GTP 50µM epoxy ATP	–	0.25
poly dC	GTP	250	
poly dT	ATP	10	
poly dA	UTP	1	
poly d(A, C)	UTP, GTP	1	

* The templates were replicated in the presence of 100µM dNTPs and
500µM of the ribonucleotides given. dNMP and AMP incorporation was
determined by measuring the incorporation of (α^{32}P)-dAMP or (α^{32}P)-AMP
into acid insoluble material.

Products of the priming reaction

The products of the priming reaction were analyzed by following
the incorporation of α-^{32}P and γ-^{32}P labeled ATP or GTP into the RNA
primer. Acid precipitation and denaturing electrophoresis of the replication products shows that the γ-phosphate group of ATP or GTP is
incorporated into high molecular weigth products (data not shown).
The length of the RNA primer was determined from assays that contained
poly(dC,T) or M13 DNA as a template, the complementary α-^{32}P labeled
NTPs at 300 µM and the complementary dNTPs at 30 µM. Following incu-

bation at $37^{O}C$, the replication products were digested with DNAse I and were then analyzed on denaturing polyacrylamide gels. Figure 1 shows that the RNA primer synthesized is 8-16 nucleotides long, if dNTPs are also present. If no dNTPs are included in the assay, RNA primers of 8-40 nucleotides length are synthesized.

Inhibition studies

A distinctive feature of DNA polymerase α is the high sensitivity to ionic strength (1). Addition of salt to the primase assay also results in a strong inhibition of the priming reaction (Table 1). 2'-3' Epoxy-ATP has been shown to be a potent inhibitor of DNA polymerase I from <u>Escherichia coli</u> and a much weaker inhibitor of DNA polymerase α and RNA polymerase (9). Our data show that the primase

A   B   C

Figure 1: Length of RNA primers as determined from denaturing polyacrylamide gels. M13mp7 (+) DNA was replicated with the 9 S enzyme in the presence of α-^{32}P-ATP, GTP, CTP, UTP (100 μM each) and in the presence (B) or absence (C) of dATP, dGTP, dCTP, dTTP (100 μM each). Following incubation at $37^{O}C$ for 1h, the DNA polymerase α was inactivated by heating at $65^{O}C$. The replication products were digested by treatment with DNAse I (50 μg/ml) for 1h at $37^{O}C$ and were then electrophoresed in 20% polyacrylamide gels containing 8 M urea. Lane A: $(rA)_{6-28}$ as a standard.

and DNA polymerizing activity of the 9 S enzyme are inhibited to the
same extent by this compound.

The primase activity is not sensitive to α-amanitin, which is
a specific inhibitor of RNA polymerase II. Thus a contamination of
the 9 S enzyme with RNA polymerase II is very unlikely.

We also have investigated the inhibitory effect of actinomycin
D on the primase reaction in an assay that contained M13mp7 (+) DNA
as template. In this assay, actinomycin D inhibits the primase by 80%,
whereas the DNA polymerizing activity is unaffected. Analysis of the
replication products on denaturing agarose gels reveals that the
specific product bands are largely suppressed, whereas the nonspecific
priming is only slightly affected.

Specificity of the priming reaction

An important question to answer refers to the sequence specificity of the priming reaction. We have used single stranded M13mp7
DNA as template and have analyzed the size distribution of the replication products on alkaline agarose gels. Distinct product bands
are detectable on the gels, which indicates that the priming events
occur mostly at specific sites on the M13 genome. A background of
nonspecific priming is however also observable. The product of replication migrates on neutral gels like the double-stranded, nicked form
of M13mp7 (data not shown). From this result it can be estimated that
5-10 priming events occur per M13 genome.

DISCUSSION

The 9 S DNA polymerase α from calf thymus contains a powerful
primase activity, that enables the enzyme to prime DNA synthesis on
single-stranded DNA templates in the presence of ribonucleoside triphosphates. The primase exhibits base selectivity, with ATP and/or
GTP being the first nucleotide to be incorporated. Initiation with
CTP and/or UTP is not possible. This holds both for initiation on
synthetic homopolymers and on natural single-stranded DNA. The primase
activity of the 9 S enzyme is thus very similar to enzymes isolated
from Drosophila and mouse cells (2, 10). These enzymes are also associated with a DNA polymerase α and synthesize primers of comparable
length.

The primase and DNA polymerizing activity enable the 9 S enzyme
to replicate single-stranded M13 DNA nearly to completion. Not more
than 10 priming events occur on this template, which points to a sequence specificity of the primase. From the actinomycin inhibition
and from the requirements for purine ribonucleotides as the first
nucleotide, we conclude that the priming occurs in helical, pyrimidine
rich regions of the M13 genome. Since priming during lagging strand

synthesis in vivo is assumed to be largely independent from sequence and structure of the strand to be copied, it remains to be shown, whether the primase activity actually is involved in the synthesis of the Okazaki fragments. However the detection of primase activity in the 9 S enzyme and the ability of this enzyme to copy long stretches of single-stranded DNA suggests that we are dealing with a real replicative entity.

REFERENCES

1.   Kornberg,A. (1980) in DNA Replication, pp 390 Freeman, San Francisco.
2.   Conoway,R.C. and Lehman,I.R. (1982) Proc. Natl. Acad. Sci. U.S.A. 79, 2523.
3.   Riedel,H.D., König,H., Stahl,H. and Knippers,R. (1982) Nucl. Acid. Res. 10, 5621.
4.   Shioda,M., Nelson,E.M., Bayne,M.L. and Benbow,R.M. (1982) Proc. Natl. Acad. Sci. U.S.A. 79, 7209.
5.   Hübscher,U. (1983) EMBO Journal 2, 133.
6.   Kaguni,L.S., Rossignol,J.M., Conoway,C.R. and Lehman,I.R. (1983) Proc. Natl. Acad. Sci. U.S.A. 80, 2221.
7.   Yagura,T., Tanaka,S., Kozu,Y., Seno,T. and Korn,D. (1983) J. Biol. Chem. 258, 6698.
8.   Grosse,F. and Krauss,G. (1981) Biochemistry 20, 5470.
9.   Abboud,M.M., Sim,J.W., Loeb,L.A. and Mildvan,A.S. (1978) J. Biol. Chem. 253, 3415.
10. Kozu,T., Yagura,T. and Seno,T. (1982) Nature 298, 180.

V
DNA Polymerases and Accessory Proteins

DNA POLYMERASE III HOLOENZYME OF ESCHERICHIA COLI:
AN ASYMMETRIC DIMERIC REPLICATIVE COMPLEX CONTAINING DISTINGUISH-
ABLE LEADING AND LAGGING STRAND POLYMERASES

C.S. McHenry and K.O. Johanson

Department of Biochemistry and Molecular Biology
The University of Texas Medical School
USA-Houston, Texas 77225

INTRODUCTION

The DNA polymerase III holoenzyme is the major replicative enzyme in Escherichia coli (for a review see ref. 1). A final definition of the enzyme's structure has not yet been accomplished, but it is apparent that it contains at least seven different subunits: α, ε, θ, β, δ, γ and τ (2-4). The holoenzyme contains a catalytic core, termed DNA polymerase III, composed of the α, ε and θ subunits. DNA polymerase III can catalyze limited synthesis in short gaps created by nuclease treatment of duplex DNA, but it is inactive in reconstituted natural Escherichia coli replication systems (3, 5).

DNA polymerase III', a complex of polymerase III and the τ subunit of holoenzyme, has been isolated (14). The most striking difference between this form of the enzyme and the core polymerase is its dimeric structure. Our laboratory has proposed that the τ subunit may serve to hold two polymerase molecules together, forming a structure upon which replication proteins could assemble creating a dimeric replicative complex. This provided enzyme structural evidence for the earlier novel proposals that the replication fork may be dimeric, permitting the concurrent coordinated replication of both strands at the replication fork (6, 7).

The β subunit of the DNA polymerase III holoenzyme, itself a dimer, has been purified to homogeneity (18). It requires the addition of DNA polymerase III* (holoenzyme minus β) to reconstitute holoenzyme activity. Holoenzyme (or DNA polymerase III* and β) can form an isolable initiation complex with primed DNA in the presence of either ATP or dATP (9-13). This reaction is blocked by anti-β IgG only if antibody is added prior to initiation complex formation;

once the complex has formed, it is resistant to antibody. This resistance is not due to the absence of β in the initiation complex but due
to its immersion in the complex, sterically preventing contact by antibody (11). Adenosine 5'-O-(3-thiotriphosphate) (ATPγS) can also
support initiation complex formation (14). It, like ATP, is hydrolyzed,
yielding ADP. However, ATPγS differentially affects two populations
of polymerase in holoenzyme preparations. We discuss these results
in the light of the dimeric polymerase hypothesis and extend this
hypothesis to include a dimeric polymerase that is asymmetric, including both a leading and lagging strand polymerase.

RESULTS AND DISCUSSION

The resistance of initiation complexes to the inhibitory action
of anti-β IgG provides a convenient assay to monitor complex formation.
Since anti-β IgG inhibits holoenzyme when it is free in solution (8,
11), any replication observed subsequent to antibody addition must
arise from preformed initiation complex. We have used this method to
determine the relationship between the amount of initiation complex
formed and the concentration of ATP or ATPγS (Table 1). One half
maximal complex formed with either 0.6 μM ATP or 0.3 μM ATPγS; complex
formation appeared maximal at 5 μM ATP and 2 μM ATPγS. In spite of the
greater efficacy of ATPγS in enabling complex formation at low concentrations, the maximum quantity of complex formed was only one half
that attained with ATP (Table 1).

Table 1.   Initiation Complex Formation in the Presence of ATP or
           ATPγS

Nucleotide added.	Concentration (μM)	Initiation complex formed[1] (fmol)
ATP	0.1	2
	0.5	10
	1	16
	2	19
	5	23
ATPγS	0.1	3
	0.5	8
	1	11
	2	12
	5	12

[1] Initiation complexes were formed as described (14) by the addition
of DNA polymerase III holoenzyme (70 units) to primed M13 Gori DNA
(60 fmol DNA circles) and incubated in the presence of the designated
nucleotide. After 5 min, the initiation complex formed was determined
by (^{3}H)dNTP incorporation after the addition of anti-β IgG (25 μg) to
block further initiation complex formation. Incubation for 30 min
instead of 5 min did not lead to increased complex formation.

Since more initiation complex can be formed with ATP than with ATPγS, we examined the effect ATP S has upon initiation complexes formed with ATP. Initiation complexes were formed with ATP, isolated by gel filtration to free them of ATP and unbound protein, and then either ATP or ATPγS was added (Table 2). The isolated initiation complex was fairly stabile in the presence of ATP; approximately 25% dissociated within 30 minutes. However, upon the addition of ATPγS the initiation complex rapidly dissociated until approximately one-half as much complex remained in the presence of ATPγS as ATP (Table 2). This effect was independent of ATPγS concentration from 1 to 100 μM ATP S. After 30 minutes complexes remaining were stabile. This demonstrated that the two-fold difference observed using ATP and ATPγS to support complex formation could also be seen in the reversal of the reaction.

Our observations suggest the existence of two populations of holoenzyme in our reaction: one that can form initiation complexes in the presence of ATPγS and one that can form complexes only with ATP and is dissociated by ATPγS.

We speculate that this functional heterogeneity arises not from two distinct enzymes in solution, but from a difference between two halves of a dimeric DNA polymerase III holoenzyme. Previous suggestions have been made regarding the role of a dimeric polymerase in the replication process (4, 6, 7, 11). We suggest an extension of these proposals to include a dimeric polymerase that is asymmetric (Figure 1).

Table 2.   Initiation Complex Dissociation in the Presence of ATPγS[1]

Time of dissociation	Complex remaining (%)	
	ATP	ATPγS
0	100	100
2	95	60
5	85	45
10	80	40
30	75	35

[1]To initiation complex (0.28 pmol DNA circles) formed in the presence of ATP and isolated by gel filtration as described (14) was added ATP or ATPγS (100 μM final). At the designated time aliquots were transferred to tubes containing anti-β IgG to inhibit free holoenzyme. After incubation (1 min, 0°C) dNTPs were added and elongated of the remaining initiation complex was permitted to proceed to completion.

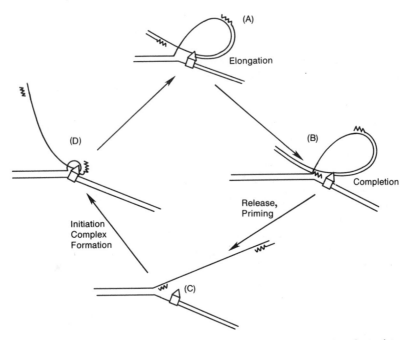

*Figure 1: Cycling of an asymmetric polymerase on the lagging strand
of a replication fork. A, a dimeric DNA polymerase concurrently replicating both strands at a replication fork. B, completion of an
Okazaki fragment on the lagging strand. Following fragment completion,
DNA is released from the lagging strand half of the asymmetric polymerase. The next primer at the fork is then synthesized. (This could
occur concurrently with the preceding steps.) C, asymmetric dimeric
polymerase associated with only the leading strand. Upon binding ATP,
the lagging strand half can form an initiation complex with the newly
synthesized primer to form (D).*

This asymmetry could have a structural basis. One half of the polymerase could contain an extra subunit* or a subunit that contains a
covalent modification. Alternatively, this asymmetry could be induced
by allosteric interactions between the polymerase halves. In any case,
we suggest that this functional asymmetry could be used within the
cell to solve the asymmetric functional problem of polymerases at the
replication fork (Fig. 1). One polymerase, the leading strand polymerase, once associated with the Escherichia coli chromosome need not
dissociate until the chromosome is completely replicated. The lagging

---

* A model for an asymmetric calf α polymerase has concurrently been
  proposed by Hübscher and Ottiger (15).

strand polymerase, however, must dissociate once per 1000 nucleotide Okazaki fragment synthesized (ca. once per second under physiological conditions). It would appear advantageous for a cell to contain a dimeric polymerase that has an exceedingly high affinity for DNA in one-half and a lesser affinity in the other. This could permit the necessary release and recycling each second during Okazaki fragment synthesis, while keeping the polymerase fixed at the replication fork. Our observation, made with an ATP analog, may be unveiling this functional asymmetry. We are optimistic that ATPγS and other analogs will prove to be useful tools in pursuing this problem in future experiments.

REFERENCES

1. McHenry,C.S. and Kornberg,A. (1982) in: The Enzymes, P.Boyer, ed., Academic Press, New York.
2. McHenry,C.S. and Kornberg,A. (1977) J. Biol. Chem. 252, 6478.
3. McHenry,C.S. and Crow,W. (1979) J. Biol. Chem. 254, 1748.
4. McHenry,C.S. (1982) J. Biol. Chem. 257, 2657.
5. Livingston,D.M., Hinkle,D.C. and Richardson,C.C. (1975) J. Biol. Chem. 250, 461.
6. Sinha,N.K., Morris,C.F. and Alberts,B.M. (1980) J. Biol. Chem. 255, 4290.
7. Kornberg,A. (1982) DNA Replication, Supplement, W.H. Freeman and Co., San Francisco.
8. Johanson,K.O. and McHenry,C.S. (1980) J. Biol. Chem. 255, 10984.
9. Wickner,W. and Kornberg,A. (1973) Proc. Natl. Acad. Sci. U.S.A. 70, 3679.
10. Wickner,S. (1976) Proc. Natl. Acad. Sci. U.S.A. 73, 3511.
11. Johanson,K.O. and McHenry,C.S. (1982) J. Biol. Chem. 257, 12310.
12. Burgers,P.M.J. and Kornberg,A. (1982) J. Biol. Chem. 257,11468.
13. Burgers,P.M.J. and Kornberg,A. (1982) J. Biol. Chem. 257, 11474.
14. Johanson,K. and McHenry,C.S. (1984) submitted for publication.
15. Hübscher,U. and Ottiger,H.P. (1984) this volume.

MAMMALIAN DNA POLYMERASE α HOLOENZYME

Ulrich Hübscher and Hans-Peter Ottiger

Department of Pharmacology and Biochemistry
University of Zürich-Irchel
Winterthurerstrasse 190
CH-8057 Zürich, Switzerland

DNA replication as a macromolecular process requires the concerted action of many enzymes and proteins either alone or in a complex (1). These include e.g. DNA polymerases, DNA polymerase accessory proteins, primase, topoisomerases, helicases, DNA-binding proteins, ribonuclease H, DNA ligase and others (1). Using the DNA of small bacteriophages of Escherichia coli and of plasmids containing the oriC as model replicons to understand the host DNA replication events, it was discovered that multienzyme systems are involved (1). The DNA elongation step alone needs DNA polymerase III, the major replicase in bacteria. DNA polymerase III functions in the form of a multipolypeptide complex called DNA polymerase III holoenzyme (2). The term holoenzyme has been introduced in Escherichia coli for a complex of proteins including a core DNA polymerase III and several auxiliary proteins, which are essential for replication of primed natural single stranded genoms (2). Many details of the structure and functions of this multipolypeptide complex have emerged during the last few years (3, 4).

The fact that replicative DNA polymerases have been conserved from bacteria to human (5) prompted us to search for a DNA polymerase α holoenzyme in mammalian tissues. In analogy to the prokaryotic holoenzyme we defined, identified and purified a similar form from freshly harvested calf thymus (6, 7). This eukaryotic holoenzyme was isolated by using in vivo like template probes such as parvoviral DNA and could functionally and physicochemically be distinguished from a corresponding homogeneous DNA polymerase α, termed the core enzyme (6, 8).

Our progress can be summarised as follows: (i) the purified DNA polymerase α holoenzyme has a defined polypeptide pattern distinguishable from the apparently homogeneous DNA polymerase α core enzyme; (ii) the primase activity appears to be part of a high molecular weight DNA polymerase α polypeptide; (iii) a protein factor has been separated from the holoenzyme and this protein can partially restore holoenzyme activity on long single-stranded (ss) templates; (iv) an early purification stage of DNA polymerase α holoenzyme was separated upon chromatography on DEAE-cellulose into four different functional forms. Besides possessing holoenzyme activity on long ss templates the following enzymatic activities copurify to this early stage with certain of these DNA polymerases. These are: DNA primase, double-stranded (ds) DNA dependent ATPase, DNA methylase, DNA topoisomerase, 3'-5' exonuclease and ribonuclease H; (v) the results favour a hypothetical model for a concerted action of leading and lagging strand DNA polymerase α holoenzymes at the replication fork.

Isolation of a DNA polymerase α holoenzyme using natural occurring templates

By using ss parvoviral DNA and singly-primed M13 DNA a complex form of DNA polymerase α from freshly harvested calf thymus can be isolated and this complex enzyme form is called DNA polymerase α holoenzyme (6). The purification procedure originally included ammonium sulfate precipitation followed by an extensive backwashing, phosphocellulose chromatography and gel filtration on Sephacryl S-200 (6). This purification procedure has now been extended to a Bio-Rex 70 and a second phosphocellulose step (7, Hübscher,U., to be published elsewhere). The final DNA polymerase α holoenzyme contains at least eleven polypeptides after electrophoresis in a SDS-polyacrylamide gel (Figure 1). The most purified DNA polymerase III holoenzyme so far achieved from Escherichia coli has ten bands after SDS-gelelectrophoresis (9). A DNA polymerase α fully active on DNase treated DNA, designated as activated DNA, was isolated by a different protocol. It resembles the holoenzyme procedure up to the phosphocellulose step. The three additional steps included DEAE-cellulose, ssDNA-cellulose and chromatofocussing (Figure 1). This DNA polymerase α is completely inactive on templates with long single-strand regions and is therefore designated as the core enzyme. It contains two high molecular weight polypeptides of approximately 125,000 and 150,000 dalton respectively, and three polypeptides in the M_r region of 50,000 to 60,000 (Figure 1). The two high molecular weight bands both contain DNA polymerase activities (10) as measured by the "SDS activity gel method" (11) and the lower M_r bands derive from the higher M_r catalytic polypeptides, most likely by proteolysis (8).

The functional distinction of a mammalian DNA polymerase α holo- and core enzyme on defined long ss DNA templates provides a basis to begin the characterisation, dissection and reconstitution of a functionally operating replicase.

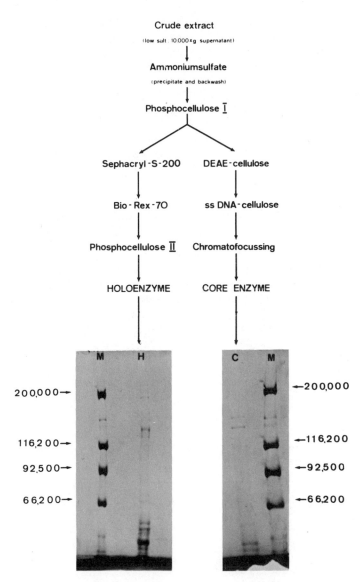

Figure 1: Purification protocol for calf thymus DNA polymerase α holo- and core enzyme. Both enzymes were electrophoresed in a 8% polyacrylamide gel. Markers were : Myosin (M_r 200,000), β-galactosidase (M_r 116,200), phosphorylase B (M_r 92,500) and BSA (M_r 66,200), H: Holoenzyme; C: Core enzyme.

The mammalian primase appears to be part of the high molecular
weight DNA polymerase α polypeptide

DNA primase was identified in purified DNA polymerase α holoenzyme and this activity could never be separated from the polymerase
by the usual chromatographic and separation techniques. Even our
apparent homogeneous DNA polymerase α core enzyme always contained
the primase (10). This raised the possibility that the primase and
the polymerase activities might reside on the same polypeptide. To
test this a homogeneous DNA polymerase α core enzyme was electrophoresed on a "SDS-activity gel" (11). DNA polymerase (tested by
including gapped DNA in the gel) and primase (tested by including
ssM13 DNA in the gel) had identical mobilities and coincided with
the high M_r catalytic subunit (125,000) of DNA polymerase α (ref.
10 and Figure 1).

Separation of a DNA polymerase α accessory protein from the
holoenzyme

A partially purified DNA polymerase α holoenzyme (6) was bound
to a hydroxylapatite column containing 50 mM Tris-HCl (pH 7.5),
5 mM potassium phosphate (pH 6.8), 5 mM DTT, 25% glycerol, 10 mM
sodium-bisulfite, 1 µM pepstatin and 1 M KCl. The bound DNA polymerase was eluted by raising the potassium phosphate concentration
in a gradient to 100 mM. The eluted DNA polymerase was unable to
replicate ss parvoviral DNA. If, however, dialysed flow-through fraction of the same column was added to the DNA polymerase a five to
ten fold stimulation was evident on long ss DNA (Table 1). This assay

Table 1: Preliminary characterisation of the DNA polymerase α
          accessory protein

Separation from holoenzyme on hydroxylapatite

Heat-labile

Freeze-thaw sensitive

Acid isoelectric point

Sedimentation coefficient of 2-3 S

Stimulation of DNA polymerase α:

activated DNA	none
primed M13 DNA	10 times
parvoviral DNA	5 times

---

was used to isolate the accessory protein. So far complete isolation has been hampered because this protein is extremely labile as well as sensitive to freezing and thawing. It appears to have an acidic isoelectric point, a sedimentation coefficient of 2-3 and a tendency to aggregate under low salt conditions.

An early purification stage of DNA polymerase α holoenzyme can be separated into four different DNA polymerase α forms upon chromatography on DEAE-cellulose

During the studies with DNA polymerase α we realised that a partially purified DNA polymerase α holoenzyme can be separated into four different forms by chromatography on DEAE-cellulose (Figure 2). This elution pattern is reproducible if, (i) extremely fresh calf thymus is used, (ii) the protease inhibitors sodium-bisulfite and pepstatin are included and (iii) potassium phosphate elution with a shallow gradient is carried out. The four peaks are designated A, B, C and D respectively, indicating the order of elution from the column.

All of these forms bear characteristics of DNA polymerase α. They all are completely inhibited by aphidicolin and N-ethylmaleimide and are insensitve to dideoxythymidinetriphosphate. No DNA polymer-

Table 2: Template specificity of the four DNA polymerase α forms and purified DNA polymerase α holoenzyme

| Template | DNA polymerase α peak | | | | Purified holoenzyme |
	A	B	C	D	
activated DNA	+	+	+	+	+
heat-denatured DNA	+	(+)	(+)	+	+
poly(dA)·(dT)$_{12-18}$ (base ratio 1:1)	(+)	+	+	+	(+)
poly(dA)·(dT)$_{12-18}$ (base ratio 3:1)	(+)	+	+	+	+
poly(dA)·(dT)$_{12-18}$ (base ratio 10:1)	(+)	(+)	+	+	+
(dA)·$_{100}$-(dT)$_{25}$ (hook)	(+)	+	+	+	+
primed M13 DNA	+	+	+	+	+
ss parvoviral DNA	+	+	+	+	+
pβG form I DNA	−	−	−	−	−
pβG form II DNA	−	−	−	−	−
pβG form V DNA	−	−	−	−	−

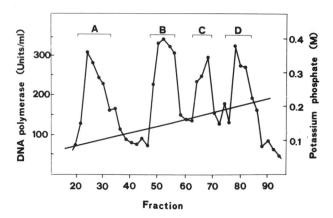

*Figure 2:   Separation of four different DNA polymerase α forms by
DEAE-cellulose chromatography. DNA polymerase α holoenzyme, partially purified by ammonium sulfate precipitation, backwash and by
chromatography on phosphocellulose (6), was dialysed, applied to a
DEAE-cellulose column, the column extensively washed and the enzymes
finally eluted with a potassium phosphate gradient. Details of this
separation step will be published elsewhere (12).*

ase β and γ or terminal deoxynucleotidyltransferase were found in
the four peaks. Every peak is able to replicate primed long ss DNA
(Table 2), as is the purified holoenzyme that has been isolated by
this criterion (6, 7). No replication is evident when ds circular
DNA's are tested as templates.

Copurification of replication enzymes with the different DNA polymerase forms

Eight additional enzymatic activities were measured in the
peaks and also in the purified DNA polymerase α holoenzyme (12). At
least six additional enzymatic activities were identified in all of
these forms (Table 3). A DNA dependent ATPase with ds supercoiled
DNA as the best effector was preferentially found in peak A. On the
other hand, primase was not identified in this peak but could be
identified in peaks B, C and D. Furthermore, the A form included DNA
topoisomerase type II, 3'-5' exonuclease and to a lesser degree
RNase H. A DNA topoisomerase type II is also measured in the three
peaks containing the primase. DNA methylase and 3'-5' exonuclease
copurified with form C and D, the latter as mentioned also with form
A. A putative primer removal enzyme, RNase H, was detected in peak
B and also to a lesser extent in peak A. Finally no DNA ligase and
poly(ADP-ribose) polymerase was found in any of these four peaks. The
purified holoenzyme, on the other hand, possesses the primase as the
only additional activity. Peaks C and D are indistinguishable by
enzymatic criteria. However, peak C is more resistant to inhibition
by salt and spermidine or to inactivation by heat (12).

Table 3: Enzymatic activities copurifying with the different forms
of DNA polymerase α and the purified holoenzyme

Enzymatic activity	| DNA polymerase α peak				holoenzyme
	A	B	C	D	
DNA polymerase α	+	+	+	+	+
DNA polymerase α holoenzyme	+	+	+	+	+
Primase					
poly(dT)-assay	−	+	+	+	+
M13 DNA-assay	−	+	+	+	+
DNA dependent ATPase					
no DNA	−	−	−	−	−
denatured DNA	(+)	(+)	−	−	−
ss φX174 DNA	(+)	−	−	−	−
ds supercoiled pAT153 DNA	+	(+)	−	−	−
DNA topoisomerase type II	+	+	+	+	−
3'-5' exonuclease	+	−	+	(+)	−
RNase H	(+)	+	−	−	−
DNA methylase	−	−	+	+	−
DNA ligase	−	−	−	−	−
poly(ADP-ribose)polymerase	−	−	−	−	−

Based on these results we propose (Figure 3) that form A is
the candidate for the leading strand replicase because it possesses
in addition to the DNA polymerase, a ds DNA dependent ATPase (possibly
a helicase), a 3'-5' exonuclease and a DNA topoisomerase type II.
Peaks B, C and D represent different DNA polymerase forms from the
lagging strand of the replication fork. The RNase H in peak B is
suggestive of a primer removal function of this DNA polymerase α
complex in analogy to Escherichia coli DNA polymerase I (1). Peaks
C and D might perform the bulk replication as DNA polymerase α holoenzymes on the lagging strand. The salt resistant C form eventually
represents a more in vivo-like complex. It has been demonstrated in
Escherichia coli that DNA polymerase III becomes resistant to physiological salt concentration assuming it is intact and activated by
ATP (13).

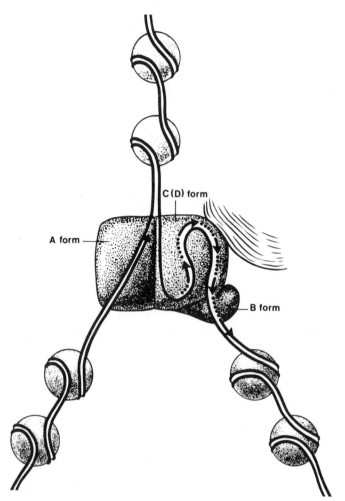

Figure 3: Hypothetical model for a coordinated function of leading and lagging strand DNA polymerase α holoenzyme at the replication fork. The stippled big complex symbolizes the entire DNA polymerase α holoenzyme and this complex might be connected to a higher order structure such as a nuclear matrix (hatched area). For clarity the chromatin is only documented by the core histones (stippled balls).

      Defined separation of DNA polymerase α forms exclusively by chromatography on DEAE-cellulose might favour the idea that these DNA polymerase α forms might function in vivo in a coordinated

fashion on the leading and the lagging strands of the replication
fork.[*]

CONCLUSIONS

The current status of calf thymus DNA polymerase α holoenzyme
is: (i) a DNA polymerase α holoenzyme can be isolated using naturally
occurring DNA templates and this holoenzyme has a distinguishable
polypeptide pattern, as well as different functional properties, from
a homogeneous DNA polymerase α core enzyme; (ii) the primase activity
appears to be part of the high molecular weight DNA polymerase α polypeptide; (iii) a DNA polymerase α accessory protein can be separated
from the holoenzyme and (iv) chromatography of a holoenzyme from an
early purification step on DEAE-cellulose separates four different
DNA polymerase α forms and they contain in a defined pattern at least
six additional DNA replication enzymes. The results favour a hypothetical model for a coordinated action of leading and lagging strand
DNA polymerase α holoenzymes at the replication fork.

ACKNOWLEDGEMENTS

This work was supported by the Swiss National Science Foundation, Grant 3.006-0.81 and by the Sandoz Stiftung zur Förderung der
medizinisch-biologischen Wissenschaften.

REFERENCES

1.   Kornberg,A. (1980, 1982 supplement) DNA replication, Freeman
        and Co. San Francisco.
2.   McHenry,C.S. and Kornberg,A. (1977) J. Biol. Chem. 252, 6478.
3.   Kornberg,A. (1984) this volume.
4.   McHenry,C.S. and Johanson,K.O. (1984) this volume.
5.   Hübscher,U., Spanos,A., Albert,W., Grummt,F. and Banks,G.R.
        (1981) Proc. Natl. Acad. Sci. USA 78, 6771.
6.   Hübscher,U., Gerschwiler,P. and McMaster,G.K. (1982) EMBO J.
        1, 1513.
7.   Hübscher,U. and Ottiger,H.-P. (1983) in: Mechanisms of DNA
        replication and recombination, Cozzarelli,N.R. ed., UCLA
        Symposia on Molecular and Cellular Biology, Alan R. Liss,
        Inc., New York Series, vol. X, 517.
8.   Albert,W., Grummt,F., Hübscher,U. and Wilson,S.H. (1982) Nucl.
        Acids Res. 10, 935.
9.   McHenry,C.S. (1982) J. Biol. Chem. 257, 2657.
10.   Hübscher,U. (1983) EMBO J. 2, 133.

---

[*] An asymmetric dimeric replicative DNA polymerase III holoenzyme
complex from <u>Escherichia coli</u> has concurrently been proposed by
McHenry and Johanson (4).

11.   Spanos,A. and Hübscher,U. (1983) in: Methods in Enzymology,
        Hirs,C.H.W. and Timasheff,S.N. eds., Academic Press,
        New York, 91, 263.
12.   Ottiger,H.-P. and Hübscher,U. (1984) submitted for publication.
13.   Burgers,P.M.J. and Kornberg,A. (1982) J. Biol. Chem. 257, 11474.

DNA POLYMERASE ACTIVITIES OF MAMMALIAN CELL-CYCLE MUTANTS AND
"WILD-TYPE" CELLS IN DIFFERENT STATES OF PROLIFERATIVE
ACTIVITY AND QUIESCENCE

E. Schneider, B. Müller and R. Schindler

Department of Pathology, University of Bern
Freiburgstrasse 30
CH-3010 Bern, Switzerland

INTRODUCTION

Of the three major DNA-dependent DNA polymerases of animal
cells, polymerase α usually exhibits high activities in proliferating cells, while activities are low during proliferative quiescence
(1, 2). On the other hand, polymerases β and γ do not seem to be
correlated with cell proliferation. The present communication is
concerned with control of DNA polymerase activities in heat- and
cold-sensitive cell-cycle mutants of a murine mastocytoma which enter
qualitatively different states of reversible proliferative quiescence
when incubated at the respective nonpermissive temperature (3, 4).
In addition, DNA polymerase activities were studied in "wild-type"
cells that are reversibly arrested in cell-cycle progression by
butyrate or dimenthyl sulfoxide (DMSO), and in variants with relative
resistance to these agents, thus permitting a comparison of DNA
polymerase activities of homogeneous cell populations in exponential
cell proliferation and in different states of proliferative quiescence.

MATERIALS AND METHODS

The clonal subline (K21) of the P-815 murine mastocytoma, the
cold-sensitive (cs, multiplying at 39.5°C, arrested at 33°C, termed
21-Fc) and heat-sensitive (hs, multiplying at 33°C, arrested at
39.5°C, termed 21-Tb) cell-cycle mutants, and the culture conditions
used have been previously described (3, 5). In addition, a butyrate-
resistant (21-B) and a dimethyl sulfoxide-resistant (21-DD) subline
were selected from K21 cells. Cell multiplication was determined by
measuring cell numbers per ml with a Coulter Counter. Relative

numbers of DNA-synthesizing cells were determined by evaluation of
autoradiographic preparations after incubation of cells for 30 min
with (^{3}H)-thymidine (0.5 µC/ml, 24 Ci/mmol).

For determination of DNA polymerase activities, approx. 10×10^6
cells were collected by centrifugation, resuspended in 0.35 ml of
low-salt buffer (10 mM Tris-HCl, pH 7.5/10 mM KCl/1.5 mM $MgCl_2$/
5 mM 2-mercaptoethanol) and frozen at -70°C. Within less than seven
days, they were thawed and centrifuged for 15 min at 30,000 x g. The
resulting supernatant (designated as low-salt extract) was preserved,
and the sediment was resuspended in 0.3 ml of high-salt buffer (20 mM
Tris-HCl, pH 7.5/200 mM KCl/1.5 mM $MgCl_2$/0.1 % triton X-100) and
homogenized by sonication. The homogenate thus obtained was designated
as high-salt extract. For the assay of polymerase α and β, 125 µl of
each of the extracts were mixed, while the polymerase γ assay was
carried out separately with both extracts. In the α and β assay, the
incorporation of (^{3}H)dTTP into acid-insoluble material in the presence
of activated calf thymus DNA as template-primer and of all four
deoxyribonucleotides was measured. In the γ assay, poly(rA)(dT)$_{12-18}$
was used as template-primer. The reaction mixture for the α assay
contained 0.5 mM 2',3'-dideoxythymidine 5'-triphosphate (ddTTP).
With purified enzymes, this resulted in more than 95% inhibition of
the polymerases β and γ, while polymerase α was only slightly inhibited. Similar results were obtained in the β assay with 20 mM NEM
and in the γ assay with 25 mM phosphate. In all three assays, a linear
relationship between enzyme concentration and incorporation of
(^{3}H)dTTP was observed. Enzyme activities are expressed in units per
10^6 cells. One unit is defined as 1 nmol of total deoxyribonucleotides
converted into acid-insoluble form during 10 min at 30°C. For use in
the α or β polymerase assay, calf thymus DNA was activated by DNase I
treatment for 15 or 9 min, respectively (6, 7).

RESULTS AND DISCUSSION

The effects of various culture conditions on cell multiplication
are illustrated in Figure 1. As previously described (Schneider et al.,
1983), K21 cells multiplied rapidly both at 33°C and 39.5°C. On the
other hand, incubation of 21-Tb cells at 39.5°C and of 21-Fc cells at
33°C resulted in near complete arrest of cell proliferation, while the
rate of multiplication of 21-Tb cells at 33°C and of 21-Fc cells at
39.5°C was similar to that of K21 cells (data not shown). In the presence of 2 mM Na butyrate, a slight decrease in the number of K21
cells was observed, while 21-B cells multiplied, though at a relatively slow rate. The addition of 2.5% DMSO to the medium caused a
marked, but not complete inhibition of K21 cell proliferation, while
21-DD cells multiplied relatively rapidly.

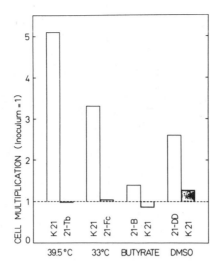

*Figure 1:   Cell multiplication of K21 and four variant cell lines
under different culture conditions. The cells were incubated during
4 days at 39.5°C, 33°C, or with 2 mM Na butyrate as indicated in the
figure. 21-DD and K21 cells were incubated for 1 day with 1% DMSO,
for 1 day with 2% DMSO and for 5 days with 2.5% DMSO.
The values for cell multiplication presented in the figure were
obtained during the last 24 hours of the 4-day or 7-day incubation
period and represent means of values from 3 parallel cultures.*

        Qualitatively similar results were obtained by autoradiographic
determination of relative numbers of DNA-synthesizing cells (labeling
indices in Table 1). At the respective permissive temperature, and
in the absence of butyrate and DMSO, between 40% and 50% of K21 and
variant cells were in S phase. At the nonpermissive temperature,
however, very few cells of the cell-cycle mutants 21-Tb and 21-Fc
were in S phase, while labeling indices of K21 cells remained essentially unchanged. Butyrate and DMSO were somewhat less effective in
decreasing relative numbers of K21 cells in S phase than exposure of
21-Tb and 21-Fc cells to the respective nonpermissive temperature,
and butyrate also caused a moderate decrease of the labeling index
of 21-B cells.

        Activities of DNA polymerases α, β and γ of cells under diffe-
rent culture conditions are summarized in Table 1. Under control conditions permitting exponential cell multiplication (left side of table), similar results were obtained for the K21 and the variant lines
with respect to the three polymerase activities, except for minor
effects of the incubation temperature (33°C vs. 39.5°C) on polymerases
β and γ. When 21-Tb and 21-Fc cells exposed to the nonpermissive temperature were compared with K21 cells (right side of table), only

Table 1: DNA polymerase α, β and γ activities and relative numbers of cells in S phase (labeling index, L.I.) of K21 and four variant cell lines under different culture conditions.

Cell line	Culture condition	L.I. %	DNA polymerase activities (units/10⁶ cells)			Culture conditions	L.I. %	DNA polymerase activities (units/10⁶ cells)		
			α	β	γ			α	β	γ
K21	33°C	44	4.14	0.32	0.77	39.5°C for 4 days	44	2.32	0.24	0.37
21-Tb		53	3.65	0.30	0.87		0.5	0.36	0.14	0.49
K21	39.5°C	48	3.58	0.51	0.45	33°C for 4 days	50	3.87	0.32	0.73
21-Fc		41	3.08	0.47	–		3	0.70	0.22	–
21-B	no buty-rate (37°C)	50	3.55	0.45	0.78	2 mM butyrate for 4 d.	25	2.33	0.37	0.51
K21		50	3.63	0.34	0.76		9	2.01	0.47	0.52
21-DD	no DMSO (37°C)	48	3.53	0.51	0.63	2.5 % DMSO for 7 days *	41	2.90	0.41	0.59
K21		50	3.43	0.33	0.78		9	1.08	0.35	0.60

All values given are means of three independent experiments.

* Cells were incubated for 1 day with 1% DMSO, for 1 day with 2% DMSO, and for 5 days with 2.5% DMSO.

small differences were detected with respect to polymerases β and γ. These two enzyme activities were also similar in K21 cells arrested by butyrate or DMSO, as compared with 21-B and 21-DD cells.

On the other hand, large differences of DNA polymerase α activities were observed: these were considerably lower in hs and cs cell-cycle mutants exposed to the respective nonpermissive temperature than in K21 cells that were multiplying under these conditions. A less pronounced decrease of polymerase α activity occurred in K21 cells arrested by DMSO, and in K21 cells arrested by butyrate, this enzyme activity was nearly the same as in 21-B cells that were proliferating slowly under this condition.

In conclusion, the results obtained confirm the notion that DNA polymerase α, but not β and γ is correlated with proliferative activity of cells. On the other hand, the changes in polymerase α activity observed in cells that had been reversibly arrested in their cell-cycle progression by four different culture conditions were smaller than corresponding changes in relative numbers of DNA-synthesizing cells. It thus appears that in these model systems, DNA polymerase α is under less strict control than the process of DNA replication.

ACKNOWLEDGEMENTS

This work was supported by the Swiss National Science Foundation.

REFERENCES

1.   Bollum,F.J. (1975) Progr.Nucl.Acid Res.Mol.Biol. 15, 109.
2.   Hübscher,U., Kuenzle,C.C. and Spadari,S. (1977) Nucl.Acid
         Res. 4, 2917.
3.   Zimmermann,A., Schaer,J.C., Schneider,J., Molo,P. and
         Schindler,R. (1981) Somat.Cell Genet. 7, 591.
4.   Zimmermann,A., Schaer,J.C., Muller,D.E., Schneider,J.,
         Miodonski-Maculewicz,N.M. and Schindler,R. (1983)
         J.Cell Biol. 96, 1756.
5.   Schneider,E., Müller,B. and Schindler,R. (1983) Biochim.
         Biophys. Acta, in press.
6.   Korn,D., Fisher,P.A. and Wang,T.S.F. (1981) Progr. Nucl.
         Acid Res. Mol. Biol. 26, 63.
7.   Spanos,A., Sedgwick,S.G., Yarranton,G.T., Hübscher,U. and
         Banks,G.R. (1981) Nucl. Acid Res. 9, 1825.

CORRELATION OF DNA POLYMERASE ACTIVITIES WITH THE INITIATION OF DNA SYNTHESIS

Angela M. Otto

Friedrich Miescher-Institut
CH-4002 Basel, Switzerland

INTRODUCTION

The regulation of mammalian cell proliferation is composed basically of two types of control: one is exerted by the interaction of the cell with its extracellular environment; the other is exerted by the genetic program expressed in the cell.

Among the extracellular components are specific growth factors as well as hormones or other non-mitogenic compounds. Quiescent Swiss mouse 3T3 cells have been used to study how growth factors, e.g. prostaglandin $F_{2\alpha}$ (PGF$_{2\alpha}$), and non-mitogenic hormones, e.g. insulin and hydrocortisone, interact with these cells to stimulate the initiation of DNA replication (for review see ref. 1). Each growth factor induces a sequence of events comprising a constant prereplicative period of about 15 hours before the onset of DNA synthesis. This period can be dissected into temporal sectors during which non-mitogenic compounds can interact with growth factor-stimulated cells to alter the rate of initiation of DNA synthesis. For example, insulin has a synergistic effect when added at any time after the initial addition of PGF$_{2\alpha}$, while hydrocortisone inhibits the initiation of DNA replication, but only when added within five hours after PGF$_{2\alpha}$.

Little is known about the intracellular program that is turned on by a specific growth factor and that regulates and coordinates the enzymatic activities involved in DNA replication. One approach to unravel biochemical events involved in executing the program is to study the activity of enzymes known to be required for DNA replication. Numerous studies have shown that the activity of DNA polymerase α changes during the cell cycle, the activity being maximal during S phase; this suggested the involvement of the α-polymerase in replicative DNA synthesis (2-5). Furthermore, it now appears that a primase

337

activity is tightly associated to the DNA polymerase α enzyme (6-8).

The questions asked in this study are: 1) Is the activity of
DNA polymerase α in Swiss 3T3 cells stimulated by PGF2α alone or together with insulin related to the rate of initiation of DNA replication? 2) Is priming activity associated to DNA polymerase α in this
experimental system? Cultures of confluent quiescent Swiss 3T3 cells
offer the advantage that they do not require adding metabolic or
mitotic inhibitors to obtain an initially synchronous population of
cells, and the rate of initiation of DNA replication can be manipula-
ted by the addition of various combinations of growth factors and
hormones. The results presented here show that the activity of DNA
polymerase α, using different DNA templates, is enhanced with the
fraction of cells in S phase. Insulin, which together with PGF2α has
a synergistic effect on the rate of initiation of DNA replication
does not enhance the specific activity of DNA polymerase α in S phase
cells above the level observed in cells stimulated by PGF$_{2\alpha}$ alone.

MATERIALS AND METHODS

Cell culture

Swiss mouse 3T3 cells (9) were maintained as previously de-
scribed (1). For experiments, 1.2×10^5 cells and 1×10^6 cells were
plated in 35 mm and 15 cm dishes, respectively, in Dulbecco-Vogt's
medium supplemented with 6% Fetal calf serum and low molecular nutri-
ents (10). Cultures were confluent and without mitotic figures 4 days
after an intermittant medium change. Cells were then stimulated by
making additions directly to the conditioned medium.

Assay for initiation of DNA synthesis in cell culture

This was carried out as previously described (1).

Assay for DNA synthesis in cell lysates

Lysates were prepared by overlaying the monolayer with a hypotonic buffer as described by Hunter and Francke (11). They were then
made isotonic with sucrose and incubated with an incorporation solution containing 2.5 μCi (^{3}H)-dTTP for 60 min at 37°C. TCA-precipitable
material was trapped on GF/C filters and counted.

Assay for priming and DNA polymerase α activities

Cells were lysed with a hypotonic buffer essentially as described (12). Aliquots of the supernatant were incubated with single
stranded M13 DNA (20 μg/assay) or activated calf thymus DNA (10 μg/
assay) in an incorporation solution containing 2 μCi (^{3}H)-dTTP for 30
min at 37°C. TCA-precipitable material was trapped on GF/C filters and
counted.

Materials

PGF$_{2\alpha}$ was a generous gift from John Pike, Upjohn Company, Kalamazoo, MI. Insulin and deoxynucleotides were from Sigma. (^{3}H)-dTTP was from New England Nuclear, Boston, MA.

RESULTS AND DISCUSSION

To be able to compare the activity of enzymes involved in DNA replication with the rate of initiation of DNA replication in a cell culture, confluent quiescent Swiss 3T3 cells were stimulated with a specific growth factor, PGF$_{2\alpha}$, and a hormone, insulin. Insulin alone under these experimental conditions does not stimulate DNA synthesis. As previously established (see Introduction) and illustrated in Fig. 1, after the addition of PGF$_{2\alpha}$ there is a prereplicative phase of about 15 hours before the cells initiate DNA synthesis at a rate depending on the stimulating conditions. Insulin has a synergistic effect with PGF$_{2\alpha}$ without affecting the length of the prereplicative phase (Figure 1).

When whole cell lysates were prepared from similar cultures stimulated by PGF$_{2\alpha}$ alone or with insulin for 23 hours, these lysates

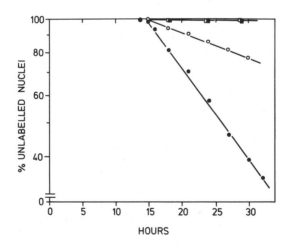

Figure 1:   Rate of initiation of DNA synthesis in Swiss 3T3 cells stimulated by PGF$_{2\alpha}$ alone or with insulin. Cells were plated in 35 mm dishes as described in Materials and Methods. (^{3}H)-Thymidine (3 µCi/ml, 1 µM) was added at the beginning until the times indicated when cells were fixed for autoradiography.
Additions: (□) none; (■) insulin (100 ng/ml); (O) PGF$_{2\alpha}$ (400 ng/ml); and (●) PGF$_{2\alpha}$ + insulin.

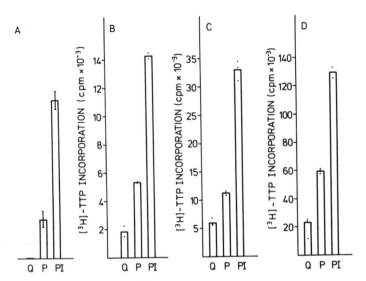

Figure 2. Comparison of priming and DNA polymerase activities with DNA synthesis in lysates. Cells plated in 15 cm dishes were stimulated with PGF$_{2\alpha}$ (500 ng/ml) alone or plus insulin (100 ng/ml) for 23 hr before preparation of lysates for DNA synthesis in whole cell lysates (A, B) and for assay of enzyme activities in supernatants (C, D) as indicated in Materials and Methods. A. Percentage of labeled nuclei, and B. Total DNA synthesis as quantified by incorporation of (^{3}H)-dTTP into acid precipitable material; C. Primase and DNA polymerase activity assayed with the phage DNA M13 single stranded template; D. DNA polymerase activity assayed with activated calf thymus DNA. Enzyme activity was quantified by incorporation of (^{3}H)-dTTP into acid precipitable material. Q: quiescent culture (no addition); P: PGF$_{2\alpha}$; PI: PGF$_{2\alpha}$ + insulin.

incorporated (^{3}H)-dTTP according to the stimulation (Fig. 2A, B). The percentage of nuclei in the lysate labeled during the 1 hour incubation was similar to the percentage of nuclei labeled in culture (compare Fig. 1 and Fig. 2A). When scintillation counting was used to quantify the incorporation of (^{3}H)-dTTP, lysates from quiescent cells had a constant background (Fig. 2B), which is not significantly increased in lysates from cells treated with insulin alone (not shown). Aphidicolin, a specific inhibitor of DNA polymerase α activity (13), when added to the assay reduced the incorporation of (^{3}H)-dTTP by stimulated lysates to background values, but it had little effect on the incorporation in quiescent lysates (not shown). These results indicate that the differences in the incorporation of (^{3}H)-dTTP into whole cell lysates reflects the stimulation of DNA replication by PGF$_{2\alpha}$ alone or with insulin.

The incorporation of (^{3}H)-dTTP into nuclei of whole cell lysates may reflect, among other things, the number of replicating sites already initiated as well as the activity of DNA polymerases. To assay for priming and DNA polymerase α activity, low salt extracts from whole cells were assayed with exogenous templates. When using circular single stranded M13 DNA, incorporation of deoxynucleotides by DNA polymerase α requires the activity of a primase to synthesize an RNA primer (6). As shown in Fig. 2C, the incorporation of (^{3}H)-dTTP into acid-precipitable material indicates that priming occurred since DNA polymerase α could utilize the resulting structure to incorporate deoxynucleotides. The differences in incorporation reflected the different fractions of cells in S phase depending on the stimulation. Likewise, when "activated" calf thymus DNA was used as a template, the incorporation of (^{3}H)-dTTP reflected the fraction of cells in S phase. The incorporation obtained with extracts from stimulated as well as quiescent cells was sensitive to aphidicolin (not shown), indicating that incorporation reflected mainly the activity of DNA polymerase α.

As these experiments show, there appears to be good correlation between the initiation of DNA replication as quantified by the per-centage of cells having entered S phase (Fig. 1) and DNA polymerase α activity (Fig. 2B,C) when comparing Swiss 3T3 cells stimulated by PGF$_{2\alpha}$ and by PGF$_{2\alpha}$ with insulin. Such a correlation was also apparent when the rate of initiation of DNA replication stimulated by PGF$_{2\alpha}$ was inhibited by the addition of hydrocortisone; the activity of DNA polymerase α then also remained at a basal level (14). These obser-vations are in agreement with other studies, using other cell types and various synchronization methods, which show that the activity of DNA polymerase α is increased in S phase cells (2-5). It can be proposed that the enhancement of DNA polymerase α activity requires some process(es) involved in the initiation of DNA replication. Whether the activity of the primase itself is also enhanced as a result of the initiation process or whether its activity is regulated by other events before initiation remains to be investigated. The reports that the primase and DNA polymerase α activities appear to be structurally associated (6-8) would favor the first possibility.

ACKNOWLEDGEMENTS

I thank Dr. Christina Smith for providing me with activated calf thymus DNA and Hans Rudolf for M13 DNA. Furthermore, I thank Drs. Luis Jimenez de Asua and Christina Smith for stimulating discus-sions. A.M.O. is a Special Fellow ot the Leukemia Society of America.

REFERENCES

1. Otto,A.M. and Jimenez de Asua,L. (1982) in: Genetic Expression in the Cell Cycle, Padilla, G.M. and Mc Carty,K.S., Sr., eds., Academic Press, New York, pp. 318-334.

2.   Chang,L.M.S., Brown,M. and Bollum,F.J. (1973) J. Mol.Biol.74,1.
3.   Chiu,R.W. and Baril,E.F. (1975) J. Biol. Chem. 250, 7951.
4.   Spadari,S. and Weissbach,A. (1974) J. Mol. Biol. 86, 11.
5.   Bertazzoni,U., Stefanini,M., Pedrali Noy,G., Giulotto,E.,
         Nuzzo,F., Falaschi,A. and Spadari,S. (1976) Proc. Natl.
         Acad. Sci. USA 73, 785.
6.   Conaway,R.C. and Lehman,J.R. (1982) Proc. Natl. Acad. Sci. USA
         79, 2523.
7.   Hübscher,U. (1983) EMBO Journal 2, 133.
8.   Kozu,T., Yagura,T. and Seno,T. (1982) Nature (London) 298, 180
9.   Todaro,G.J. and Green,H. (1963) J. Cell Biol. 17, 299.
10. O'Farrell,M.K., Clingan,D., Rudland,P. and Jimenez de Asua,L.
         (1979) Exp. Cell Res. 118,311.
11. Hunter,T. and Francke,B. (1974) J. Virol. 13, 125.
12. Jazwinski,S.M., Wang,J.L. and Edelman,G.M. (1976) Proc. Natl.
         Acad. Sci. USA 73, 2231.
13. Ikegami,S., Taguchi,T., Ohashi,M., Oguro,M., Nagano,H. and
         Mano,Y. (1978) Nature 275, 458.
14. Otto,A.M., Shmith,C., Foecking,M.K. and Jimenez de Asua,L.
         (1983) Cell Biol. Int. Rep. 7, 551.

SYNTHESIS OF CATALYTICALLY ACTIVE POLYMERASE α BY IN VITRO

TRANSLATION OF CALF RNA

Sevilla Detera-Wadleigh, Essam Karawya
and Samuel H. Wilson

Laboratory of Biochemistry
National Institutes of Health
Bethesda, MD 20205   USA

SUMMARY

The cell-free synthesis of DNA polymerase in translation mixtures containing calf thymus total and poly (A$^+$) RNA was examined
using activity gel analysis and immunobinding with a monoclonal antibody to calf thymus α-polymerase. Activity gel analysis indicated
that functional DNA polymerase catalytic polypeptides of M_r = 110,000
to 120,000 and $M_r \sim$ 68,000 were synthesized. Sucrose density gradient
centrifugation of total RNA resulted in resolution and partial purification of the mRNAs encoding these two DNA polymerase polypeptides.
Immunobinding experiments with the monoclonal antibody to calf α-polymerase confirmed that an immunoreactive polypeptide of 110 to 120
kilo-daltons had been formed in vitro. This polypeptide and the 68,000-
M_r polypeptide correspond in size to α-polymerase catalytic polypeptides observed in crude extracts of calf cells and in purified calf
α-polymerase.

INTRODUCTION

A variety of important questions can be approached through use
of cDNA probes for α-polymerase genes. Among these are questions of
the biochemical basis and functional significance of the heterogeneity
of mammalian α-polymerases (1-10). We are studying this problem by isolating and characterizing α-polymerase genes and their corresponding
transcripts.

343

Our approach to the identification of α-polymerase messenger RNA has been 1) assay of cell-free translation products for catalytically active DNA polymerase polypeptides by activity gel analysis (5, 10) and 2) immuno-binding of translation products to a monoclonal antibody against calf α-polymerase. Here we report that translation of calf thymus RNA results in synthesis of several polypeptides with DNA polymerase activity. One of these polypeptides produced a particularly strong signal in the activity gel assay and a polypeptide of the same size was immunobound by the α-polymerase monoclonal antibody. The M_r of this polypeptide, 110,000 to 120,000, is identical to that of one of the α-polymerase catalytic polypeptides in crude extracts from calf cells and in preparations of purified calf α-polymerase (5, 6, 8). This demonstration of in vitro synthesis of a catalytically active DNA polymerase indicates an approach toward unequivocal identification of cDNA probes for α-polymerase genes.

MATERIALS

Thymus glands from 14-16 week old calves were from Henry W. Stapf Co., Inc., Baltimore, MD. Properties of a rat IgM monoclonal antibody against calf DNA polymerase α (termed MC pol 2) have been described (11). Sodium dodecyl sulfate (SDS) was from Bio-Rad. Calf thymus DNA from P.L. Biochemicals was activated by DNase I treatment (12). $(\alpha-{}^{32}P)dTTP$ and $({}^{35}S)$ methionine were from Amersham. Ultrapure guanidine hydrochloride and formalin-fixed Staphylococcus aureus cells (13) were from Bethesda Research Laboratories. The cells were processed before use as recommended by the supplier

METHODS

Preparation of RNA from Calf Thymus

Thymus gland was frozen in liquid N_2 immediately after excision and stored at -80°C. The frozen tissue (20 g) was pulverized and then homogenized in 10 volumes 7M guanidine·HCl, 25 mM sodium acetate, pH 5, 0.1 M 2-mercaptoethanol. Homogenization was at 4°C for 5 min with a Polytron homogenizer and then for 3 min with a Waring Blendor. An equal volume of $CHCl_3$ was added, and the homogenate was mixed for 3 min in a Waring blendor. The resulting mixture was centrifuged at 4,500 rpm in a Beckman JS-7.5 rotor for 15 min. The $CHCl_3$ treatment was repeated four times. RNA then was obtained from the aqueous phase by ethanol precipitation and processed further according to the method described by Peterson and Roberts (14). RNA then was extracted with two volumes of a mixture of phenol:$CHCl_3$:isoamyl alcohol (24:24:1, v/v) (15) and precipitated from the aqueous phase with 0.2 M sodium acetate and two volumes of ethanol. The precipitated RNA was washed with 70% ethanol twice, lyophilized, dissolved in H_2O and stored at -196°C. Poly(A^+)RNA was obtained after two cycles of oligo(dT)-cellulose column chromatography as described by Aviv and Leder (16).

Electrophoresis of RNA

The RNA size was examined by electrophoresis in a 1.2% agarose-methyl mercury gel (17). RNA was visualized in the gel by ethidium bromide staining.

In Vitro Cell-free Synthesis

Both total RNA and poly(A$^+$)RNA were translated in a BRL reticulocyte lysate system or with a BRL wheat germ extract. Incubation was at 30°C for 2 h with the reticulocyte lysate system and 23°C for 3 h with the wheat germ extract. The translation mixtures were supplemented with calf liver tRNA at a concentration of 1.8 µg/30 µl. Reactions were in the presence of 50 µCi (^{35}S) methionine (∿800 Ci/mmol) per 30 µl translation mixture. At the end of the incubation, aliquots were taken and (^{35}S)methionine incorporation was determined by hot TCA precipitation.

The ^{35}S-labeled products were analyzed by electrophoresis in SDS-polyacrylamide gels (10%) as described by Laemmli (18). Fluorography was according to Bonner and Laskey (19). Alternatively, the gels were fixed in 10% methanol, 5% acetic acid for 30 min, washed with H$_2$O for 15 min and treated with Autofluor for 2 h. The gels were dried and autoradiography was done at -80°C using Kodak XAR-5 film with a Dupont Lightning Plus intensifying screen.

Assay for DNA Polymerase Activity in Translation Mixtures by the Activity Gel Method

Aliquots of the translation mixture were analyzed by electrophoresis on SDS-polyacrylamide gels (8% or 10%) containing 150 µg/ml of activated DNA (5, 10). The gels were washed with two changes of 50 mM Tris-HCl, pH 7.5, for 30 min at 25°C. Washing was continued at 4°C for three hours in DNA polymerase assay buffer (50 mM Tris-HCl, pH 7.5, 6 mM magnesium acetate, 40 mM KCl, 500 µg/ml BSA, 1 mM dithiothreitol, 0.01 mM EDTA, 16% glycerol). Each gel then was transferred to a flask containing 40 ml of DNA polymerase assay buffer plus 12.5 µM each of dATP, dCTP and dGTP and 1 µM (α^{32}P)dTTP(∿250µCi). Incubation was at 37°C for 24 hours. The gels were washed with two changes of 5% TCA, 1% pyrophosphate for 1 h at 25°C and then with several changes of the same solution at 4°C over a period of 40 hours. The gels were dried, and autoradiography was for 24 hours at -80°C with Kodak XAR-5 film with a Dupont Cronex Lightning Plus intensifying screen.

Immunobinding with Monoclonal Antibody to DNA Polymerase α

An aliquot (20 µl) of the translation mixture was dilluted four-fold and adjusted to give a solution containing 10 mM methionine in immunobuffer (50 mM Tris-HCl, pH 7.5, 0.15 M MaCl, 1 mM EDTA, 1% Triton X-100, 0.5% sodium deoxycholate, 0.1% SDS, 1% aprotinin and

1 mg/ml BSA). The solution was centrifuged for 30 min at 11,000 rpm
in a Sorvall HB-4 rotor. The supernatant fraction was then incubated
at 4°C for 16 h in a solution containing 100 μg/ml monoclonal antibody
(MC pol 2) in immunobuffer. After incubation the mixture was added
to packed Staphylococcus aureus cells that were precoated with rabbit
anti-rat IgM. Precoating was with 50 μl of a 10% cell suspension in
immunobuffer containing 420 μg/ml rabbit anti-rat IgM at 4°C for 1 h
with gentle shaking. The mixture was centrifuged at 3,000 rpm for
10 min, and the cells were washed once with 750 μl of immunobuffer.

Reaction between the precoated cells and IgM-translation product
complex was at 4°C for 16 hours. Cells were collected by centrifugation at 3,000 rpm at 4°C for 10 min. The supernatant fraction was
discarded, and the cells were washed, twice with 500 μl of immunobuffer containing 2.5 M KCl and once with 500 μl of 10 mM Tris-HCl,
pH 7.3. The cells were collected and the immune complexes dissociated
by incubation for 5 min at 25°C in 50 mM Tris-HCl, pH 7.0, 2% SDS,
6 M urea, 5% β-mercaptoethanol (20). The mixture was centrifuged in
an Eppendorf microfuge for 1.5 min, and the supernatant fraction was
analyzed on a 10% SDS-polyacrylamide gel. Fluorography was conducted
as described above. An identical immunobinding experiment was conducted using a nonimmune monoclonal IgM antibody (11) as a control.

Sucrose Density Gradient Centrifugation of Total RNA

Total RNA was fractionated on a 5-20% linear sucrose density
gradient containing 0.1 M NaCl, 10 mM Tris-HCl, pH 7.5, 1 mM EDTA.
Approximately 1.8 mg of total RNA was applied to the gradient in a
nitrocellulose tube. Centrifugation was in a Beckman SW 27 rotor at
25,000 rpm at 4°C for 20 hours. Fractions of 1 ml were collected from
the bottom of the tube. Portions of the fractions were diluted 10-
fold with H_2O and absorbance at 260nm was measured. Pools of five
fractions each were made and diluted two-fold with H_2O; RNA was recovered by ethanol precipitation, washed with 70% ethanol, lyophilized,
and dissolved in autoclaved glass distilled H_2O. All RNA fractions
were tested for translation in vitro.

RESULTS

The poly(A$^+$)RNA preparation obtained from calf thymus was
examined by electrophoresis in a 1.2% agarose-methyl mercury gel.
Most of the RNA molecules were larger than 28S, suggesting the presence of hnRNA. It was evident also that 18S rRNA had not been completely removed after two cycles of oligo(dT)-cellulose chromatography.

Total and poly(A$^+$)RNA were translated in vitro using a rabbit
reticulocyte lysate system, and to determine whether DNA polymerases

had been synthesized, these translation mixtures were analyzed by a modification (10) of the activity gel method described by Spanos et al. (5) (Figure 1). Endogenous DNA polymerase catalytic polypeptides were observed at $\sim$70,000-M_r and 40,000-M_r (Fig. 1, Lane 1). These bands are identical in M_r to those of α-polymerase catalytic polypeptides, respectively (10). Upon addition of total RNA to the translation mixtures, an activity band between $\sim$110,000-and 120,000-M_r was detected (Fig. 1, Lane 2). This molecular weight corresponds to that of one of the α-polymerase polypeptides in highly prurified enzyme preparations from calf thymus (6, 8) and in crude extracts from calf and a variety of other mammalian cells (6, 8, 10).

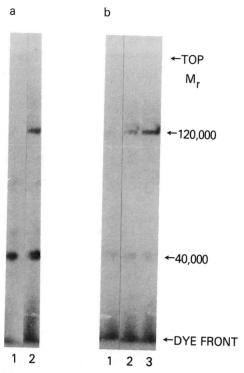

Figure 1: Autoradiogram showing results of activity gel analysis of in vitro translation products in the reticulocyte lysate system. Translation was conducted with or without exogenous mRNA; 80 μM unlabeled L-methionine was present. Aliquots of the translation mixtures, 10 μl, were analyzed using a 10% activity gel. Each sample was mixed with a heterogeneous protein mixture (10). Panel a shows results with calf total RNA. Lane 1, control incubation, no exogenous mRNA. Lane 2, incubation in the presence of 7 μg of calf thymus total RNA. Panel b, results with poly(A^+)RNA. Lane 1, control incubation, no exogenous mRNA. Lane 2, incubation in the presence of 0.9 μg poly(A^+)RNA. Lane 3, incubation in the presence of 2.7 μg poly(A^+)RNA. Details of translation, activity gel analysis and autoradiography are described under Methods.

With the translation mixture containing poly(A$^+$)RNA, an activity band
was observed between 110,000- and 120,000-M$_r$ (Fig. 1, Lanes 2 and 3);
a stronger signal was obtained with a higher amount of poly(A$^+$)RNA
(Fig. 1, Lane 3). These results suggest that sufficient DNA polymerase
catalytic polypeptide had been formed in the translation mixture to
enable detection by activity gel analysis and that α-polymerase messenger RNA was enriched in the poly(A$^+$) fraction. Analysis of translation mixtures incubated with rabbit globin mRNA did not reveal a
band in the ∿120,000-M$_r$ region of the gel (not shown), indicating
that some type of polynucleotide activation of an endogenous DNA
polymerase probably was not responsible for the 110,000-M$_r$ to
120,000-M$_r$ band observed in the translation mixtures containing calf
RNA.

In our experience, variation was observed in the intensity of
the 110,000 to 120,000-M$_r$ activity band designated in Figure 1. Intensity ranged from essentially identical to control incubations (without RNA) to much stronger than that shown in Figure 1b. This variation
probably was due to differences among RNA preparations. Strong signals
also could be obtained after fractionating the total RNA preparation
by sucrose density gradient centrifugation (Figure 2, panel A). Pooled
RNA fractions were translated in vitro and the products were examined
by both electrophoresis of ^{35}S-labeled proteins and activity gel analysis. The profiles of ^{35}S-labeled translation products (Fig. 2, panel
B) indicated that the various size populations of RNA encoded proteins
of different sizes. Activity gel analysis indicated a dramatic enrichment of the mRNAs that encode synthesis of the 110,000 to 120,000-M$_r$
and 68,000-M$_r$ polypeptides, respectively, in the pooled RNA fractions
4 and 6 (Fig. 2, panel C). RNA molecules in these fractions correspond
to approximately 18S and 7S, respectively. In all the translation
mixtures, endogenous DNA polymerase β activity at 40,000-M$_r$ was observed.

In vitro synthesis of α-polymerase polypeptides was evaluated
immunologically using a rat IgM monoclonal antibody to calf α-polymerase (11). A rabbit reticulocyte translation mixture with calf total
RNA was probed with this antibody and with a nonimmune rat IgM. Specific immunoprecipitation of ^{35}S-labeled polypeptides was observed
(Fig. 3); these polypeptides correspondend to M$_r$ = ∿110,000, ∿118,000
and ∿155,000.

DISCUSSION

In the present study, the cell-free synthesis of DNA polymerase
catalytic polypeptides was evaluated using activity gel analysis. The
results indicated the formation of two active polypeptides of M$_r$ =
110,000- to ∿120,000 and ∿68,000, respectively, that are putative
α-polymerases. This finding is important because unequivocal demonstration of in vitro synthesis of the enzyme can not be based solely

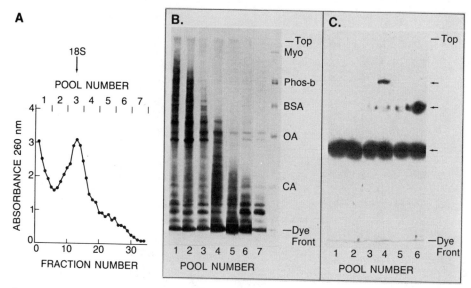

Figure 2: Enrichment of α-polymerase mRNA by sucrose density gradient centrifugation of calf thymus total RNA. Panel A, calf thymus total RNA (1.8 mg) was fractionated in a 5-20% linear sucrose gradient using a Beckman SW27 rotor. Fractions of 1 ml were collected and absorbance at 260 nm was measured after a ten-fold dilution of each fraction. The top of the gradient is at the right. Pools of 5 fractions each are indicated. Panel B, the RNA pools were translated in vitro in a reticulocyte lysate system containing 50 μCi of ^{35}S-methionine (∿800 Ci/mmol) in a 30 μl incubation mixture. Aliquots (1 μl) were analyzed by electrophoresis on a 10% polyacrylamide-0.1% SDS gel. The profiles of ^{35}S-labeled translation products are presented. Lanes 1, 2, 3, 4, 5, 6 and 7: translation reactions made in the presence of RNA pools 1 through 7, respectively; Lane 8: ^{14}C-labeled protein markers, M_r:myosin, 200,000; phosphorylase b, 100,000 and 92,000; BSA, 66,000;ovalbumin, 43,000; carbonic anhydrase, 30,000. Panel C, the RNA pools were translated in the presence of 85 μM L-methionine and the products were analyzed on a 10% activity gel. Aliquots (25 μl) of the translation mixtures were mixed with 9 μl of a denatured heterogeneous protein mixture (10) and sample buffer, heated for 3 min at 37°C and applied onto the gel. The autoradiogram of ^{32}P-labeled products is presented. Lanes 1, 2, 3, 4, 5 and 6, translation reactions in the presence of RNA pools 1, 2, 3, 4, 5 and 6, respectively. Each 30 μl translation mixture contained ∿16 μg of RNA. The activity bands are denoted with arrows and correspond to M_r = 110,000, 68,000 and 40,000. With unfractionated total RNA the signals obtained in the 110,000-120,000-M_r and ∿68,000-M_r regions were approximately 5% and 0.5%, respectively, without exogenous RNA, was identical to Lane 1. Details of the electrophoresis, activity gel analysis, fluorography and autoradiography are given under Methods.

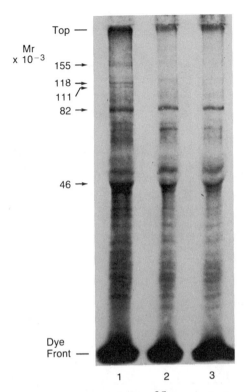

Figure 3: *Specific immunobinding of* ^{35}*S-labeled polypeptides trans-*
lated in vitro using α-polymerase monoclonal antibody. Lanes 1 and 2
are fluorograms of immunobound ^{35}*S-labeled polypeptides. Lane 1, a*
20 μl aliquot of a translation mixture containing 40 μg total RNA
per 60 μl translation mixture was reacted with 20 μl of a 2 mg/ml
monoclonal rat IgM antibody. Lane 2, same conditions as in Lane 1,
except that 10 μg non-immune rat IgM was used in the immunobinding.
Lane 3, same conditions as in Lane 1, except that rabbit antirat IgM
was the only antibody used. Translation reactions were conducted in
the presence of 100 μCi of ^{35}*S-methionine per 60 μl total traslation*
mixture. The immunobound products were analyzed in a 10% polyacryla-
mide-SDS gel. Details of the translation, immunobinding, electropho-
resis and fluorography are described under Methods.

upon the widely used approach of immunoprecipitation with a specific
antibody. This follows because there is no generally recognized subunit structure of the enzyme, as various species and preparations of
calf thymus DNA polymerase α have been shown to consist of multiple
polypeptides with M_r's of ∿200,000. Further, the primary structure
of the enzyme is unknown, ruling out identification by partial primary
structure determination. Hence, at the present time, it appears that

the most reliable method for identification of in vitro translation products must depend upon the DNA polymerase catalytic activity itself of the newly formed peptide molecule.

The routine solution assay of DNA polymerase activity in the translation lysate is complicated by the presence of endogenous polymerases. However, by activity gel analysis, newly formed DNA polymerase polypeptides can be distinguished from endogenous enzymes by 1) the appearance of a band at a unique M_r, such as 110,000- to 120,000-M_r and 2) the mRNA-dependent enhancement of a specific endogenous band. In addition to the advantage of specificity, the activity gel assay appears to be more sensitive than (^{35}S)methionine incorporation combined with specific immunoprecipitation. This is reasonable because the number of (^{32}P)-dNMP residues incorporated per enzyme molecule appears to be much higher (21) than the number of methionine residues per enzyme molecule. It should be recognized, however, that the activity gel assay, can not detect DNA polymerase subunits that are inactive as individual proteins. Also, some enzyme species active in routine solution assay are not active in the gel assay, perhaps because they are not able to refold properly after removal of SDS. In spite of these limitations, activity gel analysis appears to be a reliable method for identification of cDNA clones homologous with mRNA encoding the 110,000- to 120,000-M_r and 68,000 M_r catalytic polypeptides. In this regard, the enrichment of α-polymerase mRNAs by sucrose gradient centrifugation appears to be of particular value. First, the intensity of the activity gel signal is enhanced. Second, we believe that the mRNAs encoding these catalytic polypeptides are rare, comprising less than 0.1% of the total mRNA population, and it is likely that the number of α-polymerase cDNA clones in a cDNA library could be increased by using the enriched mRNA fractions.

The results from immunoprecipitation with an α-polymerase specific monoclonal antibody gave further evidence supporting the in vitro synthesis of the 110,000 to 120,000-M_r α-polymerase polypeptide. It is noteworthy that the ∿155,000-M_r and the ∿110,000- 120,000-M_r polypeptides shared a common antigenic determinant, yet the 155,000-M_r polypeptide did not produce a signal in the activity gel. This is consistent with earlier results on immunoprecipitation of polyptides from crude extracts of calf cells (11). We have not yet evaluated whether the ∿68,000 M_r polypeptides could be a truncated version of the 110,000 to 120,000-M_r polypeptide or whether these polypeptides are structurally related to putative higher M_r α-polymerase polypeptides (8, 11). Further studies using specific cDNA probes should clarify any structural relationships between these α-polymerase polypeptides and the higher M_r polypeptides identified by use of the monoclonal antibody (11).

The presence of a large amount of endogenous β-polymerase activity prevented our evaluation of the synthesis of this enzyme.

However, with β-polymerase the approach of specific immunoprecipitation of a 40,000 dalton polypeptide appears sufficient since it has
been established that mammalian β-polymerase is a polypeptide of this
mass (22).

ACKNOWLEDGEMENTS

        We thank Drs. Thomas Sargent and Bruce Paterson for helpful
discussions.

REFERENCES

1.   Matsukage,A., Sivarajan,M. and Wilson,S.H. (1976)
          Biochemistry 15, 5305.
2.   Momparler,R.L., Rossi,M. and Labitan,A. (1973) J. Biol. Chem.
          248, 285.
3.   Holmes,A.M., Hesslewood,I.P. and Johnston,I.R. (1974)
          Eur. J. Biochem. 43, 487.
4.   Mechali,M. Abadiedebat,J. and DeRecondo,A.M. (1980)
          J. Biol. Chem. 255, 2114.
5.   Spanos,A., Sedgwick,S.G., Yarranton,G.T. and Hübscher,U. (1981)
          Nucleic Acids Res. 9, 1825.
6.   Hübscher,U., Spanos,A., Alberts,W., Grummt,F. and Banks,G.R.
          (1981) Proc. Natl. Acad. Sci. U.S.A. 78, 6771.
7.   Grosse,F. and Krauss,G. (1981) Biochemistry 20, 5470.
8.   Albert,W., Grummt,F., Hübscher,U. and Wilson,S.H. (1982)
          Nucleic Acids Res. 10, 935.
9.   Filpula,D., Fisher,P.A. and Korn,D. (1982) J. Biol. Chem.
          257, 2029.
10.   Karawya,E. and Wilson,S.H. (1982) J. Biol. Chem. 257, 13129.
11.   Krawya,E., Albert,W., Swack,J., Fedorko,J., Minna,J.D. and
          Wilson,S.H. (1983) J. Biol. Chem., submitted.
12.   Schlabach,A., Fridlender,B, Bolden,A. and Weissbach,A. (1971)
          Biochem. Biophys. Res. Commun. 44, 879.
13.   Kessler,S.W. (1975) J. Immunol. 115, 1617.
14.   Paterson,B.M. and Roberts,E.B. (1981) in: Gene Amplification
          and Analysis (Chirikjian,J.G. and Papas,T.S.,eds),
          Elsevier/North Holland, New York, 2, pp. 417.
15.   Goodman,H.M. and MacDonald,R.J. (1979) in: Methods in Enzymology
          (Wu,R., ed), Academic Press, New York, 68, pp. 75.
16.   Aviv,H. and Leder,P. (1972) Proc. Natl. Acad. Sci. U.S.A.
          69, 1408.
17.   Bailey,J.M. and Davidson,N. (1976) Anal. Biochem. 70, 75.
18.   Laemmli,U.K. (1970) Nature, 227, 680.
19.   Bonner,W.M. and Laskey,R.A. (1974) Eur. J. Biochem. 46, 83.
20.   Lustig,A., Padmanaban,G. and Rabinowitz,M. (1982) Biochemistry
          21, 309.

21.   Karawya,E., Swack,J.S. and Wilson,S.H. (1983) Anal. Biochem.,
       in press.
22.   Tanabe,K., Yamaguchi,M., Matsukage,A. and Takahashi,T.
       (1981) J. Biol. Chem. 256, 3098.

INTERACTION OF DNA ACCESSORY PROTEINS WITH DNA POLYMERASE β OF THE NOVIKOFF HEPATOMA

R.R. Meyer, D.C. Thomas, T.J. Koerner and D.C. Rein

Department of Biological Sciences
University of Cincinnati
Cincinnati OH 45221, USA

INTRODUCTION

Replication and repair require the concerted action of a DNA polymerase with several accessory proteins. In prokaryotes the interaction of such proteins in replication has been well documented (1). With DNA repair, less is known, although with the recent progress of the uvr system of Escherichia coli we can anticipate a better understanding of excision-repair in bacteria (2). In eukaryotes there is a wealth of information on polymerases, but little progress has been made with mechanistic studies of accessory proteins for replication or repair.

Several years ago our laboratory began a search for eukaryotic accessory proteins which interact with the DNA polymerases. Two approaches were taken. The first involved searching for factors which enhance the activity of the polymerase (3, 4). Because DNA polymerase-β was available in a homogeneous form (5), we concentrated our efforts on this enzyme. This led to the finding of several factors (3, 4) and the characterization of two of them as an exonuclease (DNase V) which forms a complex with the DNA polymerase (6) and a novel single-stranded DNA-binding protein (SSB-48) (7). The second approach involved searching for specific classes of accessory proteins and then examining their interaction with the polymerase. This led to the finding of five distinct DNA-dependent ATPases, one of which (ATPase III), was shown to stimulate DNA polymerase-β (8). We summarize here some of the properties and interactions of ATPase III and SSB-48 with DNA polymerase-β.

RESULTS AND DISCUSSION

We have recently described the purification of three accessory proteins for DNA polymerase (6-8). Binding protein SSB-48 exists as a monomer of 48,000 daltons in solution. This globular protein has several novel features which distinguish it from all other binding proteins previously described. Unlike these others, SSB-48 binds weakly to DNA. It dissociates from DNA at 120-150 mM NaCl as indicated by its elution from single-stranded DNA (ssDNA)-cellulose columns (7). In addition, treatment of φX174 ssDNA-SSB-48 complexes with 0.15 M NaCl before preparation for electron microscopy (Koerner and Meyer, unpublished data) revealed no SSB-48 bound to the DNA. Further evidence comes from studies of the competition for DNA of DNA-dependent ATPases and nucleases with single-stranded DNA binding proteins (9). DNA-binding proteins such as that of Escherichia coli (SSB) bind very tightly to ssDNA and can readily displace DNA-dependent ATPases. Figure 1 shows a loss of ATPase activity with increasing concentrations of Escherichia coli SSB. However, no loss of ATPase activity is seen with SSB-48, indicating that it can not displace the ATPases from the effector even when SSB-48 is in excess. Similarly, nucleases such as S1 are strongly inhibited by the Escherichia coli SSB (9) but are unaffected by SSB-48 (data not shown).

A second feature of SSB-48 is its ability to bind cooperatively to DNA. The evidence for cooperativity is two-fold. First, there is a sigmoidal response when measuring the stimulatory activity of SSB-48 with DNA polymerase (7). Secondly, when φX174 ssDNA is incubated with subsaturating amounts of SSB-48 and then prepared for electron microscopy, only two types of molecules are seen: those fully coated and those completely lacking SSB-48. No intermediate, partially coated forms are detected. These studies also show that SSB-48 promotes a condensation of the DNA by a factor of 2.7. These properties, cooperative binding and strand condensation, are similar to those of the strongly binding SSB of Escherichia coli. Only one eukaryotic DNA-binding protein, that of Ustilago maydis (10), has been shown to bind cooperatively to DNA.

Finally, SSB-48 is capable of destabilizing a double helix at 37°C (ref. 7, Figure 2). There are several interesting features about this reaction: it occurs at physiological temperatures, and this destabilization occurs in A-T-rich regions. DNA can be rendered sensitive to nuclease S1 up to about 50% (7). However, poly(dG)·poly(dC) is resistant, but poly(dAdT) is easily rendered S1-sensitive (Figure 2).

SSB-48 was first identified (3, 7) by its ability to stimulate DNA polymerase-β (Table 1). This protein most likely exerts its action on the template rather than on the polymerase, since direct protein-protein interactions can not be demonstrated. Normally DNA synthesis with polymerase-β reaches a plateau in 1-2 hours. In the presence

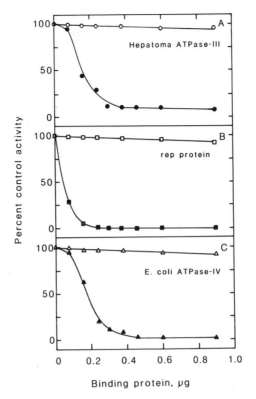

*Figure 1:   Comparison of the effect of SSB-48 (open symbols) and
Escherichia coli SSB (closed symbols) on DNA-dependent ATPases. Assay
conditions for Novikoff hepatoma ATPase III (40 units, panel A),
Escherichia coli rep ATPase (250 units, panel B) and Escherichia coli
ATPase IV (250 units, panel C) have been described (8, 9, Meyer et
al., in preparation). Reactions contained 115 pmole (nucleotide) of
G4 ssDNA. SSB-48 or SSB was added at the concentration given in the
figure, and the tubes were incubated for 20 min at 37°C.*

of SSB-48 there is an increase in the initial velocity and a marked
prolongation of linear synthesis. Indeed, we have shown such linearity
continues for at least 45 hrs (7). Since SSB-48 can both destabilizing a double helix and bind to ssDNA, these properties suggest two
possible mechanisms which could explain its stimulatory effect.
Since the DNA polymerase prefers a short gap, SSB-48 could promote
strand displacement from a nick or short gap in advance of polymerization. Alternatively it could bind to the template strand placing
it in a conformation promoting polymerization and active recycling
of the polymerase. To test this, model substrates were prepared with
poly(dA) as template to which varying amounts of oligo(dT)$_{15}$ were
annealed as primers. The average gap size was controlled by the ratio

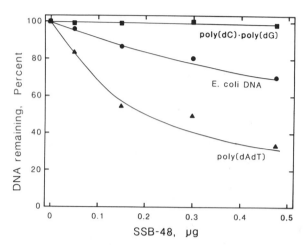

Figure 2:   Helix destabilization activity of SSB-48. DNA unwinding assays were measured as described (7) using tritiated substrates and increasing concentrations of SSB-48 as indicated in the figure. After incubation of 3 nmol (nucleotide) of DNA or polynucleotide with SSB-48 for 20 min at 37°C, excess nuclease S1 was added and after an additional 20 min incubation at pH 4.6, the DNA was precipitated to determine the amount of substrate rendered single-stranded as indicated by S1 nuclease sensitivity.

of primer to template. If strand displacement is the mode, we would predict that stimulation would be greatest on smaller gap sizes templates. Conversely, if template binding is the mode, we would predict that stimulation would be greatest on long gapped substrates. In one such experiment gap sizes of 0, 35 and 280 nucleotides showed a 2-, 3- and 10-fold stimulation of the rate of reaction, respectively, by the binding protein. This supports the interpretation that SSB-48 places the template in a favorable conformation for the polymerase (7).

In extracts of the tumor there are five distinct DNA-dependent ATPases (8), all of which have now been purified to near homogeneity (Thomas and Meyer, unpublished data). ATPase III has been chosen for further study, since it is capable of stimulating DNA polymerase-β (8) as shown in Table 1. This ATPase stimulates the polymerase about 10-fold on nicked DNA as substrate. The stimulation is greatly reduced in the presence of the ATP analogues ATPγS, ADP or α, β-methylene ATP, which are inhibitors of ATPase III (8). On gapped DNA there is virtually no effect of the ATPase on the polymerase. This is probably due to the presence of sufficient productive sites on this substrate for synthesis. It is interesting that in the presence of both ATPase III and SSB-48 there is a further enhancement of stimu-

Table 1:   Effect of ATPase III and SSB-48 on the activity of
           DNA polymerases *

Additions	DNA polymerase activity			
	gapped DNA		nicked DNA	
	pmoles	% control	pmoles	% control
DNA-polymerase α				
None	15.5	100	3.5	100
+ ATPase III	8.2	53	5.3	151
+ SSB-48	18.1	117	6.4	183
+ ATPase III + SSB-48	7.8	50	7.9	226
DNA polymerase β				
None	53.4	100	1.2	100
+ ATPase III	57.0	107	12.5	1040
+ SSB-48	84.9	159	4.5	375
+ ATPase III + SSB-48	93.1	174	20.1	1680
+ ATPase III + ATPγS			2.9	241
+ ATPase III + ADP			2.0	167
+ ATPase III + α, β-methylene-ATP			0.4	33

* The following concentrations of reactants were used: DNA polymerase-
α, fraction VIII, 0.1 units; DNA polymerase-β, fraction VI, 0.05
units; ATPase III, fraction VIII, 80 units; SSB-48, fraction IV,
0.15 μg; ATPγS and ADP, 0.8 mM, α, β methylene-ATP, 4.0 mM. Incubation was for 60 min at 37°C. Gapped DNA was prepared by treatment
of calf thymus DNA with DNase I and nicked DNA consisted of high
molecular weigth calf thymus DNA containing a few nicks per molecule.

lation greater than the additive effect of each protein alone, suggesting a possible interaction.

In contrast to the marked stimulation of DNA polymerase-β, these
two proteins have a minimal effect on the rat hepatoma DNA polymerase-α (Table 1).

Model for the interaction of DNA polymerase and accessory proteins
in DNA repair

DNA polymerase-β has been implicated in DNA repair (1). The
following observations are relevant to any model for the interaction
of the polymerase and accessory proteins in this process: (a) DNA
polymerase-β alone prefers a short gapped substrate; (b) this enzyme
forms a complex with DNase V, a 3'→5' and 5'→3' exonuclease; (c)
SSB-48 and ATPase III each act upon the template, not upon the polymerase; (d) these proteins together stimulate synthesis best from
nicks, thus suggesting gap formation by a weak helicase activity;
(e) SSB-48 stimulates the polymerase better on large rather than
small gaps, suggesting stabilization of the template strand. Given
the properties of this DNA polymerase and the accessory proteins,
the scheme shown in Figure 3 is suggested as one of several possible
models for the role of these proteins in DNA repair. Using UV-photo-
dimers as a model lesion, the reaction is initiated by incision on
the 5' dimer site by an endonuclease which can recognize strand
distortion. Assuming that the mechanism of action of the rat repair
endonuclease is similar to that of the UV-specific endonucleases of
bacteriophage T4 or Micrococcal luteus, this will result in an

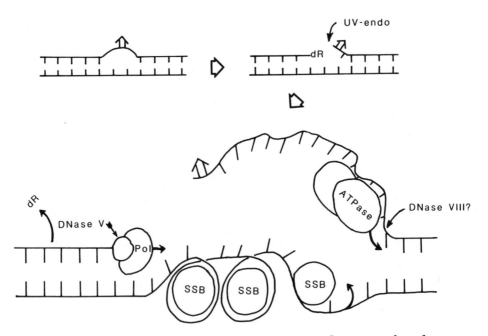

Figure 3: Model for the interaction of DNA polymerase-β and accessory proteins in DNA repair. For explanation see text.

apyrimidinic site on the 3' side and a dangling dimer on the 5' side
of the incision. Deoxyribonuclease V is capable of excising the
3'-deoxyribose (dR), thereby restoring the 3'OH site to primer competence (Small, Meyer and Rein, manuscript in preparation). We visualize ATPase III and SSB-48 coordinating destabilization of the
helix with resulting primer strand displacement (Figure 3). A weak
helicase activity postulated for ATPase III may be aided by SSB-48
pulling the template strand into a condensed conformation. The gap
is then filled by the polymerase. Removal of the displaced strand is
achieved by another nuclease such as DNase VIII which has the desired
specificity (11). Repair is completed by DNA ligase. Work in progress
is directed at testing this and other models for the repair of DNA
in vitro using purified mammalian enzymes.

ACKNOWLEDGEMENTS

This work was supported by grants NP277C from the American
Cancer Society and CA17723 from the National Institutes of Health.

REFERENCES

1.   Kornberg,A. (1980) "DNA Replication", W.H. Freeman Co.,
        San Francisco.
2.   Sancar,A. and Rupp,W.D. (1983) Cell 33, 249.
3.   Probst,G.S., Stalker,D.M., Mosbaugh,D.W. and Meyer,R.R.
        (1975) Proc. Natl. Acad. Sci. USA 72, 1171.
4.   Mosbaugh,D.W., Stalker,D.M. and Meyer,R.R. (1977)
        Biochemistry 16, 1512.
5.   Stalker,D.M., Mosbaugh,D.W. and Meyer,R.R. (1976)
        Biochemistry 15, 3114.
6.   Mosbaugh,D.W. and Meyer,R.R. (1980) J. Biol. Chem.
        255, 10239.
7.   Koerner,T.J. and Meyer,R.R. (1983) J. Biol. Chem.
        258, 3126.
8.   Thomas,D.C. and Meyer,R.R. (1982) Biochemistry 21, 5060.
9.   Meyer,R.R., Glassberg,J., Scott,J.V. and Kornberg,A. (1980)
        J. Biol. Chem. 258, 2897.
10. Banks,G.R., Spanos,A., Kairis,M.V. and Molineux,I.J. (1981)
        J. Mol. Biol. 151, 321.
11. Pedrini,A.M. and Grossman,L. (1983) J. Biol. Chem.
        258, 1536.

PLASMID CLONING OF DNA POLYMERASE I IN <u>ESCHERICHIA COLI</u> AND

<u>SACCHAROMYCES CEREVISIAE</u>

Ad Spanos and Steven Sedgwick

Genetics Division
National Institute for Medical Research
Mill Hill
London NW7 1AA, Great Britain

INTRODUCTION

The <u>Escherichia coli</u> <u>pol</u>A$^+$ gene, which encodes DNA polymerase I, was first cloned on bacteriophage λ (1). More recently a mutant <u>pol</u>AI gene, which makes a negligible amount of active protein, was cloned on λ. (2) and then transferred to a multicopy plasmid for determination of the <u>pol</u>Al nucleotide sequence (3). Also, a sequence encoding the Klenow fragment polymerase I has been cloned on a plasmid expression vector (4). However, attempts by ourselves and others (1) to construct multicopy plasmids with wild-type <u>pol</u>A$^+$ genes were uniformly unsuccessful, presumably because high levels of DNA polymerase are lethal. Thus, despite the advantages to biochemical studies gained from these cloned sequences, it remains difficult to study the biological effects of a cloned polymerase gene. The products from the above cloning procedures are either inactive, incomplete, produced in lethal amounts, or are carried by λ vectors which lack the practical versatility of plasmids, and which are unsuitable for introduction into organism other than <u>Escherichia coli</u>.

This communication describes attempts to circumvent these problems by cloning the <u>pol</u>A$^+$ gene on plasmid pHSG415 (5). Two features of this vector were important in chosing it for cloning the <u>pol</u>A$^+$ gene. Firstly, it had a low enough copy number to avoid the apparently lethal effects of excessive DNA polymerase I activity which foiled earlier attempts at <u>pol</u>A$^+$ cloning with multicopy plasmids. Secondly, it could transform <u>pol</u>A mutant hosts, permitting complementation tests between host and plasmid genes and allowing assays of plasmid-encoded DNA polymerase to be made against low levels

363

of endogenous activity. polA⁺ plasmids can also be modified for transformation of other organisms to test whether unknown deficiencies in
DNA metabolism can be complemented by any of the well characterized
actvities offered by cloned DNA polymerase I. As a first step in this
approach, shuttle plasmids have been constructed which are capable
of transforming the yeast, <u>Saccharomyces cerevisiae</u>, and which direct
the synthesis of active DNA polymerase I therein.

RESULTS AND DISCUSSION

Construction of polA⁺ plasmids

λ <u>polA</u>⁺ <u>cI</u>857 <u>nin</u>⁵ <u>Qam</u>73 <u>Sam</u>7, stock NM964 (2) was used as a
source of a 5.1 kbp <u>Hind</u> III fragment which encodes the <u>polA</u>⁺ gene
(1). This fragment was ligated into the <u>Hind</u> III site of pHSG415.
The insertion was always in the orientation shown in Figure 1, which,
from the work of Joyce et al. (3), is transcribed in the opposite

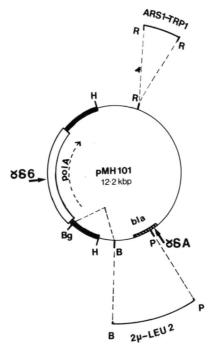

Figure 1: A map of <u>polA</u>⁺ plasmid pMH1O1 and its derivatives. Vector
pHSG415 DNA, —— ; 5.1 kbp <u>Hind</u> III fragment encoding the <u>polA</u>⁺ gene,
▭, and flanking sequences, ▬. Sites of transposition by γδ are
shown →. Broken lines indicate sites of in vitro deletion, or insertion of either yeast 2 μ or <u>ars</u>1 segments (see text). Abbreviations:
B, <u>Bam</u>Hi; Bg, <u>Bgl</u>II; P, <u>Pst</u>I; R, <u>Eco</u>RI. DNA methodology was described
elsewhere.

direction to the interrupted kanamycin-resistance gene. A representative plasmid, pMH101, was chosen for further attention.

Restoration of Pol⁺ phenotype by pMH101

Preliminary half-plate UV-irradiations of cross-streaked transformants showed that pMH101 conferred UV-resistance to the UV-sensitive, polAI, strain, JG112 (6). Subsequent quantitative experiments showed that pMH101 restored the UV-resistance of JG112 to wild-type levels (Figure 2). Similarly, the MMS-sensitivity of this mutant strain was reversed by transformation with pMH101 (data not shown). Resistant, polA⁺, Escherichia coli JG113 showed no additional increase in UV-survival after transformation with pMH101 (Figure 2b). The UV-sensitivity of polAI bacteria is attributed to a lack of 5'→3' polymerase and 3'→5' exonuclease; 5'→3' exonuclease activity is retained. However, complementation tests in bacteria with other polA mutations showed that pMH101 could complement deficiencies in all the activities of DNA polymerase I. For example, polAex1 and polAex2 mutants (7), which retain substantial polymerase activity (Table 1), but lack 5'→3' exonuclease (9) displayed wild-type UV-resistance when transformed with pMH101. pMH101 also complemented the radiation-sensitivity of temperature sensitive polA12 mutants (10) which at 42° lack all three of DNA polymerase I's activities (11). pMH101 also provided the DNA polymerase I needed for growth of red gam mutants of λ (12)

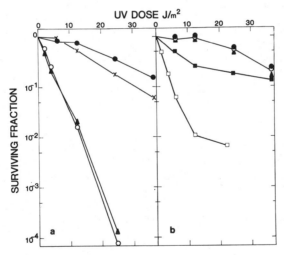

Figure 2: Radio-protective effects of pMH101 and its derivatives. The UV-survival of polAI JG112, panel a), and polA⁺ JG113, panel b), containing no plasmid, o ; pMH101, ● ; pHSG415, ▲ ; or pMH191, X. Panel b) also shows JG112 transformed with pMH101 bla::γδ, ■ ; or pMH101 polA::γδ, □ . Culture methods and irradiation conditions were as previously described (8).

Table 1:   Relative DNA polymerase I activity in pMH101 and pHSG415
           transformants of wild-type and polA-defective
           Escherichia coli

Strain	Percentage of wild-type activity	
	Plasmid	
	pHSG415	pMH101
polA⁺	98	288
polA1	<2	269
*polA12	<2	193
polA ex 1	90	246
polA ex 2	80	274

*Extracts preincubated at 42°C for 20 minutes before assay.

Crude cell extracts were prepared from cells grown at 32°C in 0.05
M Tris/HCl, pH 7.5, 1 M EDTA and 10% glycerol, using a French
pressure cell (14). Enzyme assays (see ref. 11) contained 10 mM
n-ethylmaleimide with activated calf thymus DNA as template. Incubation was for 20 min at 37°C. Enzyme levels are expressed as a
percentage of the activity in the untransformed wild-type strain.

and for replication of pBR322 in polA mutations (data not shown).
In view of the earlier problems of cloning the polA⁺ gene, particular
care was taken to show that genuine polA⁺ plasmid transformation had
occurred. Transformation with a plasmid-encoded polA⁺ gene was confirmed by curing experiments, Southern hybridization, and γδ transposition mutagenesis (13) (Figures 1 and 2). Selection of a sub-popu-
lation of polA⁺ revertant cells during transformation of polAI Escherichia coli was furthermore excluded by finding that polA⁺ plasmid,
and vector plasmid DNA, transformed Escherichia coli polAI with equal
efficiency.

Expression of a plasmid-encoded polA⁺ gene in Escherichia coli

     DNA polymerase I activity increased 2 to 3 times above the
normal wild-type level when polA⁺, polAI, polA12, polAex1 and polAex2
bacteria were transformed with pMH101 (Table 1). No increase in activity was detected after transformation with pHSG415 vector DNA.

Similarly, transposition of the $\gamma\delta$ sequence into the putative $\underline{polA}^+$ region of pMH101 eliminated any plasmid contribution to DNA polymerizing activity. However, polymerase I was made, though at lower levels, after transpositional inactivation of the bla gene of pMH101 (Table 2). The reduced amount of DNA polymerase I derived from pMH101 $\underline{bla}::\gamma\delta$ may be due to an apparent instability of the transposed plasmid, coupled with its lower copy number, estimated at approximately 1 per cell.

        DNA polymerase bands in SDS-polyacrylamide gels were visualized by a functional in situ assay for DNA polymerases (14, 15). After electrophoresis, SDS was washed from the gel and protein renaturation occured. The position of active DNA polymerases was determined by the incorporation of appropriately radio-labelled substrates and autoradiography. An extract of a polAI mutant transformed with pMH101 was analysed with this technique and a pattern of polymerase bands similar to those obtained with wild-type control extracts was seen (Figure 3). Therefore, the intact 109 kD polymerase band and its

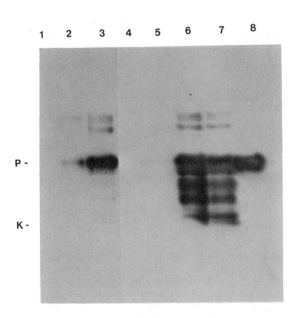

Figure 3: Autoradiograms of in situ DNA polymerase assays after SDS-polyacrylamide gel electrophoresis of crude extracts of wild-type and $\underline{polAI}$ strains transformed with pHSG415 and pMH101 ($\underline{polA}^+$). Lanes 1 to 7 contained 250 μg of extract of the following combinations of hosts and plasmids: 1, $\underline{polAI}$ + pMH101 $\underline{polA}::\gamma\delta$; 2, $\underline{polAI}$ + pMH101 $\underline{bla}::\gamma\delta$; 3, $\underline{polAI}$ + pMH101; 4, $\underline{polAI}$ (no plasmid); 5, $\underline{polAI}$ + pHSG415; 6, wild-type + pMH101; 7, wild-type + pHSG415. Lane 8 contained 10 units of pure $\underline{Escherichia\ coli}$ polymerase I. Molecular weight markers P and K refer to 109 kD polymerase I and its 76 kD Klenow fragment.

Table 2:   Relative DNA polymerase I activity in extracts of polAl
           Escherichia coli JGll2 containing the plasmid indicated

Plasmid	Percentage of wild-type activity
No plasmid	<2
pHSG415	<2
pMHlOl	256
pMHlOl pol::γδ	<2
pMHlOl bla::γδ	15
pMH191	189

Enzyme assays were as described in Table 1. Enzyme levels are
expressed as a percentage of the activity in the untransformed
wild-type strain.

proteolytic products are produced by both chromosomal and plasmid
polA$^+$ genes.

     The amounts of activity detected in Figure 3 reflect the levels
of DNA polymerase in Tables 1 and 2. Intense bands were seen with
wild-type extracts (lanes 6 and 7) and with extracts of polAI bacteria
transformed with pMHlOl (lane 3). Less activity was detected in
pMHlOl bla::γδ transformants (lane 2). No activity was detected in
ectracts of polAI bacteria with no plasmid (lane 4), pHSG415 (lane
5) or pMHlOl polA::γδ (lane 1).

Introduction of plasmid-encoded polA$^+$ gene into Saccharomyces cerevisiae

     Derivatives of pMHlOl and pHSG415 were constructed which
harbour the arsl of 2 μ yeast replicative sequences. The arsl, and
the overlapping selective TRPl sequence, were introduced by insertion
of a 1.55 kbp EcoRl fragment from YRp7 (16) into the unique EcoRl
site of pMHlOl and pHSG415 to give plasmids pMHl51 and pMHl61 (Fig.
1). 2 μ DNA, and its linked LEU selective marker, were introduced
on a 7.5 kbp Pstl-BamHl fragment of YEpl3 (17). This replaced the
smaller Pstl-BamHl segment of pMHlOl and pHSG415 to give plasmids
pMH3Ol and pMH3ll (Fig.1). Yeast which has been transformed (18) to
tryptophan or leucine prototrophy with these plasmids were screened
for plasmid-encoded β lactamase (19), and subjected to Southern
hybridization analysis, to confirm the presence of extrachromosomal

plasmid DNA. In addition, plasmid DNA preparations from transformed yeast were able to retransform polAI <u>Escherichia coli</u> to UV-resist-ance and <u>ars1</u>-based plasmids were readily lost from yeast by growth under non-selective conditions.

Expression of a plasmid encoded polA[+] gene in <u>Saccharomyces</u> cerevisiae

Cell extracts from untransformed and transformed yeast were first assayed for DNA polymerase activity in an assay with optimal conditions for DNA polymerases (20). There was a 2-fold increase in activity caused by either <u>ars1</u> or 2 μ-based polA[+] plasmids but polymerization levels were not changed by transformation with pMH161 and pMH311 plasmids lacking the polA[+] gene (data not shown).

Other assays preferentially reduced endogenous yeast DNA poly-merases so that plasmid-encoded DNA polymerase I was the primary activity detected. This was achieved by reducing the concentration of radioactive dTTP in the assay and/or by adding inhibitors with a preferential action towards yeast DNA polymerases. When yeast DNA polymerase activity in extracts of untransformed cells was assayed in the presence of 0.33 μM dTTP, instead of 50 μM dTTP, only 20-30% of its activity was detected. Extracts of yeast transformed with either pMH151 (Figure 4) or pMH311 (data not shown) showed a 3- to 4-fold increase in DNA polymerase with this assay, compared with

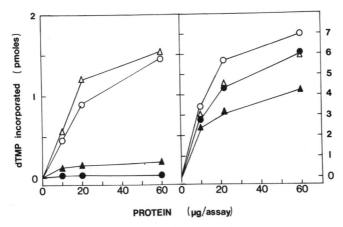

Figure 4: Inhibition of polymerase activity by PCMB and aphidicolin in extracts of pMH161 and pMH151 transformed yeast cells. Extracts of yeast Ml-2B pMH161, panel a) and Ml-2B pMH151, panel b) were prepared and assayed as described by Wintersberger and Wintersberger (20), except that mercaptoethanol and DTT were omitted for deter-mining PCMB inhibition (21) and 0.3 M dTTP was used in all assays.
o , no PCMB; ● , 500 μM PCMB; △ , no aphidicolin, ▲ , 100 μg/ml aphidicolin.

extracts of control cells transformed with pMH161 or pMH311. The additional activity was shown to be DNA polymerase I since it was not inhibited by aphidicolin, p-chloromercurobenzoate and n-ethyl maleimide (21, 22) (Figure 4b). These compounds did, however, inhibit more than 90% of the DNA polymerase activity in extracts of pMH161 and pMH311 transformed yeast which contain only yeast DNA polymerase (Figure 4a). With PCMB the extracts of polA$^+$ transformed yeast frequently retained much higher activity and on some occasions even incorporated more radioactive precursors than non-PCMB treated controls. We suspect that yeast extracts contain a low molecular weight, PCMB-sensitive inhibitor of DNA polymerase I, especially because a 10- to 20-fold increase in its activity in extracts is seen after Sephadex G200 column chromatography. Conversely, an equivalent reduction in activity of pure DNA polymerase I was found when yeast crude extract was added to an assay mix.

In situ assays of DNA polymerase activity, employing low dTTP concentrations and PCMB, were also made after SDS-PAGE of yeast crude extracts (Figure 5). Polymerase activity was only detected in extracts of yeast transformed with the polA$^+$ plasmid, pMH151. It was absent in untransformed yeast or in yeast transformed with shuttle vector pMH161. Activity co-migrated with purified polymerase I and Klenow fragment. Compared with Escherichia coli crude extracts, however, a much greater proportion of activity was present as the Klenow fragment and intermediate proteolytic fragments. This may be partly due to the absence of protease inhibitors during sample preparation. Thus, an analysis of enzyme prepared under conditions which minmize endogenous proteolysis is required before conclusions can be made about the in vivo yeast form of DNA polymerase I. However, it is interesting that the different proteases of the two species do produce some common sized intermediary degradation products of DNA polymerase I.

Two possible promoter sequences of the polA gene have been tentatively identified by Joyce et al. (3) with Pribnow boxes at -28 to -22, and at -150 to -144. Derivatives of polA$^+$ plasmids were made in which one of these sites was removed by deleting the cloned sequences preceeding a BglII site at bp -99 and extending to a BamHI site in the flanking vector DNA (Figure 1). In Escherichia coli this deletion did not prevent expression of the polA$^+$ gene on plasmid pMH191 (Table 2), nor did it prevent radio-protection (Figure 2), indicating that at least one polA$^+$ promoter is between bp -99 and the structural gene. However, in yeast this deletion eliminated synthesis of DNA polymerase I, indicating that the two organisms use different features of the polA$^+$ plasmids for expression of the polA$^+$ gene. Is has thus far been impossible to invert the cloned polA$^+$ gene to determine whether the elements needed for yeast expression are in the cloned sequence or the DNA of the vector flanking the cloning site.

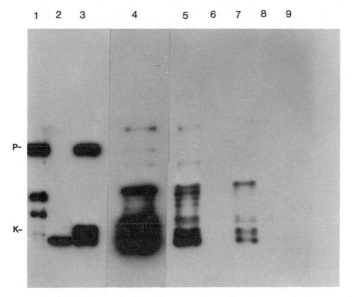

Figure 5: Autoradiogram of in situ DNA polymerase assays after SDS-
polyacrylamide gel electrophoresis of yeast transformed with pMH151
(<u>polA</u>$^+$) and pMH161 (control). Extracts were prepared as described in
Figure 4. Gel assays were as previously described, except mercaptoethanol was omitted and 500 µM PCMB was added. The autoradiogram was
exposed for 3 hrs for lanes 1 to 4 which contained: 1, 10 units partially purified DNA polymerase I; 2, 5 units Klenow fragment (both
PL Biochemicals Inc.); 3, 20 units pure DNA polymerase I stored 1
year at -20°; 4, 500 µg of pMH151 transformed yeast extract. Lanes
5 to 9 were exposed for 30 min and contained: 5, sample 4; 6, no
sample; 7, 200 µg pMH151 transformed yeast extract; 8, 500 µg pMH161
transformed yeast extract; 9, 500 µg extract of untransformed yeast.
P and K indicate the position of DNA polymerase I and its Klenow
fragment.

The introduction of active <u>Escherichia coli</u> DNA polymerase I
into yeast permits complementation analyses of mutational and drug-
induced deficiencies in yeast DNA metabolism. These experiments are
underway. More generally, the low bacterial copy number yeast shuttle
plasmids may be useful in inter-species cloning of other genes whose
excessive expression is lethal to <u>Escherichia coli</u>.

ACKNOWLEDGEMENTS

We are very grateful to Drs. N.Murray and P.Strike for gifts of
phages and plasmids, and for advice about their use. Aphidicolin was
a kind gift of Dr. A.H.Todd of I.C.I. Pharmaceuticals Division.

Geoffrey Yarranton, Geoffrey Banks and Robin Holliday contributed
with helpful advice and critical reading of the manuscript. Charles
White gave valuable assistance with yeast transformation.

REFERENCES

1.   Kelley,W.S., Chalmers,K. and Murray,N.E. (1977)
          Proc. Natl. Acad. Sci. USA 74, 5632.
2.   Murray,N.E. and Kelley,W.S. (1979) Mol. Gen. Genet. 175, 77.
3.   Joyce,C.M., Kelley,W.S. and Grindley,N.D.F. (1982)
          J. Biol. Chem. 257, 1958.
4.   Joyce,C.M. and Grindley,N.D.F. (1982) Proc. Natl. Acad. Sci.
          USA 80, 1830.
5.   Timmis,K.N. (1981) in: Genetics as a Tool in Microbiology,
          eds. Glover,S.W. and Hopwood,D.A., Cambridge University
          Press, pp. 49.
6.   Gross,J. and Gross,M. (1969) Nature 224, 1166.
7.   Konrad,E.B. and Lehman,I.R. (1974) Proc. Natl. Acad. Sci. USA
          71, 2048.
8.   Sedgwick,S.G. and Yarranton,G.T. (1982) Mol. Gen. Genet.
          185, 93.
9.   Uyemura,D., Eichler,D.C. and Lehman,I.R. (1976) J. Biol. Chem.
          251, 4085.
10.   Monk,M. and Kinross,J. (1972) J. Bacteriol. 109, 971.
11.   Uyemura,D. and Lehman,I.R. (1976) J. Biol. Chem. 251, 4078.
12.   Zissler,J., Signer,E.R. and Schaeffer,F. (1971) in: The
          Bacteriophage Lambda, ed. Hershey,A.D., Cold Spring
          Harbor Laboratories, Cold Spring Harbor, New York,
          pp. 455.
13.   Guyer,M.S. (1978) J. Mol. Biol. 126, 347.
14.   Spanos,A., Sedgwick,S.G., Yarranton,G.T., Hübscher,U. and
          Banks,G.R. (1981) Nucleic Acids Res. 9, 1825.
15.   Spanos,A. and Hübscher,U. (1983) Methods in Enzymol. 91, 263.
16.   Struhl,K., Stinchcomb,D.T., Scherrer,S. and Davis,R.W. (1979)
          Proc. Natl. Acad. Sci. USA 76, 1035.
17.   Broach,J.R., Strathern, J.N. and Hicks,J.B. (1979) Gene 8, 121.
18.   Hinnen,A., Hicks,J.B. and Fink,G.R. (1978) Proc. Natl. Acad.
          Sci. USA 75, 1929.
19.   Chevallier,M.R. and Aigle,M. (1978) FEBS Letts. 108, 179.
20.   Wintersberger,U. and Wintersberger,E. (1970) Eur. J. Biochem.
          13, 11.
21.   Wintersberger,E. (1974) Eur. J. Biochem. 50, 41.
22.   Sugino,A., Kojo,H., Greenberg,B.D. and Brown,P.O. (1981)
          The Initiation of DNA Replication, Academic Press Inc.

STRUCTURAL AND FUNCTIONAL PROPERTIES OF DNA POLYMERASE α

FROM CALF THYMUS

Friedrich Grummt, Waltraud Albert, Gerd Zastrow
and Andrea Schnabel

Institut für Biochemie, Universität Würzburg
Röntgenring 11
D-8700 Würzburg, Germany

INTRODUCTION

DNA polymerase α, the enzyme most widely believed to catalyze chromosomal DNA replication in animal cells, is typically purified as a high molecular weight complex of multiple polypeptides, as revealed by SDS-polyacrylamide gel electrophoresis (1-11). The question of precise subunit structure of this enzyme and the structure-function relationship of the individual polypeptides are subjects of our current research. We report here on the association of several functional properties -besides the catalytic DNA polymerizing activity- with the α-polymerase complex isolated from calf thymus, i.e. Ap_4A binding, amino acyl tRNA synthetase, primase and $3' \rightarrow 5'$ exonuclease-(proofreading-) activities.

Purification of scheme for bovine α-polymerase

The enzyme from calf thymus extracts was prepared as essentially described (7). Modifications were: 1) 2 mM benzamidine hydrochloride, 10 µg/ml of soybean trypsin inhibitor, and 100 µg/ml β-aminopropionitrile fumarate as protease inhibitors were present during the purification procedure. 2) The polymerase activity containing fractions of the DEAE-cellulose step were applied on a hydroxyapatite (Clarkson Biochem. Co.) column (2 ml) equilibrated with 10 mM potassium phosphate, pH 6.9, washed with 10 mM potassium phosphate, pH 6.9, 7 mM 2-mercaptoethanol, 20% glycerol and eluted with a 10 ml gradient from 10 to 300 mM potassium phosphate. The enzyme was eluted at about 180 mM potassium phosphate and was then applied on native polyacrylamide gels as described (7). The purification scheme is shown in Table 1. A 24000 fold enrichment in enzymatic activity was reached

Table 1: Purification of DNA polymerase α from calf thymus

	Volume ml	Protein concentration mg/ml	Activity U	Specific activity U/mg	Purification - fold
Crude extract	350	25	$11 \cdot 10^3$	1.25	-
Phosphocellulose Chromatography	180	1.4	$200 \cdot 10^3$	800	640
DEAE-sephacel chromatography	73	0.48	$112 \cdot 10^3$	1500	1840
Hydroxyapatite chromatography	50	0.2	$48 \cdot 10^3$	4800	3840
Nondenaturing polyacrylamide gel electrophoresis	0.3	0.008	74	30000	24000

by this isolation procedure and the enzyme was purified to apparent
homogeneity as determined by staining of the protein after nondenaturing gel electrophoresis.

Structural properties

The purified α-polymerase from calf thymus contained an abundant
118.000-M_r polypeptide as well as five lower molecular weight polypeptides with M_r 54.000, 56.000, 57.000, 58.000 and 64.000. One of
the low molecular weight polypeptides (M_r 57.000 or 58.000) is lost
on the hydroxyapatite purification step. Tryptic peptide mapping
indicated that the 118.000-M_r polypeptide shared extensive primary
structure homology with 57.000-, 58.000- and 64.000-M_r polypeptides
and some limited homology with 54.000- and 56.000-M_r polypeptides
(12). These results suggest the existence of a common precursor with
molecular weight >140.000.

Catalytic activity during chain elongation

To determine in which polypeptide of the heterooligomeric
α-polymerase the polymerizing activity resides we have applied a
technique enabling to detect DNA polymerase activity in situ after
Na Dod SO_4 gel electrophoresis (13). These experiments revealed that
the high M_r (118-120.000) polypeptide is responsible for chain elongation in preparations of purified DNA polymerase α from calf thymus
(12,13). Since one possible interpretation of the results of our
tryptic peptide mapping predicted the existence of an α-polymerase
polypeptide with M_r > 140.000 we examined a crude homogenate of calf
thymus with the in situ gel technique (13). This analysis revealed
the presence of a relative abundant DNA polymerase polypeptide with
an M_r of approximately 200.000 and several other bands with M_r greater
than the 120.000-M_r band exhibiting catalytic activity in purified
α-polymerase. It remains to be elucidated whether this high molecular
weight forms of catalytically active polymerase represents the polypeptides in DNA replication in living cells or precursor molecules.

Association with aminoacyl-tRNA synthetases

Recently, it was shown that human and wheat germ α-polymerases
are intimately associated with a tryptophanyl-tRNA synthetase activity
(14,15). Calf thymus α-polymerase at various purification steps also
charges mixed yeast tRNA if assayed with amino acids from a tritium-
labelled mixture. Competition experiments with an excess of individual
unlabelled amino acids revealed that exclusively histidyl-tRNA
synthetase activity is copurifying with catalytic α-polymerase activity. Whereas tryptophanyl-tRNA synthetase activity is completely
separated from DNA polymerase α at the very first purification step,
i.e. chromatography on phosphocellulose (Figure 1A), histidyl-tRNA
synthetase appears tightly associated with α-polymerase activity
during purification on phospho- and DEAE-cellulose as well as on

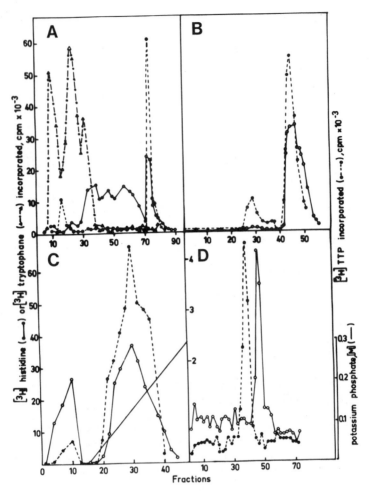

Figure 1: *Separation of tryptophanyl-tRNA synthetase (△—●—●—△) and
histidyl-tRNA synthetase activities (○——————○) from DNA from DNA polymerase activity (●– – – ●) during the purification of DNA polymerase
α from calf thymus. A: phosphocellulose column; B: DEAE-Sephacel
column; C: hydroxyapatite column; D: polyacrylamide gelelectrophoresis
under nondenaturing conditions.*

hydroxyapatite (Figure 1 A-C). Electrophoresis on polyacrylamide gels
under nondenaturing conditions, however, leads to quantitative separation of both activities (Figure 1D). Our data suggest that the
coincidence of histidyl-tRNA synthetase and α-polymerase activities
during purification by ion exchange or hydroxyapatite chromatography
might rather be fortuitous than a reflection of physical association.
This assumption was further supported by the observation that neither
the catalytic α-polymerase activity was affected by amino acids or
tRNA nor was that of the histidyl-tRNA synthetase influenced by

dXTPs and DNA. In case that both activities are physically associated the interaction does not withstand electrophoresis under native conditions.

Ap_4A-binding activity

By equilibrium dialysis an Ap_4A binding activity was shown to be present in calf thymus DNA polymerase α (7). A similar activity was later also found in α-polymerase preparations from HeLa-cells (14). By affinity labeling one of the low M_r-polypeptides (57.000) has been shown to be the Ap_4A-binding constituent of DNA polymerase α. First preliminary results from our laboratory demonstrated that this subunit is separated from the catalytic subunit during chromatography on hydroxyapatite. This loss of the Ap_4A-binding activity is accompanied by a drastic reduction of Ap_4A being a preferential substrate of the priming reaction using ssM13-DNA as a template (see below).

Primase activity

DNA polymerase α from calf thymus catalyzes ribonucleoside triphosphate-dependent DNA synthesis on single-stranded circular M13 DNA templates. The rNTP-dependent primase activity copurified on phosphocellulose-, DEAE-cellulose and hydroxyapatite columns and comigrated during electrophoresis in nondenaturing polyacrylamide gels with DNA polymerase α (Figure 2). The primase-dependent DNA synthesis is not inhibited by α-amanitin. Analysis of (^{32}P)-dAMP-labelled products by alkaline agarose gel electrophoresis showed that 0.1-1.5-kilobase DNA fragments were synthesized. Using all 4 NTPs alone, the priming products consist of oligoribonucleotides of four main size ranges between 9 to 12 nucleotides in length. With all 4 dNTPs in addition to 4rNTPs in the reaction mixture, the oligonucleotide products are primarily 10-12 residues long. The primase products are alkali labile and resistant against DNase I.

Besides the mixture of all 4rNTPs or ATP alone the priming activity can utilize Ap_4A as substrate. At low concentration (0.1 mM) Ap_4A is a preferential substrate of primase-dependent DNA synthesis. If Ap_4A is used as priming substrate it becomes covalently incorporated into nascent DNA chains. The utilization of Ap_4A as priming substrate is lost, however, after purification of α-polymerase on hydroxyapatite columns. Besides that, the hydrxyapatite-purified α-polymerase preparations cannot be affinity-labelled with periodate-oxidized (^{3}H)-Ap_4A any longer. From these results we suppose that the Ap_4A-binding subunit is involved in the incorporation of Ap_4A into nascent DNA chains during the priming reaction and that either two different priming mechanisms are carried out by calf thymus α-polymerase, one utilizing rNTPs and the other one using Ap_4A or that the primase has two different substrate sites, one for rNTPs and one for Ap_4A.

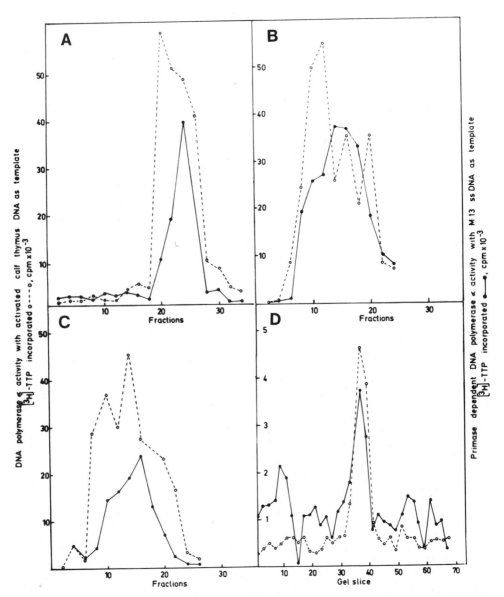

Figure 2: Copurification of DNA polymerizing and primase activities during the isolation procedure for DNA polymerase α from calf thymus (o——o). DNA polymerizing activity using activated calf thymus DNA as template, (●——●) rNTP-dependent activity using ssM13 DNA as template. A: phosphocellulose column; B: DEAE-Sephacel column; C: hydroxyapatite column; D: polyacrylamide gelelectrophoresis under nondenaturing conditions.

3'→5' exonuclease (proofreading-) activity associated with
α-polymerase

      Association of exonuclease activity with lower eukaryote DNA
polymerases (16-21) as well as with DNA polymerase δ from calf thymus
(22) and a subspecies of murine α-polymerase (6) is already well
known. The discovery of 3'→5' exonuclease activity in our calf thymus
DNA polymerase α is of particular interest in view of the important
role of DNA polymerase associated exonucleolytic activity in proofreading in prokaryotes. The DNA substrates used in the exonuclease
assay were either mismatching (poly dA: poly T with (^{32}P)-α-dAMP at
the 3' end of poly T) or base pairing (poly dA: poly dT with (^{32}P)-
α-dTMP at the 3' end of poly dT). α-polymerase fractions from the
DEAE-cellulose step of the purification scheme catalyzed release of
the ultimate residue of the 3' end from both mismatching and base-
pairing substrats in the absence of the 4 dNTPs. The release from
the annealed end was blocked if TTP, the substrate appropriate for
chain elongation was added to the reaction mixture not however in
the presence of the other three dNTPs (Figure 3). The releasing
activity was not impaired by the presence of substrates for chain
elongation if DNA with mismatching 3' ends was used. In this case
release and incorporation of the cognate substrate occurred in parallel
(Figure 4). The ^{32}P released was in the form of dAMP or dTMP, no
release of oligonucleotides was detected.

      This 3'→5' exonuclease activity was lost from the α-polymerase
complex upon purification on hydroxyapatite. So far experiments to
isolate the exonuclease from hydroxyapatite column fractions and
study it in complementation assays with α-polymerase remained unsuccessful.

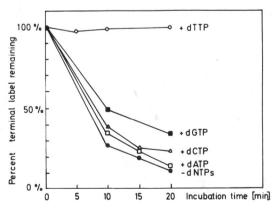

*Figure 3: Kinetics of 3'→5' exonuclease activity associated with
α-polymerase from calf thymus. Release of base pairing end label was
assayed in the presence of the base pairing substrate (dTTP), in the
presence of the noncognate substrates for primer extension (dGTP,
dCTP, dATP) or in the absence of dNTPs.*

F. GRUMMT ET AL.

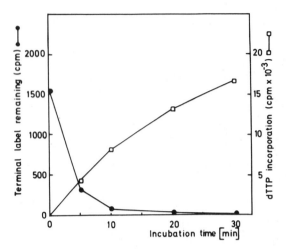

Figure 4:   Kinetics of 3'→5' exonuclease (●——●) and DNA polymerizing (□——□) activities of DNA polymerase α from calf thymus. Poly dA: poly dT with (^{32}P)-d-dAMP at the 3' end of poly dT was used as a mismatching template and TTP as substrate for polymerization was added.

REFERENCES

1.   Holmes,A.M., Hesslewood,I.P. and Johnston,I.R. (1976)
        Eur. J. Biochem. 62, 229.
2.   Hesslewood,I.P., Holmes,A.M., Wakeling,W.F. and Johnston,I.R.
        (1978) Eur. J. Biochem. 84, 123.
3.   Banks,G.R., Holloman,W.K., Kairis,M.V., Spanos,A. and
        Yarranton,G.T. (1976) Eur. J. Biochem. 62, 131.
4.   Fischer,P.A. and Korn,D. (1977) J. Biol. Chem. 252, 6528.
5.   Banks,G.R., Boezi,J.A. and Lehman,I.R. (1979)
        J. Biol. Chem. 254, 9886.
6.   Chen,Y.C., Bohn,E.W., Planck,S.R. and Wilson S.H. (1979)
        J. Biol. Chem. 254, 11678.
7.   Grummt,F., Waltl,G., Jantzen,A.-M., Hamprecht,K., Hübscher,U.
        and Kuenzle,C.C. (1979) Proc. Natl. Acad. Sci. USA
        76, 6081.
8.   McKunc,K. and Holmes,A.M. (1979) Nucleic Acids Res. 6, 3341.
9.   Méchali,M., Abadiedebat,J. and De Recondo,A.M. (1980)
        J. Biol. Chem. 255, 2114.
10. Villani,G., Sauer,B. and Lehman,I.R. (1980) J. Biol. Chem.
        255, 9479.
11. Grosse,F. and Krauss,G. (1981) Biochem. 20, 5470.
12. Albert,W., Grummt,F., Hübscher,U. and Wilson,S.H. (1982)
        Nucleic Acids Res. 10, 935.
13. Hübscher,U., Spanos,A., Albert,W., Grummt,F. and Banks,G.R.
        (1981) Proc. Natl. Acad. Sci. USA 78, 6771.

14. Rapaport,E., Zamecnik,P.C. and Baril,E.F. (1981)
    Proc. Natl. Acad. Sci. USA 78, 838.
15. Castronejo,M., Fournier,M., Gatius,M., Gander,J.C.,
    Labonesse,B. and Litvak,S. (1982) Biochem Biophys. Res.
    Commun. 107, 294.
16. Helfman, W.B. (1973) Eur. J. Biochem. 32, 42.
17. Crear,M. and Pearlman,R.E. (1974) J. Biol. Chem. 248, 3123.
18. McLennan,A.G. and Kier,H.M. (1975) Biochem. J. 151, 239.
19. Banks,G.R. and Yarranton,G.T. (1976) Eur. J. Biochem. 62, 143.
20. Chang,L.M.S. (1977) J. Biol. Chem. 252, 1873.
21. Wintersberger,E. (1978) Eur. J. Biochem. 84, 167.
22. Byrnes,J.J., Downey,K.M., Black,V.L. and So,A.G. (1976)
    Biochemistry 15, 2817.

VI
DNA Helicases and DNA Topoisomerases

FUNCTIONS OF DNA HELICASES IN THE DNA METABOLISM OF ESCHERICHIA COLI

Mahmoud Abdel-Monem, Helen M. Arthur[1], Inga Benz,
Hartmut Hoffmann-Berling, Ursula Reygers, Anita Seiter
and Gisela Taucher-Scholz

Max-Planck-Institut für medizinische Forschung
Abteilung Molekulare Biologie
Jahnstrasse 29
6900 Heidelberg, Federal Republic of Germany

[1]The University of Newcastle upon Tyne
Department of Biochemistry
Ridley Building
Newcastle upon Tyne NE1 7RU, United Kingdom

DNA helicases catalyze the separation of double-stranded DNA into single strands using the energy of ATP hydrolysis. The four helicases which have been found in Escherichia coli are listed in Table 1. These enzymes are the helicases I, II, III and the helicase specified by the rep gene. As the table shows the four proteins differ considerably with respect to M_r, number of molecules per cell and, in particular, mechanism of action. Each of the helicases requires a region of single-stranded DNA to initiate unwinding, and each of the four enzymes unwinds DNA unidirectionally relative to the chemical polarity of the DNA strand to which it is bound. The direction of unwinding depends on the nature of the helicase.

Relevant properties of the four helicase proteins and current ideas regarding their modes of action have recently been reviewed (1). The biochemical aspect of helicase action will thus be described only briefly.

Helicase I is a very large, fibrous molecule, actually the largest soluble protein detectable in Escherichia coli K12. The helicase, which readily forms aggregates and which according to electron microscopy binds in clusters to DNA, seems to unwind DNA

Table 1: DNA Helicases of Escherichia coli

Helicase	Gene	M_r (kd)	Molecules/cell	Mode of action	SSB required	Direction of action
I	traI(F)	180	500	processive	no	5'→3'
II	uvrD	75	>5000	DNA binding	no	5'→3'
III	*)	2x20	~20	processive	yes	5'→3'
rep	rep	68	50	processive	yes	3'→5'

*) as yet unidentified

by migrating as a group of 70 or more molecules along one of the DNA strands. Unwinding occurs in the 5'-to 3'-direction of the bound DNA strand (i.e. opposite to the direction of nucleotide polymerase action). This processive mode of action implies that, for unwinding a long and a short DNA molecule, the same amount of helicase protein per DNA molecule is required.

Helicase II, a globular protein of moderate size, seems to unwind DNA by binding in a stoichiometric amount to one of the DNA strands. Consistent with this interpretation is the fact that this enzyme is present in a high number of copies per cell. Helicase II unwinds DNA in the same direction as does helicase I.

rep Helicase was detected because single-stranded DNA phages such as ϕX174 and the filamentous phages fd and M13 make use of this enzyme for replicating their double-stranded Replicative Form DNA molecules. Results obtained in several laboratories, in particular that of A. Kornberg, suggest that rep helicase unwinds DNA processively. The helicase seems to translocate on the bound DNA strand in the 3'-to 5'-direction, opposite to the direction of the helicases I and II. It is estimated that there is one rep helicase molecule per replication fork. To completely dissociate the strands of a DNA duplex, rep helicase requires the presence of Escherichia coli single strand binding protein (SSB) in an amount stoichiometric with the unwound DNA, and electron microscopy shows that the unwound DNA is saturated with this protein. Furthermore, rep helicase is the only helicase for which the amount of ATP hydrolyzed during unwinding has been determined. It amounts to two molecules of ATP hydrolyzed per base-pair unwound.

Few publications concerning helicase III have appeared. This enzyme is thought to act essentially like rep helicase except for the direction of action which seems to be opposite to that of the rep enzyme. Helicase III requires the presence of SSB for maximum DNA denaturation.

The four helicases, especially the enzymes I, II and rep, are the most prominent DNA dependent ATPases in an Escherichia coli K12 extract.

For several years only the gene for rep helicase was known (2, 3). Furthermore, the rep enzyme was the only Escherichia coli helicase with a known function, i.e. that in the replication of ϕX174 and fd DNA. Recently the genes encoding the helicases I and II have been identified. Considerations regarding the biological functions of these enzymes have thus become possible. Such speculations, which also affect our understanding of the physiological role of the rep helicase, will be presented here. We will disregard helicase III as nothing is yet known concerning the biological role of this enzyme.

Helicase I has recently been shown to be the product of traI, one of the 20 or 30 so-called transfer genes of the Escherichia coli F sex plasmid (4). These transfer genes, which form an operon on the F factor DNA, specify functions needed for transferring DNA between conjugating Escherichia coli cells (5). Most of their products (which include, for instance, the protein of F pili) are bound to the bacterial membrane. Helicase I is one of the few products of a transfer gene which is found in the cytoplasm of artificially lysed, male Escherichia coli cells.

Exchange of DNA between conjugating bacteria requires transfer of one of the F factor strands with its 5'-end first from a donor to a recipient cell. A site- and strand-specific nuclease, apparently the product of the transfer genes Y and Z (6), seems to nick the F factor DNA in a position near the transfer region. We assume that this nicking of the F factor DNA at the origin of transfer is followed by action on the DNA of helicase I, i.e. that helicase I serves to unwind the F factor DNA for transfer. Since the helicase unwinds DNA in the 5'-to 3'-direction of the bound DNA strand, it would proceed along the DNA strand to be transferred. Transfer of the F factor DNA is accompanied by DNA synthesis: both the F factor strand which remains in the donor and that which is transferred to the recipient are converted, through the synthesis of complementary DNA strands, to double-stranded DNA. Since transfer can be uncoupled from transfer-induced DNA synthesis the driving force of the transfer reaction is not derived from replication (7, 8). Furthermore, since helicase I is found predominantly in the cytoplasmic fraction of male Escherichia coli cells we find it difficult to believe that the energy required for transferring F factor DNA through the membranes of conjugating bacteria derives from action of the helicase. Transfer-induced synthesis of F factor DNA can be distinguished from vegetative replication of the F factor. Transfer-induced DNA synthesis and vegetative replication of the F factor DNA proceed from different origins on the F factor. Moreover, traI deficient F factor, which cannot be transferred, is still capable of replicating. Helicase I is thus not needed for replicating the F factor. It would be of interest to know wheter the recently discovered single strand binding protein of the F factor (9) is involved in any of these reactions.

In contrast to helicase I the second helicase of Escherichia coli is encoded by a chromosomal gene, namely uvrD. This correlation of helicase II with uvrD was established by molecular cloning studies. As the results of these studies showed, selection for uvrD complementing activity (10-12) and selection for a gene product with the antigenic character of helicase II (13) yield DNA clones of the same type as judged on the basis of restriction analysis. Biochemical studies on the products of the cloned DNA, the application of antibody against helicase II to the product of uvrD (14, 15) and complementation assays performed with the DNA encoding helicase II (13) have meanwhile confirmed that helicase II is the product of uvrD.

uvrD mutations are characterized by wide-ranging pleiotropy. The mutants are sensitive to ultraviolet (UV) light, ionizing irradiation and alkylating agents. Unlike incision mutants of the uvrA, B, C type the uvrD mutants are capable of incising damaged DNA. DNA breaks in these mutants, however, are closed only slowly (16). uvrD mutants are further characterized by hyperactivity in genetic recombination (17, 18), a mutator phenotype (19, 20), and a defect in the correction of mismatched DNA bases (21, 22). Recent in vitro studies have shown that, in Escherichia coli mismatch-repair of DNA bases requires recognition not only of the mismatched bases but also that of a methylation signal which may be several hundred nucleotides apart on the DNA (22). It is tempting to speculate that the coordination of this identification of two widely separated signals is mediated, in some way, by the helicase. It should be noted that mutations in uvrD are allelic with those in uvrE, recL, mutU and pdeB (23).

The identification of helicase II as the product of a uvr gene came to us somewhat as a surprise as our previous studies had revealed that, by applying antibody against helicase II to crude subcellular Escherichia coli systems, it is possible to inhibit the replication of Escherichia coli DNA as well as that of the DNAs of phage λ and plasmid ColE1 (24). For elongation synthesis the two latter DNA species make use of bacterial enzymes. Specificity of the antibody effect was indicated by the fact that the inhibition did not extend to the replication of fd DNA which depends on rep helicase, as already mentioned.

On the other hand, it was also noted in these studies that the replication of Escherichia coli DNA and λ DNA is not completely inhibited by antibody raised against helicase II - as if some complementing activity in the replication system were capable of substituting for the defective helicase. In retrospect we realize that there are parallels between these results and those of others who have studied the growth properties of rep mutants. Although rep mutation does not abolish chromosomal replication it reduces its rate (25). It should be added that rep mutants are slightly more sensitive to UV-irradiation than are normal cells (26).

Our studies gained a new aspect when it was found, by P1 transduction, that uvrD mutation is apparently incompatible with rep mutation. This result, originally obtained for the classical mutations uvrD502 and rep3 (R. Moses, unpublished results), is consistent with data obtained recently for two other mutations, namely uvrD210 and rep71 (H. Arthur, unpublished results). Using these mutations we have not succeeded in transducing a rep defect to a uvrD cell and a uvrD defect to a rep cell. We thus suspect that, in the replication of Escherichia coli DNA, it might be rep helicase which substitutes for a defective helicase II while helicase II substitutes for a defective rep.

Following this idea we have proposed, as a working hypothesis (27), that the chromosomal fork of Escherichia coli is shifted through the action of both helicase II and rep helicase. Since the two helicases act in different directions we expect them to be bound to opposite sides of the fork, namely rep helicase to the leading parental strand and helicase II to the lagging parental strand (Figure 1). uvrD⁻ cells and rep⁻ cells have no serious difficulties in replicating their chromosomes; accordingly, both helicases would be capable of moving the fork on their own. Loss of the activities of both uvrD and rep, however, would be critical. These considerations might hold for the DNA of λ as that of Escherichia coli.

This hypothesis, i.e. the assumption that helicase II and rep helicase fulfil redundant functions at the chromosomal fork, is not implausible if one assumes that the cell might be prepared to deal with occasional disruption of the fork. Disruption would occur if the advancing fork meets with a single strand interruption in the template duplex - for instance an incision break resulting from incomplete repair of a UV-lesion. As a consequence of this encounter one of the arms would be disconnected from the fork, together with the helicase bound on the damaged side of the replicating duplex. In the case of fork breakage it might be essential to continue replication, i.e. to complete at least one of the daughter chromosomes to provide a continuous template for the subsequent round of chromosomal replication.

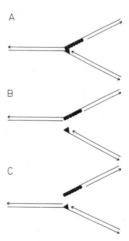

Figure 1: A. Hypothetical scheme of the chromosomal replication fork of Escherichia coli. The proteins shown are helicase II (●) and rep helicase (▲). Arrows indicate 5'-to 3'-direction of DNA. B. State after disconnection of the leading arm from the fork. C. State after disconnection of the lagging arm.

Since the helicase released from the fork fails to find single-stranded DNA beyond the break it would be unable to re-initiate. An additional helicase bound to the intact side of the disrupted fork, however, might be capable of guiding the replication process through the region of DNA damage. By displacing single-stranded DNA beyond the break this enzyme would create an initiating site for the detached helicase. As consequence, a replication complex including both helicases would be re-constituted beyond the break.

One can express these considerations also in a negative form. We believe that if the fork were formed with only one helicase (as is presumably the case with uvrD⁻cells and rep⁻cells) it would risk permanent arrest at a single strand break in the template duplex. Unwinding would terminate if an interruption in the template duplex deprives the fork of the arm which carries the helicase.

Accordingly, uvrD⁻ and rep⁻ mutants would suffer from replication defects. As these defects would affect predominantly the replication of damaged DNA these mutants are phenotypically repair-defective. The fact that the sensitivity to DNA damage is higher in uvrD⁻ cells than in rep⁻cells could be explained by assuming that unlike rep helicase, helicase II is used also for repairing primary DNA lesions. The recently discovered role of the uvrD function in the correction of mismatched DNA bases is consistent with this interpretation as is the fact that the uvrD function is inducible. Treatment of cells with nalidixic acid or mitomycin C, typical SOS inducing agents, increases the level of helicase II ATPase up to fivefold (15) and also increases the level of β-galactosidase when the lac operon is fused to the uvrD gene (28).

On the other hand, because of the asymmetry inherent in the organization of a replication fork the roles of helicase II and rep helicase at the fork are not strictly comparable. Moreover, the mechanisms of action of the two helicases are different, as already mentioned. It is therefore conceivable that the problems posed by a disrupted fork are different depending on whether the leading arm or the lagging arm is disconnected from the fork and whether disruption occurs in a uvrD⁻cell or a rep⁻cell. Interestingly enough, all effects we can conceive of would disfavour a uvrD⁻cell relative to a rep⁻cell (27).

We will try to test these predictions by isolating temperature-sensitive uvrD and rep mutants and by testing these mutations in combination with a defect for what we believe to be the complementing helicase. Unfortunately, no temperature-sensitive mutants with sufficiently unambiguous lesions have thus far been found. Nevertheless, genetic analysis seems presently to be the most promising means to prepare for further biochemical studies on the Escherichia coli helicases.

REFERENCES

1.   Geider,K. and Hoffmann-Berling,H. (1981) Ann. Rev. Biochem.
         50, 233.
2.   Denhardt,D.T., Dressler,D.H. and Hathaway,A. (1967)
         Proc. Natl. Acad. Sci. U.S.A. 57, 813.
3.   Tessman,I., Fassler,J.S. and Benneth,D.C. (1982) J. Bact.
         151, 1637.
4.   Abdel-Monem,M., Taucher-Scholz,G. and Klinkert,M.-Q. (1983)
         Proc. Natl. Acad. Sci. U.S.A. 80, 4659.
5.   Manning,P.A. and Achtman,M. (1979) in: "Bacterial Outer
         Membranes", M. Inouye, ed., Wiley & Sons, Inc., New York.
6.   Everett,R. and Willets,N. (1980) J. Mol. Biol. 136, 129.
7.   Sarathy,P.V. and Siddiqi,O. (1973) J. Mol. Biol. 78, 443.
8.   Kingsman,A. and Willets,N. (1978) J. Mol. Biol. 122, 287.
9.   Kolodkin,A.L., Capage,M.A., Golub,E.I. and Low,K.B. (1983)
         Proc. Natl. Acad. Sci, U.S.A. 80, 4422.
10.   Oeda,K., Horiuchi,T. and Sekiguchi,M. (1982) Nature 298, 98.
11.   Arthur,H.M., Bramhill,D., Eastlake,P.B. and Emmerson,P.T.
         (1982) Gene, 19, 285.
12.   Maples,V.F. and Kushner,S.R. (1982) Proc. Natl. Acad. Sci. U.S.A.
         79, 5616.
13.   Taucher-Scholz,G. and Hoffmann-Berling,H. (1983)
         Eur. J. Biochem. 137, 573.
14.   Hickson,J.D., Arthur,H.M., Bramhill,D. and Emmerson,P.T.
         (1983) Mol. Gen. Genet. 190, 265.
15.   Kumura,K., Oeda,K., Akiyama,M., Horiuchi,T. and Sekiguchi,M.
         (1983) in: Cellular responses to DNA damage, Friedberg,
         E.C. and Bridges,B.R., eds. UCLA Symposia on Molecular
         and Cellular Biology, New Series, Volume 11, Alan R. Liss,
         Inc., New York, in press.
16.   Rothman,R.H. and Clark,A.J. (1977) Mol. Gen. Genet. 155, 267.
17.   Arthur,H.M. and Lloyd,R.G. (1980) Mol. Gen. Genet. 180, 185.
18.   Lloyd,R.G. (1983) Mol. Gen. Genet. 189, 157.
19.   Smirnov,G.B., Filkova,E.V. and Skavronskaya,A.G. (1972)
         Mol. Gen. Genet. 118, 51.
20.   Siegel,E.C. (1981) Mol. Gen. Genet. 184, 526.
21.   Nevers,P. and Spatz,H.C. (1975) Mol. Gen. Genet. 139, 233.
22.   Lu,A.L., Clark,S. and Modrich,P. (1983) Proc. Natl. Acad. Sci.
         U.S.A. 80, 4639.
23.   Kushner,S.R., Shaepherd,J. Edwards,G. and Maples,V.F. (1978)
         in: "DNA Repair Mechanisms", P.C. Hanawalt, E. Friedberg
         and C.F. Fox, eds., Academic Press, New York.
24.   Klinkert,M.-Q., Klein,A. and Abdel-Monem,M. (1980) J. Biol.
         Chem. 255, 9746.
25.   Lane.H.F.D. and Denhardt,D.T. (1975) J. Mol. Biol. 97, 99.
25.   Calendar,R., Lindqvist,G., Sironi,G. and Clark,A.J. (1970)
         Virology 40, 72.

27.   Taucher-Scholz,G., Abdel-Monem,M. and Hoffmann-Berling,H.
         (1983) in: "Mechanisms of DNA Replication and Recombina-
         tion", UCLA Symposia on Molecular and Cellular Biology,
         New Series, Volume 10, N.R. Cozzarelli, ed., Alan R. Liss,
         Inc., New York, in press.
28.   Arthur,H.M. and Eastlake,P.B. (1983) Gene, in press.

ESCHERICHIA COLI DNA GYRASE

Elisha Orr, Heinz Lother[1], Rudi Lurz[1] and Elmar Wahle[2]

Department of Genetics
Leicester University
Leicester LE1 7RH, U.K.

[1]Max-Planck-Institut für Molekulare Genetik
Ihnestrasse 63-73
D-1000 Berlin 33, FRG

[2]Zoologisches Institut der Universität Münster
Badestrasse 9
D-4400 Münster, FRG

INTRODUCTION

The role of DNA gyrase -a type II topoisomerase- had been spe-
culated about for a long time before the discovery of the enzyme. In
1963, J. Cairns (1) pointed out that the Escherichia coli chromosome
is a closed circular double strand DNA molecule. It has consequently
become apparent that a replication machinery which has to unwind the
two strands would generate a positive swivel in front of the advancing
replication fork. A topoisomerase type II could, therefore, release
this tension. The second problem relating to DNA gyrase was the observation that coumarins, e.g. novobiocin, and drugs like nalidixic
acid inhibit the B and the A subunits of gyrase respectively. Previous studies have already shown that these drugs also block DNA
replication in vivo. It is therefore not too surprising that the role
of this enzyme seemed clear when it was discovered by M. Gellert and
Colleagues at the N.I.H. (2). This picture however, turned out to be
more complicated when further studies revealed that the two families
of drugs inhibit many other cellular processes, all dependent on the
DNA template. DNA gyrase has now become a most promiscuous protein,
participating in different molecular pathways such as replication,
transcription, transposition and recombination (for review see 3).

Inactivation of DNA gyrase can affect cellular events in two ways. It can either inhibit directly reactions which require double strand nicking-closing activity e.g. DNA decatanation, or it can change the superhelical density of the DNA which might further affect some cellular activities. We hoped that the isolation of bacterial mutants defective in gyrase activity would differentiate between these two mechanisms of action and perhaps enable us to clarify whether the inhibitory effects of the drugs, mentioned above, reflect a genuine need for gyrase in vivo.

RESULTS

Characterisation of mutants

The most important feature of a given mutant is to find out whether the activity of the concerned enzyme is defective both in vivo and in vitro. Figure 1 demonstrates that 5 min incubation at the restrictive temperature is sufficient to partially relax all form I pBR322 DNA in strain LE316 (gyrB ts), in vivo. The same result has

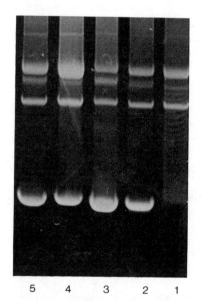

5   4   3   2   1

Figure 1: The effect of temperature on plasmid DNA in vivo. Strains LE 316 (gyrB ts) and LE234 (gyrB⁺) carrying plasmid pBR322 were grown at 30°C and transferred to 42°C. Plasmid DNA was extracted at various times and analysed by 0.8% agarose gel electrophoresis.
1- LE234 at 30°C; 2- LE234, 5 min at 42°C; 3- LE234, 30 min at 42°C; 4- LE316 at 30°C; 5- LE316, 5 min at 42°C.

been recently described by Lockson and Morris (4). Likewise, a super-infection of strain LE316λ⁺ by phage λ followed by a temperature shift, results in its inability to supercoil the λ DNA (Gellert, personal communication), consistent with the failure of λ to repli-cate in this strain (5).

Incubation of supercoiled ColEl DNA in the mutant cell free extract leads to a rapid relaxation (within 5 min) of the plasmid form I DNA (Figure 2) (6). This relaxation occurs even at permissive temperatures, indicating that the enzyme is unstable once the mutant cells are lysed.

Finally, DNA gyrase from LE316 has been purified to homogeneity by affinity chromatography on novobiocin sepharose. Both the ATPase (Orr, unpublished) and the supercoiling activities of the enzyme are completely abolished at temperatures above 37°C (Figure 3). The mutant cell free extract contains a 45K protein which reacts with anti-gyrase B serum. It is yet unclear whether this fragment is a degra-dation product of the B subunit, or is a "nonsense" fragment encoded by the mutated gene.

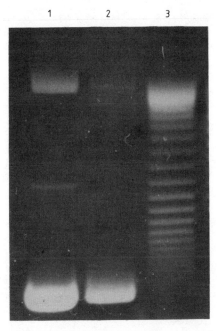

Figure 2: ColEl DNA in cell free extract. ColEl DNA was incubated at 30°C in cell free extract prepared from strains LE316 (gyrB ts) and LE234 (gyrB⁺). Plasmid DNA was re-extracted with phenol, after 5 min incubation, and analysed by 0.8% agarose gel electrophoresis. 1- ColEl DNA, no extract; 2- LE234 extract; 3- LE316 extract.

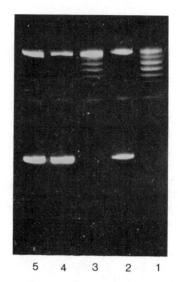

5   4   3   2   1

Figure 3: The effect of temperature on the activity of DNA gyrase. DNA gyrase from strains LE316 (gyrB ts) or LE234 (gyrB⁺) was purified by affinity chromatography on novobiocin-sepharose and incubated with relaxed ColEl DNA (prepared with Escherichia coli topoisomerase I). After 30 min incubation the DNA was analysed by 0.8% agarose gel electrophoresis. 1- template DNA, no enzyme; 2- LE316 enzyme at 30°C; 3- LE316 enzyme at 38°C; 4- LE234 enzyme at 30°C; 5- LE234 enzyme at 38°C.

DNA replication

Three classes of conditional lethal mutants have been analysed for their ability to sustain DNA replication at the non-permissive temperature. These are mutants defective in the B subunit (5, 6), in the A subunit (6, 7), or in both the A and the B subunits of DNA gyrase (Orr, unpublished). All these mutants are defective in initiation of DNA replication. DNA chain elongation, however, does not seem to be substantially affected in all these strains (Orr, unpublished). Recently a new gyrB ts mutant, allegedly defective in DNA chain elongation in vivo, has been described (8). Unfortunately, this mutant has not been properly characterised as being in gyrB. Moreover, none of the essential experiments which make it possible to determine whether a mutant is defective in DNA chain elongation or not (see ref. 9) have been carried out.

The inactivation of DNA gyrase could lead to an abortive termination of DNA replication. To clarify this point, we pulse-labelled the DNA in strain LE316 with radioactive thymidine at the non-permissive temperature. The DNA was then analysed on sucrose gra-

dients. All radioactive material could be chased to a high molecular
weight DNA.

The failure of ColEl to replicate in extracts prepared from
strain LE316, and the need for gyrase in the initiation of oriC in
vitro (10), further underline the importance of gyrase in this event.
In order to differentiate between the need for gyrase either during
initiation or during the other steps of ColEl replication, we added
gyrase inhibitors or anti-gyrase IgG after the initiation of ColEl
DNA had been established. The latter can be monitored by its sensitivity to rifampicin. Under these conditions, there is sufficient
DNA replication and the products are covalently closed monomer molecules (to be published elsewhere).

Effects of gyrase inactivation on gene expression

Several papers have suggested that DNA gyrase might modulate
gene expression both in vivo and in vitro. We have reinvestigated
the effect of gyrase inactivation in strain LE316 (gyrB ts) on the
activities of various promoters. The result of one such an experiment
using the lac promoter is shown in Figure 4. At the non-permissive
temperature, the expression of this promoter is not differentially
affected when it is present either on the chromosome or cloned into
a plasmid (pBR322). None of the other promoters tested in the mutant
strain, except the lacIq mutant promoter, was affected in the same
way that they were affected by gyrase inhibitors (Wahle, Müller and
Orr, submitted; see ref. 3 for the effects of gyrase inhibitors on
transcription).

The activities of some promoters in strain LE316 could however
be stimulated depending on their location on the chromosome or their
structure. The tnaA promoter is activated only when present at the
oriC region of the chromosome. Other promoters like the mutant lacIq
or the non-functional wild-type bgl promoter can be relatively stimulated upon inactivation of gyrase (Wahle, Müller and Orr, submitted). We have already reported that an extract prepared from the
gyrB mutant to promote a coupled transcription-translation does not
produce the same results as a wild-type extract treated with gyrase
inhibitors (11).

Purification of DNA gyrase

To understand better the interaction of DNA gyrase with its
substrates, we purified large quantities of the enzyme by affinity
chromatography on novobiocin sepharose. Subsequent studies with the
purified enzyme, and with anti-subunit A or B IgG enabled us to draw
several conclusions reading the enzyme.

1.   In contrast to previous published data (12, 13), DNA gyrase
      is present in thousands of molecules per cell. This puts doubts

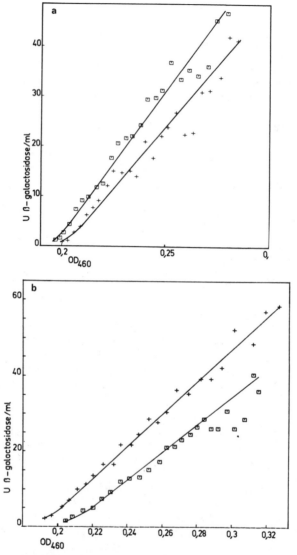

Figure 4: *The activity of the* lac *promoter. LE316 (gyrB, lac+) and*
LE234 (lac+) were grown at 31°C with glycerol as a carbon source. IPTG
was added to a concentration of 1 mM and the cultures were either left
at 31°C (a) or transferred to 42°C (b). β-galactosiodase activity was
measured every 2 min after induction. Crosses - LE234; squares - LE316.

in the necessity for regulating the respective genes in vivo
(14). It is also quite clear that there is an equal number of
the two subunits in the cell.

2.   The ν protein which has been reported as a new topoisomerase
        (13, 15), is most likely a degradation artefact of the gyrB
        subunit arising during the purification procedure and is not
        present in intact cells (Figure 5). This protein reacts with
        anti-gyrase B IgG (Orr, unpublished). The appearance of this
        fragment (Figure 5) and many others by prolonged storage,
        probably reflects the instability of the gyrase B protein du-
        ring purification.
3.   Each of the gyrase subunits has an independent activity, in
        contrast to previous reports (16, 17; see also 3). The B sub-
        unit has a DNA independent ATPase (18) while the A subunit is
        a DNA binding protein (Lother, Lurz and Orr, submitted).
4.   The reconstitution of holoenzyme in 25 mM hepes pH 8.0, 100 mM
        KCl is poor and requires at least 5-10x more gyrase B than A.

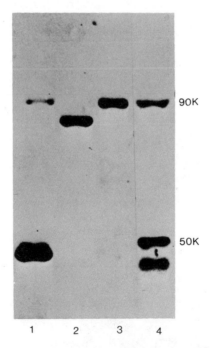

*Figure 5: The activity of anti-Escherichia coli gyrase B serum.
Bacterial cell free lysates were electrophorised on 10% (SDS) acrylamide gel and blotted on to a nitrocellulose paper (Western blotting).
The paper was incubated with rabbit anti-Escherichia coli gyrase B
serum and with goat anti-rabbit fluorescin IgG. 1. LE316 (gyrB)
lysate; Culture was grown at 30°C; 2. B.subtilis lysate; 3. Escherichia coli lysate prepared freshly from exponentially grown culture
at 37°C; 4. As in 3 but lysate was stored on ice for 5 hours after
lysis.*

Interaction of gyrase with DNA

          In an attempt to understand the interaction of gyrase with DNA,
we studied the binding of the enzyme to the Escherichia coli minichromosome pOC51 (19), containing oriC. Several workers have previously suggested that gyrase might recognise specific sequences on DNAs
(for review see 3). This was mainly based on the double strand cleavage of DNA in the presence of the enzyme, oxolinic acid and detergent. We first tried this approach. A similar treatment of the DNA
produced only one linear fragment. The fragment was purified and
treated with either of the restriction enzymes SmaI or XhoI which
cut the minichromosome only once. Figure 6 demonstrates that several
DNA bands appear when the products are re-run on agarose gel. Thus,
although gyrase cleaves the minichromosome once under our experimental conditions, it can do so at many different sites. This result was
confirmed by electron microscopy studies. Gyrase can bind through the
activity of the A protein to many sites on the minichromosome. Nevertheless, only one, located at 235 bp (20) (Figure 7) was a prominent
and a reproducible site in all the experiments carried out under
optimal conditions. DNA gyrase binds at oriC when the DNA is either
supercoiled, relaxed or linear (Figure 8). This indicates that the
enzyme recognises a sequence rather than a structure on the DNA

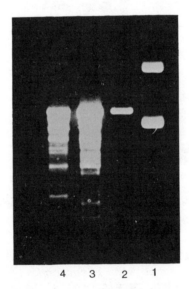

Figure 6: Cleavage of pOC51 DNA. pOC51 (lane 1) was incubated with
purified gyrase and oxolinic acid. Linearisation of DNA was induced
by SDS and proteinase K and the linear product was purified from
agarose gel. The linear DNA (lane 2) was digested by either Xho I
(lane 3) or SmaI (lane 4) and analysed by 1.5% agarose gel electrophoresis.

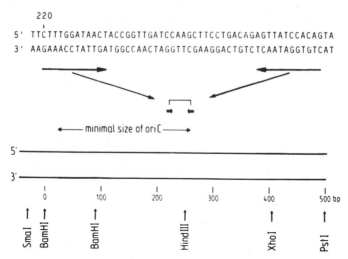

Figure 7: Location of the preferential gyrase binding site within oriC as identified by electron microscopy. The DNA sequence of the estimated gyrase binding centered around bp 235 is shown. The most prominent repeats thought to be involved in gyrase recognition are indicated by arrows. Its location within oriC is indicated by arrows connected by a bracket. The extent of the minichromosomal oriC region and some endonuclease restriction cut sites in oriC are given below.

(Lother, Lurz and Orr, submitted). The binding to this sequence only, can not be detected in the presence of oxolinic acid, the ATP analogue β-γ-imido ATP, or when B. subtilis DNA gyrase replaces the Escherichia coli enzyme (Figure 8).

The immunological properties of different gyrases

To further elucidate the differences between various DNA gyrases, we used serum raised against Escherichia coli A or B subunits. The purified B. subtilis subunit B fully reacts with the anti- Escherichia coli subunit B serum when analysed by either Western blotting (21) or by solid phase radioimmunoassay (22). No cross reactivity however, could be detected between the B. subtilis subunit A and the anti Escherichia coli subunit A serum by the Western blotting technique. A limited crossreactivity (15%) could be demonstrated by the solid phase radioimmunoassay method (Lother, Lurz and Orr, submitted).

DISCUSSION

The studies with Escherichia coli DNA gyrase have suffered from a considerable number of drawbacks during the recent years. Some of

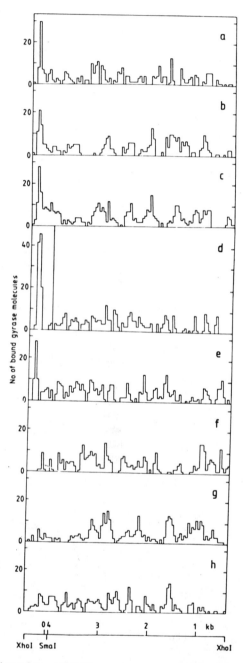

Figure 8: *Localisation of DNA gyrase on pOC51 DNA. pOC51 DNA was incubated with purified gyrase at 30°C for 10 min. Incubation was stopped by the addition of glutaraldehyde. Samples containing circu-*

lar DNA were split after the removal of glutaraldehyde by chromato-
graphy on sepharose 4B. Half the DNA was linearised with the restric-
tion enzyme XhoI and half with SmaI. In each experiment about 300
molecules with 1.0 - 2.0 DNA gyrase molecules bound on each were mea-
sured. Orientation of measured molecules and determination of peak
positions were primarily obtained by both computer program and manual-
ly, comparing (to best fit) DNA molecules cut independently with XhoI
and SmaI. All histograms (1 interval = 41 bp) shown are from DNAs
which were cut with XhoI for electron microscopic preparations. The
orientation of gyrase complexes on the histograms was however deter-
mined by DNA molecules cut independently by XhoI and SmaI. DNA gyrase
was bound to supercoiled DNA (a) to covalently closed relaxed (b)
and to linear DNA (c). In histogram (d) DNA was cut with XhoI and
SmaI into two fragments for the assay. Histograms of both fragments
are shown joined together. For evaluation of the small fragment only
130 molecules were used. Histogram 3e shows an identical experiment
as in (a), but without ATP, and in (f) ATP was replaced by β-γ-imido
ATP. Oxolinic acid was added to the binding assay at a concentration
of 200μg/ml in (g). B. subtilis subunits A and B instead of Escheri-
chia coli enzyme were bound to supercoiled DNA in histogram (h).

the past observations have been hastily put into the literature without sufficient follow up. Worse than that, several researchers have
tried very hard to fit their data into the postulated role of DNA
gyrase (see introduction). DNA supercoiling seems to be a necessary
component for almost every biological activity dependent on DNA.
Nevertheless, two sets of experiments confirm that inactivation of
DNA gyrase by drugs is not comparable to the temperature inactivation
of the enzyme; those related to DNA replication and to transcription.
The reason for this discrepancy has been suggested previously (5, 7).
Both novobiocin and oxolinic acid enhance the formation of stable
gyrase-drug-DNA complexes.
These complexes (in some of which the enzyme might be covalently
linked to DNA) could undoubtedly present obstacles for travelling
enzymes involved in either DNA replication or transcription. Furthermore, nalidixid acid has another significant effect on DNA. It induces DNA breakage in vivo which is probably capable of altering
biological activity of the template. Although this cleavage seems to
be mediated by gyrase, it is different from the inhibitory effect of
drug on the supercoiling activity. The latter is blocked by much
higher concentrations of nalidixic or oxolinic acids (23). The hypothesis that a metabolised form of these compounds is responsible for
gyrase inactivation in vivo, is not plausible. The low concentrations
of nalidixic acid which affect cellular growth in vivo, do not seem
to induce plasmid relaxation (23).

So far, there is only one activity which could reflect the need
for the enzyme in vivo. This is its interaction with oriC. The pre-
ferred binding of gyrase to a sequence within oriC may indicate an

active role of the enzyme during the initiation of DNA replication,
analogous to the proposed function of the T4 topoisomerase type II
during the initiation of T4 replication (24). The inhibitory effect
of gyrase inactivation on the initiation of DNA replication in vivo,
in the temperature-sensitive gyrase mutants, strongly support this
idea. The site specific cleavage induced by oxolinic acid has provided a lot of work for many scientists (see ref. 3). It is however
unlikely to reflect the specificity sought in the binding of gyrase
to the template DNA.

Finally, the differences between the DNA binding proteins, the
A subunits, of various microorganisms reinforce the importance in
balancing the supercoiling density of DNA prior to its replication.
A failure of subunit A to recognise efficiently a particular replicon would lead to its relaxation and loss from the host.

ACKNOWLEDGEMENTS

We wish to thank all our colleagues who encouraged us by ex-
pressing faith in our work. We are especially endebted to Prof. I.B.
Holland, Prof. R.H. Pritchard, Prof. K. Müller, Dr. T. Trautner,
Dr. W. Messer and last but not least to Dr. J. Pratt. It is a pleasure
to thank Beate Rueckert for her excellent assistance in the electron
microscopy experiments and Dr. T. Harrison for his assistance with
the immunological experiments. This work was supported by grant
No. G80/OO84/OCB(E.O) from the Medical Research Council (U.K.) and
by a short term EMBO awards (E.W.; E. O).

REFERENCES

1.   Cairns,J. (1963) J. Mol. Biol. 6, 308.
2.   Gellert,M., Mizuuchi,K., O'Dea,M.H. and Nash,H.A. (1976)
           Proc. Natl. Acad. Sci. USA 73, 3872.
3.   Gellert,M. (1981) Ann. Rev. Biochem. 50, 879.
4.   Lockson,D. and Morris,D.R. (1983) Nucl. Acids Res. 11, 2999.
5.   Orr,E., Fairweather,N.F., Holland,I.B. and Pritchard,R.H.
           (1979) Mol. Gen. Genet. 177, 103.
6.   Orr,E. and Staudenbauer,W.L. (1981) Mol. Gen. Genet. 181, 52.
7.   Kreuzer,K.N. and Cozzarelli,N.R. (1979) J. Bacteriol. 140, 424.
8.   Filutowicz,M. and Jonczyk,P. (1983) Mol. Gen. Genet. 191, 282.
9.   Pritchard,R.H. and Zaritsky,A. (1970) Nature 226, 126.
10.   Fuller,R.S., Kaguni,J.M. and Kornberg,A. (1981) Proc. Natl.
           Acad. Sci. USA 78, 7370.
11.   Pratt,J.M., Boulnois,G.J., Darby,V., Orr,E., Wahle,E. and
           Holland,I.B. (1981) Nucl. Acids Res. 9, 4459.
12.   Higgins,N.P., Peebles,C.L., Sugino,A. and Cozzarelli,N.R.
           (1978) Proc. Natl. Acad. Sci. USA 75, 1773.
13.   Brown,P.O., Peebles,C.L. and Cozzarelli,N.R. (1979)
           Proc. Natl. Acad. Sci. USA 76, 6110.

14. Menzel,R. and Gellert,M. (1983) Cell 34, 105.
15. Gellert,M., Fisher,L.M. and O'Dea,M.H. (1979) Proc. Natl. Acad. Sci. USA 76, 6289.
16. Sugino,A., Higgins,N.P., Brown,P.O., Peebles,C.L. and Cozzarelli,N.R. (1978) Proc. Natl. Acad. Sci. USA 75, 4838.
17. Higgins,N.P. and Cozzarelli,N.R. (1982) Nucl. Acids Res. 10, 6833.
18. Staudenbauer,W.L. and Orr,E. (1981) Nucl. Acids Res. 9, 3589.
19. Lother,H., Buhk,H.J., Morelli,C., Heimann,B., Chakraborty,T. and Messer,W. (1981) ICN-UCLA symp. Mol. and Cell. Biol. 22, 57.
20. Meyer,M., Beck,E., Hansen,F., Bergmans,H.E.C., Messer,W., von Meyenburg,K. and Schaller,H. (1979) Proc. Natl. Acad. Sci. USA 76, 580.
21. Towbin,H., Staehelin,T. and Gordon,J. (1979) Proc. Natl. Acad. Sci. USA 76, 4350.
22. Micheel,B., Karsten,U. and Friebach,H. (1981) J. Immunol. Meth. 46, 41.
23. Kano,J., Miyashita,T., Nakamura,H., Kuraki,K., Nagata,A. and Imamoto,F. (1981) 13, 173.
24. Liu,L.F., Liu,C.C. and Alberts,B.M. (1979) Nature 281, 456.

A UNIQUE ATP-DEPENDENT DNA TOPOISOMERASE FROM TRYPANOSOMATIDS

Joseph Shlomai, Anat Zadok and Dale Frank

The Kuvin Centre for the Study of Infectious and
Tropical Diseases
The Hebrew University-Hadassah Medical School
Jerusalem, Israel 91010

ABSTRACT

Crithidia fasciculata DNA topoisomerase (22) has been purified
to near homogeneity from trypanosomatid cell extracts. The purified
enzyme catalyzes the reversible interconversion of monomeric duplex
DNA circles and catenanes in an ATP dependent reaction. Reversible
catenane formation is affected by the ionic strength and is dependent
upon the action of a crithidial DNA binding protein, which could be
substituted for the polyamine spermidine. Covalently sealed DNA
circles are specifically used as substrates for decatenation. Nicking,
but not relaxation per se, inhibits network decatenation and has
little or no effect on catenane formation. The catalytic properties
of this enzyme and its potential role in the prereplication release
and post replication reattachment of kDNA minicircles are discussed.

INTRODUCTION

Kinetoplast DNA (kDNA) is a unique DNA structure present in
the single mitochondrion of flagellate protozoa of the order
Trypanosomatidae. It consists of thousands of small duplex DNA rings
heterogenous in their nucleotide sequences, known as minicircles
(0.8-2.5 kb pairs each, depending on the species), and a few larger
DNA circles (20-40 kb pairs) known as maxicircles. Minicircles and
maxicircles are organized as a three dimensional network and are
interlocked to form a catenane (reviewed in references 1-3). It has
been suggested (3-5) that kDNA minicircles are not replicated while
attached to the network. Instead, they are released from the network

409

and replicate as free minicircles through a Cairns type mechanism
(6-8). The resulting progeny DNA minicircles which are nicked, reat-
tach to the network periphery. The network increases in size until
it doubles and then segregates into two daughter networks. Such a
model for kDNA network replication assumes an enzymatic system which
releases minicircles from the network prior to their replication and
subsequently reattaches the progeny minicircles to the DNA network
(3). Enzymatic activities capable of interconverting monomeric circles
and catenanes have been described in various systems including extracts of Xenopus laevis oocytes (9, 10), T4-infected Escherichia
coli cells (11, 12), Drosophila embryos (13), Escherichia coli and
M. luteus (14-18) HeLa cells (19) and Yeast (20) (reviewed by Gellert
in reference 21). These enzymes, classified as type II DNA topoisomerases, can catalyze the topological passage of one double stranded
DNA segment through another, presumably by introducing a transient
enzyme bridged double-stranded break on one of the crossing DNA
segments (12, 17, 16, 13). As a consequence of such a mechanism,
type II DNA topoisomerases can catenate and decatenate covalently
closed DNA circles.

        Considering the catenane nature of kDNA networks, the require-
ment for the prereplication release of replicating kDNA minicircles
from the network, and their subsequent post replication reattachment
to it, we searched for an enzyme in trypanosomatids which can carry
out such "release and reattachment" functions. It was presumed (22)
that such an enzyme would be capable of catalyzing the catenation
and decatenation of double stranded DNA circles, while discriminating
between newly replicated kDNA minicircles as a substrate for catenation and mature ones as the exclusive substrate for decatenation. We
discuss here some of the properties of the ATP dependent DNA topoisomerase from the trypanosomatid Crithidia fasciculata, and its potential role in the course of kinetoplast DNA replication.

Reversible formation of DNA networks is catalyzed by an enzyme
purified from Crithidia cell extracts

        Fractionated Crithidia cell extracts were found to support the
efficient conversion of double stranded DNA circles into huge DNA
aggregates (22). This activity was almost undetectable in crude cell
extracts. However, in cell supernatants made virtually free of DNA,
cell membranes and large insoluble particles (Fraction I), formation
of huge DNA complexes could be detected. Further purification of
this enzymatic activity by ammonium sulfate fractionation (at 40-60%
of saturation at 0°C, Fraction II), and a subsequent step of DEAE-
cellulose chromatography (Fraction III) provided an enzyme fraction
in which a linear and sensitive catenation assay could be carried
out. This enzyme fraction could also catalyze the efficient decatenation of natural kDNA networks into the monomeric minicircle subunits
(Figure 1). Further purification by ion exchange chromatography
on phosphocellulose and BioRex-70 columns (Fractions IV and V, respec-

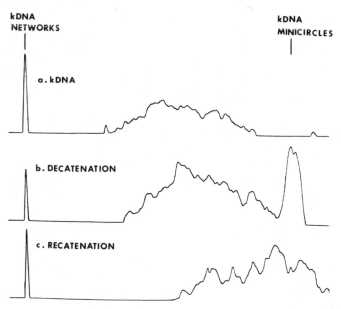

Figure 1: *Reversible decatenation displayed by Crithidia DNA topoiso-merase. 0.7 µg of kDNA (a) was incubated under standard decatenation assay conditions (in 25 µl reaction mixtures containing 20 mM Tris-HCl pH 8.3, 15% (v/v) glycerol, 10 mM MgCl$_2$, 0.5 mM EDTA, 1 mM DTT, 30 µg/ml BSA, 2 mM ATP, and 140 mM KCl; reactions were incubated at 30°C in the pressence of 0.37 units of enzyme). Reactions were stopped by the addition of EDTA to 10 mM, and the DNA products were phenol extracted (b). Minicircles resulting from the decatenation reaction were incubated with the enzyme as above under standard catenation assay conditions (as in (a) except that the concentration of KCl was adjusted to 20 mM, and 7.5 mM spermidine was added), to produce the product in (c). To the phenol extracted DNA solution of these reac-tions, were added: SDS, glycerol and bromophenol blue to 1%, 10% (v/v) and 0.01% respectively. Reaction products were electrophoresed in 1% agarose gels at constant voltage (1V/cm) at room temperature in 40 mM Tris-acetate buffer pH 7.8, 5 mM sodium acetate, 1 mM EDTA and 1 µg/ml ethidium bromide. Quantitation of catenation and decate-nation was by microdensitometry of negative photographs of the DNA bands fluorescence induced by UV light illumination.*

tively) purified the enzyme 6000 fold over the crude cell extract, and a final step of affinity chromatography on Cibacron Blue-Sepharose (Fraction VI) brought the enzyme preparation to apparent homogeneity. Electrophoresis in SDS-polyacrylamide gels under denaturing and re-ducing conditions yields, while using the sensitive silver stain procedure (23), a single polypeptide band of approximately 60,000 daltons. The exact subunit structure of the enzyme and its precise

native molecular weight are currently under study. A detailed description of the purification procedure and the physical characterization of <u>Crithidia</u> topoisomerase will be published elsewhere
(Shlomai and Frank, in preparation).

<u>DNA networks generated by the crithidial enzyme are interlocked
catenanes</u>

     DNA networks generated by the crithidial enzyme were found
to be huge DNA aggregates, which upon electrophoresis failed to
migrate in 1% agarose gels and in electron micrographs resembled
the natural kinetoplast DNA networks found in trypanosomatids. The
possibility that these networks were formed through protein-DNA
associations could be excluded, since treatments with proteolytic
enzymes, detergents or phenol did not affect the network structure
(22). Excluding this possibility, one can consider linkages between
the duplex DNA circles by one or two DNA strand exchanges or by the
formation of linear tandem array of joined DNA molecules. Alternatively, a topological, rather than covalent, linkage between the duplex
DNA rings to form an interlocked catenane could also be considered
(18, 13). The results of our experiments to address this question
have been recently published (22).

     In approaching this question we made use of the observation
that no extensive sequence specificity is involved in the reversible
formation of DNA networks by the crithidial (22), as well as by other,
DNA topoisomerases. Since no homology between the duplex DNA rings
involved was required, φX RF DNA circles could be linked to pBR322
plasmid DNA to form a heterogeneous mixed DNA network. We have digested the DNA networks using EcoRI or BamHI which cleave at a single
site pBR322 (but not φX RF DNA), XhoI which has one cleavage site
in φX DNA (but none in pBR322) and PstI which cleaves both pBR322
and φX DNA at a single site (22).

     The results of these analysis support the topological model,
for the following reasons: (a) No joined structures which would support a model of covalent linking through a single strand exchange
were found; (b) Linear DNA oligomers which were expected if the two
species of DNA were joined in a linear tandem array, were not observed; (c) Linkage of the DNA circles through double strand exchanges
could be eliminated because of the lack of homology of the DNA molecules involved; (d) Digestions of the heterogenous DNA networks using
a restriction endonuclease which introduces a single cut in each of
the rings of one of the DNA species in the networks, yield one unit
length linear molecules of this DNA species. However, such a digestion
also results in the quantitative release of free duplex DNA circles
of the other DNA species in the network. Based on these results we
have suggested that DNA circles within the networks formed in vitro
are linked topologically, rather than covalently to yield an interlocked catenane.

Catenane formation requires the action of a crithidial DNA binding protein

Catenation by DNA topoisomerases is promoted by a variety of polycations including spermidine, histones, and DNA binding proteins (17, 13, 12, 24, 19, 20). Polycations enhance the aggregation of DNA segments (25-30), which presumably faciltates the juxtaposition of DNA required for catenane formation (17, 29). DNA catenation catalyzed by Crithidia cell extract supernatants (Fraction I), the ammonium sulfate fraction (Fraction II) or the DEAE-cellulose enzyme preparation (Fraction III), was found to be independent of an externally added DNA-aggregating factor. However, upon further purification, by the removal of a double stranded DNA binding protein from the enzyme preparation, the catenation reaction becomes absolutely dependent upon the addition of an alternative aggregating factor (22). Such a dependence on DNA binding proteins or histones was also reported for other eukaryotic topoisomerases (13, 19, 20). Catenation by the crithidial enzyme could be supported by the polyamine spermidine, at optimal levels of 5-10 mM.

As was likewise reported for other eukaryotic and prokaryotic DNA topoisomerases (10, 13, 17-20, 30), generation of DNA networks and their decatenation by Crithidia topoisomerase, is affected by the ionic strength. The optima for catenation and decatenation were measured at the potassium chloride concentrations of 20 mM and 140 mM respectively (22), reflecting the effect of the monovalent ions on the aggregation state of the DNA substrate (30).

Reversible decatenation by Crithidia DNA topoisomerase is ATP dependent

The role of ATP in the catalytic activity of type II DNA topoisomerases has not yet been entirely clarified. However, all the type II DNA topoisomerases known require ATP for their catalytic activities. Reversible decatenation by Crithidia DNA topoisomerase was likewise found to be ATP dependent (22) (Figure 2, 3). None of the other common ribonucleoside triphosphates assayed (up to 2 mM) could support the reaction. Deoxy-ATP could substitute for ATP, but was much less efficient (Table 1).

The quantitative hydrolysis of ATP which was observed with less purified enzyme preparations (22) is most likely the result of a contaminating ATPase activity. Pure enzyme preparations (Fraction VI) display a very low ATPase activity (about 0.2% of that observed with Fraction III) (Figure 4). It still has to be determined if the residual ATPase activity is intrinsic to Crithidia ATP dependent topoisomerase.

The nonhydrolyzable β,γ imido and β,γ methylene analogues of ATP, as well as the ATP-γ-S derivative could not support the reaction

DNA NETWORKS   DNA CIRCLES

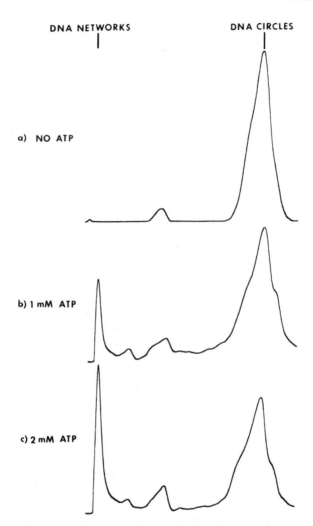

a) NO ATP

b) 1 mM ATP

c) 2 mM ATP

*Figure 2:   ATP dependence of catenation reaction. 0.6 μg pBR322 DNA
was incubated under catenation assay conditions for 1 hr in the
presence of 0.15 units enzyme (Fraction VI), in the absence of ATP
(a) and in the presence of 1 mM (b) and 2 mM (c) of the nucleotide.
Reaction conditions, gel electrophoresis, and microdensitometry were
as described in the legend to Figure 1.*

(Table 1). However, both the β,γ imido and the ATP-γ-S analogues are
bound by the enzyme, since these analogues inhibit the reaction in
the presence of ATP (Shlomai and Frank, to be published).

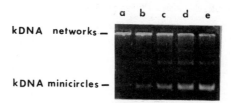

Figure 3: Decatenation dependence on ATP. 0.7 μg kDNA was incubated under decatenation assay conditions for 1 hr in the absence of ATP (a) and in the presence of 0.25 mM (b), 0.50 mM (c), 1.0 mM (d) and 2.0 mM (e) of ATP. Reaction conditions, gel electrophoresis and microdensitometry were as described in the legend to Figure 1.

The spatial arrangements of the carbon-phosphate bonds in the β,γ ATP analogues differ from those of the corresponding phosphonate linkages in ATP. Therefore their effects on the reaction are instructive in understanding the steric requirements for the interaction of phosphate groups with the ATP binding site on the crithidial topoisomerase. It was found that the more acute P-C-P bond angle (117°) of the ATP methylene analogue (31), could not satisfy these requirements and this derivative could neither support the reaction (Table 1), nor inhibit it (not shown). However, the more sterically similar β,γ imido derivative, in which the P-N-P bond form an angle of 127.2° (in comparison to 128.7° of the P-O-P bond angle) (31), could be bound by the enzyme and inhibited the reaction in the presence of ATP. It should be emphasized, that using stoichiometric levels of the enzyme, the reaction catalyzed by <u>Crithidia</u> topoisomerase could not be stimulated by the bound nonhydrolyzable β,γ ATP derivative, even for a single cycle of activity.

A different example of apparent competition on the ATP binding site of the enzyme is demonstrated by the effect of novobiocin on the reaction. This antibiotic drug has no obvious similarity to ATP and yet was found to inhibit most of the ATP dependent topoisomerases (32-34, 10, 13). We have measured a reduction of 50% in the rate of catenation or decatenation catalyzed by the crithidial enzyme in the presence of 0.1 mM of the drug and 0.5 mM of ATP. Under these conditions the rate of the reaction decreases by only 15% when the ATP concentration is raised to 2 mM (22) (Table 1). These values are similar to those reported for the other eukaryotic ATP-dependent DNA topoisomerases. As has been suggested for DNA gyrase, this might indicate competition between the inhibitory drug and ATP for the same site on the enzyme molecule.

Table 1: Nucleotide specificity of <u>Crithidia</u> topoisomerase

Nucleotide	Relative rates of reaction
ATP	100
APP(CH$_2$)P	<5
APP(NH)P	<5
ATP-γS	<5
ATP (2 mM) + Novobiocin (0.1 mM)	85
ATP (o.5 mM) + Novobiocin (0.1 mM)	50[a]
CTP	<5[b]
GTP	<5[b]
UTP	<5[b]
dATP	48[b]

a) A relative rate of decatenation in the presence of 0.5 mM ATP.
b) These values are from Shlomai and Zadok (1983) (22).

Nucleotide specificity of Crithidia *topoisomerase. Decatenation
reactions were carried out as described in the legend to Figure 1.
except that ATP was substituted in the reaction mixtures with 2 mM
of each of CTP, GTP, UTP, dATP (Sigma), or with up to 4 mM of each
of the analogues: Adenylyl-imidodiphosphate (AMPP(NH)P), Adenylyl
(β,γ,-methylene)-diphosphonate (AMPP(CH$_2$)P), and Adenosine-5'-O-
(3 thiotriphosphate) ATP-γ-S) (Boehringer), under catenation standard
assay conditions. Gel electrophoresis and quantitation by microdensitometry was as described in the legend to Figure 1. Values represent
relative rates of reaction using the rate measured in the presence
of 2 mM ATP as 100%.*

A strict preference for covalently sealed DNA rings as substrate for
decatenation is displayed by Crithidia topoisomerase

     It has been reported that <u>Escherichia coli</u> and <u>M. luteus</u> DNA
gyrases as well as T4 DNA topoisomerase unlink nicked catenated or
knotted DNA inefficiently in comparison to the unknotting and decatenation of the covalently sealed forms (12, 16). Such substrate
preference, if displayed by the trypanosomatidial topoisomerase,
might provide this enzyme with the specificity required for discriminating between the nicked newly replicated kDNA minicircles (35-38)

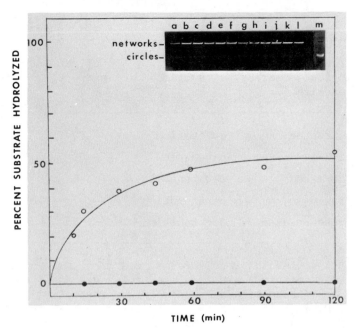

Figure 4:   Dependence on ATP versus ATPase activity of Crithidia
*topoisomerase. 0.6 µg pBR322 DNA was incubated under catenation
assay conditions with 0.4 units of either Fraction III enzyme preparation (DEAE-cellulose enzyme) (o) (a-f, in the inserted picture),
or Fraction VI enzyme preparation (Blue Sepharose enzyme preparation)
(o) (g-l, in the inserted picture). At the time intervals indicated,
samples were withdrawn for gel electrophoresis (insert) and for ATPase
activity analyses. The ATPase assay measures the production of labelled ADP from (^{3}H) ATP; Assays were carried out under standard catenation assay conditions except that (2,8-^{3}H) ATP (50 Ci/mmole, Amersham)
was used. Aliquots of 2 µl were applied to polyethyleneimine-cellu-
lose strips (0.6 x 6 cm, Brinkman) together with unlabeled ATP, ADP
and AMP markers. The strips were developed with 1 M formic acid,
0.5 M LiCl at room temperature, dried and examined with UV light to
locate and cut out the ATP, ADP and AMP spots. Radioactivity was
determined in scintillation fluid without further elution. Gel
electrophoresis was as described in the legend to Figure 1. Lane* m
represents the untreated pBR322 DNA marker.

and the covalently sealed mature ones. Such discrimination would be
anticipated (3) since each kDNA minicircle is replicated only once
in every generation (8, 35).

We have recently reported (22) on the effects of nicking and
relaxation of DNA circles on the catalytic catenation and decatenation by Crithidia topoisomerase. Table II summarizes the results of

Table 2: Effect of DNA nicking on topoisomerase reaction

DNA	Treatment prior to topoisomerase reaction	Topoisomerase reaction	% of DNA in networks	% of DNA in monomers
pBR322	untreated	catenation	72	28
pBR322	nicked	catenation	78	22
kDNA	untreated	decatenation	<5	100
kDNA	nicked	decatenation	97	<5
kDNA	nicked-closed	decatenation	<5	100

Effect of DNA nicking on topoisomerase reaction. 0.7 µg of either pBR322 or kDNA networks were incubated for 2 hrs under the standard catenation or decatenation assay conditions respectively, in the presence of 0.37 units of Crithidia topoisomerase. Reactions were stopped and subjected to gel electrophoresis and microdensitometry as described in the legends to Figure 1. Values represent extents rather than rates of reaction measured.

these experiments. It was found that the crithidial enzyme efficiently catenates nicked, as well as covalently sealed, duplex DNA circles (Table 2), and decatenates covalently sealed ones (Figure 5, Table 2). However, no significant decatenation could be observed with nicked DNA networks as a substrate (Table 2, Figure 5). It could also be concluded, that it is not the topological relaxation of the nicked DNA circles per se which causes the change in their efficiency as a substrate for decatenation. DNA networks which were "nicked-closed" by type I DNA topoisomerase were found to be as efficient substrates for decatenation as are the untreated networks (Table 2, Figure 5) (22).

DISCUSSION

It has been suggested that the circular DNA products might segregate at the end of a round of replication by mechanisms which prevent the formation of catenanes (39-41), and thus avoid the need for their resolution. Alternatively, it is possible that DNA catenanes are normally produced in the course of replication of circular DNA molecules and are unlinked at the end of a round of replication through the action of DNA topoisomerases (21). A rather clear demonstration of the involvement of catenation and decatenation in the process of DNA replication is presented in the course of the repli-

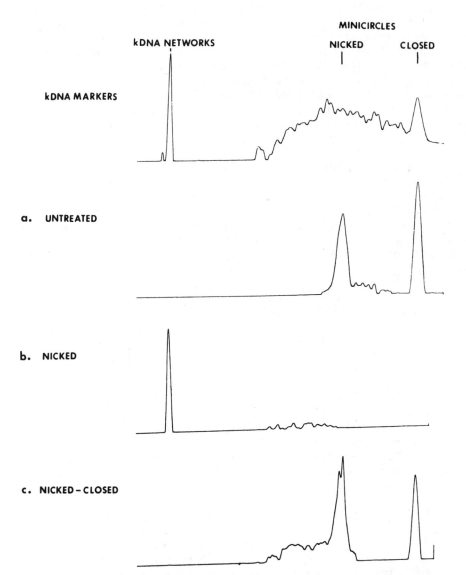

Figure 5: Decatenation of nicked DNA substrate by Crithidia topoisomerase. 0.7 µg of kDNA networks, either untreated (a), or nicked (using Crithidia nicking endonuclease (22)) (b) or nicked-closed using calf thymus topoisomerase I (c), were incubated with 0.37 units of Crithidia topoisomerase under decatenation assay conditions. Reaction conditions and gel electrophoresis were as described in the legend to Figure 1.

cation of kinetoplast DNA (kDNA) networks. Since kDNA minicircles
are replicated as free DNA circles, they have to be released from
the catenane prior to their replication and subsequently must be
topologically reattached to it once their replication has been terminated. We have previously discussed in detail the properties which
make the Crithidia ATP dependent topoisomerase a potential candidate
for this "release and reattachement" function (22). The lack of decatenation activity observed with nicked DNA networks is compatible
with the model that nicking provides the signal for discriminating
between newly replicated minicircles as a substrate for catenation
and mature ones as a substrate for decatenation (3-5).

   Crithidia fasciculata ATP dependent topoisomerase resembles
the eukaryotic type II DNA topoisomerases and T4 DNA topoisomerase
in several of its properties. It catalyzes the reversible catenation
of duplex DNA circles and the relaxation of negatively supercoiled
DNA, but fails to supercoil relaxed DNA circles under the assay conditions employed. As was likewise reported for these topoisomerases,
catenane formation by the crithidial enzyme is dependent upon the
presence of a DNA aggregating protein in the reaction mixture. This
crithidial double stranded DNA binding protein can be substituted
for by other polycations such as the polyamine spermidine. Osheroff
and Brutlag (42) and Hsieh (43) have observed the production of DNA
catenanes by stoichiometric levels of Drosophila DNA topoisomerase
in the absence of an externally added DNA condensing agent. Such
enzyme induced condensation of DNA was not observed in our studies
using Crithidia topoisomerase.

   The role of ATP and its hydrolysis in the reactions catalyzed
by the eukaryotic type II DNA topoisomerases has not yet been entirely
clarified. The role of ATP in these reactions is often explained by
assuming that the reactions catalyzed by all type II enzymes may be
similar to that of DNA gyrase, with the exception of their inability
to wrap DNA with the orienting twist required for supercoiling (13).
It was observed (34) that a single cycle of DNA supercoiling could
be produced by DNA gyrase in the presence of the nonhydrolyzable
β,γ -imido derivative of ATP. It was also suggested that the limited
relaxation of DNA by T4 type II topoisomerase in the presence of
ATP-γ-S represents an analogous reaction (11). Hsieh and Brutlag (13)
and Osheroff et al. (44) have suggested that ATP binding to the
Drosophila type II topoisomerase is sufficient to induce a double
strand DNA passage event, but the enzyme turnover requires its hydrolysis. Unlike these reports we could not detect any catenation or
decatenation by Crithidia topoisomerase in the presence of the nonhydrolyzable analogues of ATP. However, whether or not the hydrolysis
of ATP plays any role in the course of these reactions is still to
be determined.

   Crithidia ATP dependent topoisomerase is a potential candidate
for participation in the replication and the segregation of a rather

complex DNA topological structure (22). During this process, DNA minicircles are released from the central area of a DNA network and are subsequently reattached to some specific loci in the network periphery (36). At this stage, we do not understand the mechanism by which a DNA topoisomerase recognizes special zones in a DNA network. However, it is clear that such specific requirements might be imposed upon the enzyme by the system which carries out the replication of kDNA and its segregation.

ACKNOWLEDGEMENTS

We gratefully acknowledge Dianne Casuto for her excellent assistance during various stages of this work and Dr. A. Friedman from the Department of Genetics, Hebrew University for his help and advice with the EM studies. This work was supported by a grant from the United States-Israel Binational Science Foundation (BSF), Jerusalem, and by the Fund for Basic Research administered by the Israel Academy of Sciences and Humanities.

REFERENCES

1. Simpson,L. (1972) Int. Rev. Cytol. 32, 139.
2. Borst,P. and Hoeijmaker,J.H.H. (1979) Plasmid 2, 20.
3. Englund,P.T. (1980) in: Biochemistry and Physiology of Protozoa, eds. Levandowsky,M. and Hunter,S.H. (Academic Press, New York, N.Y.) 2nd edition, vol. 4, pp. 334.
4. Englund,P.T. (1978) Cell 14, 157.
5. Englund,P.T. (1979) J. Biol. Chem. 254, 4895.
6. Brock,C., Delain,E. and Riou,G. (1972) Proc. Natl. Acad. Sci. USA 69, 1642.
7. Wesley,R.D. and Simpson,L. (1973) Biochim. Biophys. Acta 319, 237.
8. Manning,J.E. and Walstenholme,D.R. (1976) J. Cell. Biol. 70, 406.
9. Gandini Attardi,D., Martini,G., Mattocia,E. and Tocchini-Valentini,G.P. (1976) Proc. Natl. Acad. Sci. USA 73, 554.
10. Baldi,M.I., Benedetti,P., Mattocia,E. and Tocchini-Valentini, G.P. (1980) Cell 20, 461.
11. Liu,L.F., Liu,C.-C. and Alberst,B.M. (1979) Nature 281, 456.
12. Liu,L.F., Liu,C.-C. and Alberts,B.M. (1980) Cell 19, 697.
13. Hsieh,T. and Brutlag,D. (1980) Cell 21, 115.
14. Brown,P.O. and Cozzarelli,N.R. (1979) Science 206, 1081.
15. Cozzarelli,N.R. (1980) Science 207, 953.
16. Mizuuchi,K., Fisher,L.M., O'Dea,M.H. and Gellert,M. (1980) Proc. Natl. Acad. Sci. USA 77, 1847.
17. Kreuzer,K.N. and Cozzarelli,N.R. (1980) Cell 20, 245.
18. Tse,Y. and Wang,J.C. (1980) Cell 22, 269.

19. Miller,K.G., Liu,L.F. and Englund,P.T. (1981) J. Biol. Chem.
         256, 9334.
20. Gioto,T. and Wang,J.C. (1982) J. Biol. Chem. 257, 5866.
21. Gellert,M. (1981) Ann. Rev. Biochem. 50, 879.
22. Shlomai,J. and Zadok,A. (1983) Nucleic Acids Res. 11, 4019.
23. Oakley,B.R., Krisch,D.R. and Morris,N.R. (1980) Anal. Biochem.
         105, 361.
24. Brown,P.O. and Cozzarelli,N.R. (1981) Proc. Natl. Acad. Sci.
         USA 78, 843.
25. Wilson,R.W. and Bloomfield,V.A. (1979) Biochemistry 18, 2192.
26. Gosule,L.C. and Schellman,J.A. (1978) J. Mol. Biol. 121, 311.
27. Osland,A. and Kleppe,K. (1977) Nucleic Acid Res. 4, 685.
28. Widom,J. and Baldwin,R.L. (1980) J. Mol. Biol. 144, 431.
29. Manning,G.S. (1978) Q. Rev. Biophys. 11, 179.
30. Krasnow,M.A. and Cozzarelli,N.R. (1982) J. Biol. Chem.
         257, 2687.
31. Yount,R.G., Babcock,D., Ballantyne,W. and Ojala,D. (1971)
         Biochemistry 10, 2484.
32. Gellert,M., O'Dea,M.H., Itoh,T. and Tomizawa,J. (1976)
         Proc. Natl. Acad. Sci. USA 73, 4474.
33. Mizuuchi,K., O'Dea,M.H. and Gellert,M. (1978) Proc. Natl.
         Acad. Sci. USA 75, 5960.
34. Sugino,A., Higgins,N.P., Brown,P.O., Peebles,C.L. and
         Cozzarelli,N.R. (1978) Proc. Natl. Acad. Sci. USA
         75, 4838.
35. Simpson,L., Simpson,A.M. and Wesley,R.D. (1974) Biochem.
         Biophys. Acta 349, 161.
36. Simpson,A.M. and Simpson,L. (1976) J. Protozool. 23, 583.
37. Benard,J., Riou,G. and Soucier,J.-M. (1979) Nucleic Acids
         Res. 6, 1941.
38. Englund,P.T., DiMaio,D.C. and Price,S.S. (1977) J. Biol. Chem.
         252, 6208.
39. Meinke,W. and Goldstein,D.A. (1971) J. Mol. Biol. 61, 543.
40. Gefter,M.L. (1975) Ann. Rev. Biochem. 44, 45.
41. Tomizawa,J. (1978) in:"DNA Synthesis: Present and Future",
         ed. M.Kohiyama, I. Molineux, pp. 797, Plenum Press,
         New York.
42. Osheroff,N. and Brutlag,D.L. (1983) in:"Mechanisms of DNA
         Replication and Recombination" UCLA Symposia on Molecular
         and Cellular Biology, New Series, Vol. 10, N.R.Cozzarelli,
         ed., Alan R. Liss Inc. New York.
43. Hsieh,T.S. (1983) J. Biol. Chem. 258, 8413.
44. Osherhoff,N., Shelton,E.F. and Brutlag,D.L. (1983) J. Biol.
         Chem., in press.

REGENERATING RAT LIVER TOPOISOMERASE II: PURIFICATION OF THE

ENZYME AND CATENATION OF DNA RINGS

Gilles Mirambeau, Catherine Lavenot[1]
and Michel Duguet

Laboratoire d'Enzymologie des Acides Nucleiques
Université Pierre et Marie Curie
96, Bd. Raspail
75006 Paris, France

[1]Unité d'Enzymologie
Institut de Recherches Scientifiques sur le Cancer
BP n⁰ 8
94800 Villejuif, France

INTRODUCTION

In search for the biological functions of these enzymes, several eukaryotic type II topoisomerases have been isolated from various organisms including Trypanosomatids (1), Yeast (2), Drosophila (3,4), Xenopus (5), Hela cells (6) and Rat liver (7). Comparison with their prokaryotic counterparts (8) suggested their possible involvement in several genetic processes such as DNA replication, recombination, transposition and repair. In addition, several lines of evidence indicated that topoisomerases are closely associated with chromatin (9, 10) and might play a role in the structural changes occuring within chromatin (condensation/decondensation).

In a previous work, we have isolated a type II topoisomerase from regenerating rat liver nuclei and found that the activity of the enzyme was modulated as a function of regenerating time, with a maximum activity 40 hours after hepatectomy (S phase). This finding strongly suggested direct or indirect involvement of the enzyme in DNA replication (7). Here, we describe the purification of topoisomerase II to near homogeneity and some of its properties in vitro. In particular, we have analyzed the different factors necessary to perform the catenation of supercoiled or relaxed DNA rigns in vitro.

423

MATERIALS AND METHODS

DNA

   Supercoiled pBR 322 DNA was isolated from Escherichia coli
strain HB 101 by using the procedure of Clewell and Helinski (11)
and purified in a CsCl-ethidium bromide density gradient. Relaxed
pBR 322 DNA was prepared by incubation of supercoiled DNA with rat
liver topoisomerase I (our phosphocellulose pool) in the relaxation
mixture described below. Kinetoplast DNA was a generous gift of Dr.
G. Riou (Institut Gustave Roussy, Villejuif).

Other products

   Rat liver HMG1 protein was a gift of Dr. C. Bonne (IRSC, Villejuif). Histone H1, ATP, ethidium bromide and novobiocin were from
Boehringer (Mannheim). Coumermycin A1 and agarose were from Sigma.
Acrylamide and Hydroxylapatite (biogel HTP) were from Biorad, Phosphocellulose P11 from Whatman. Calf thymus native DNA cellulose was
prepared according to the method of Herrick and Alberts (12). All
other chemicals were of the highest grade commercially available.

Topoisomerases assays

   ATP-independent relaxation of supercoiled DNA was used as an
assay for topoisomerase I. The 30 μl reaction mixture contained 10 mM
Tris-HCl, pH 7.9, 0.2 mM EDTA, 0.5 mM DTT, 0.5 μg pBR 322 DNA, 200 mM
NaCl, and 2 μl of the fraction to be assayed. The reaction was carried
out from 30 min at 37°C and then stopped by adding 1% final SDS,
0.25 mg/ml bromophenol blue, 15% sucrose.

   The 30 μl reaction mixtures for topoisomerase II assays contained in all cases 50 mM Tris-HCl, pH 7.9, 25 mM $MgCl_2$, 0.5 mM DTT,
30 μg/ml bovine serumalbumine, 1 mM ATP. pBR322 DNA (0.9 μg), supercoiled or relaxed, was used for the catenation reactions in the presence of 43 mM KCl and 0.3 μg histone H1/μg DNA. ATP-dependent relaxation assay was the same, except that histone H1 was omitted.
Kinetoplast DNA (0.4 μg) was the DNA substrate for decatenation, with
a KCl concentration raised to 85 mM and histone H1 omitted.

   After incubation at 30°C for 30 min, the reactions were terminated as indicated in the case of topoisomerase I. Reaction products
were analyzed by 1% agarose gel electrophoresis in Tris-Phosphate
buffer. Electrophoresis was run at 1V/cm for 15h. The gel was stained
with Ethidium bromide before photography under UV illumination. Scanning of the negative was made with a Joyce-Loebl densitometer.

Nuclei isolation and PEG supernatant

Purified nuclei were isolated from 40h regenerating rat liver (topoisomerase II maximum activity) and a high salt nuclear extract was made and treated with polyethylene glycol (7). The PEG supernatant (11.2 mg protein) obtained from 25 rats was dialyzed against 200 mM potassium phosphate, pH 7.4, 1 mM EDTA, 10 mM β mercaptoethanol, 10% glycerol and 1 mM PMSF added immediately before use, and refered as fraction II.

Phosphocellulose chromatography

Fraction II (14 ml) was loaded onto a phosphocellulose column (1 cm x 5 cm) equilibrated with the same buffer. After loading, the column was washed and eluted with a 400 ml linear gradient from 0.2 to 0.6 M potassium phosphate. Topoisomerase II eluted slightly ahead of topoisomerase I: therefore, a part of topoisomerase II activity was eliminated in order to reduce topoisomerase I contamination. Fractions of interest were pooled and dialyzed against 0.1 M potassium phosphate, pH 7.0 mM β mercaptoethanol, 0.5 mM EDTA, 20% glycerol, 1 mM PMSF to give fraction III (8 ml).

Hydroxylapatite chromatography

Fraction III was loaded onto a hydroxylapatite column (0.5 x 2 cm) equilibrated with the same buffer. After the column has been rinced, it was eluted with a 10 ml gradient from 0.1 to 0.7 M phosphate. Topoisomerase II came off as a peak centered at 0.42 M potassium phosphate. Active fractions were pooled and dialyzed against 40 mM Tris-HCl, pH 7.4, 10 mM β mercaptoethanol, 0.5 mM EDTA, 20% glycerol, 1 mM PMSF, to give fraction IV (3 ml).

DNA cellulose chromatography

Fraction IV was loaded onto a calf thymus native DNA cellulose column (0.5 x 0.75 cm) equilibrated with fraction IV buffer. The column was washed and then eluted with four 2.5 ml steps of 0.1, 0.2, 0.4 and 1.0 M KCl in the same buffer. The topoisomerase II activity eluted in the 0.4 M step. Active fractions were pooled (fraction V, 0.45 ml). A 0.1 ml aliquote was analyzed on a 7.5% acrylamide gel. The remaining fraction was adjusted to 100 µg/ml bovine Serumalbumin and dialyzed against fraction IV buffer plus glycerol raised to 50%. In these conditions, fraction V was stable for 3 months at -20°C.

Other methods

SDS polyacrylamide gels were run according to the method of Laemmly (13) and stained according to the silver ultrasensitive method of Switzer and coworkers (14).

RNA polymerase 160,000 subunit, β Galactosidase (130,000), Phosphorylase A (94,500), Catalase (60,000) and Aldolase (40,000) were used as molecular weight markers. Densitometric profiles of the gels were made by using a Joyce-Loebl densitometer. Protein concentration was determined by the method of Bradford (15).

RESULTS AND DISCUSSION

Purification of rat liver type II topoisomerase

Since both type I and type II topoisomerases are present in a nuclear extract from rat liver (7), it is necessary to distinguish between these enzymes by using specific assays. Besides the ATP-independent relaxation of a supercoiled DNA (topoisomerase I), three of the assays currently performed for topoisomerase II activity are illustrated in Figure 1. (i) ATP-dependent relaxation (lane 2); (ii) ATP-dependent formation of catenanes from monomeric circles (lane 3); (iii) ATP-dependent decatenation of the giant networks of kinetoplast DNA (lane 5). Detailed procedures for these assays are given in the experimental section and in reference (7). Decatenation has been prefered over relaxation and catenation to monitor the purification of the type II topoisomerase: it is strictly dependent on ATP, does not interfere with topoisomerase I nor it requires additional factor(s), and gives a unique final product.

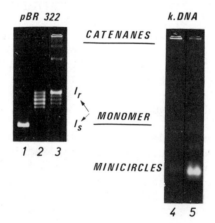

Figure 1: Reactions catalyzed by topoisomerase II: agarose gel analysis. 0.9 μg negatively supercoiled DNA (I_s), lane 1, was incubated with 5 units of rat liver topoisomerase II in the ATP-dependent relaxation standard mixture (lane 2); relaxed DNA (I_r) was obtained. The same reaction plus 10 μg/ml histone Hl catenated about 40% monomer (lane 3). 0.4 μg kinetoplast DNA (lane 4) was also incubated with 5 units of rat liver topoisomerase II in the decatenation reaction mixture: networks were converted into free minicircles (lane 5).

In a standard preparation, 100 g of regenerating rat liver were homogenized at 0°C in the presence of protease inhibitors. Detailed purification of the nuclei and preparation of fraction II (polyethylene glycol supernatant) were given in a previous work (7).

Fraction II was further purified on a phosphocellulose column as described in Figure 2 and experimental section. Topoisomerase I, followed by relaxation, and topoisomerase II, followed by decatenation, came off as partially separated peaks at 0.34 and 0.39 M respectively in a linear phosphate gradient. However, when catenation of a supercoiled DNA is used as an assay, the active fractions appeared as a broad peak overlapping topoI and topoII (Figure 2).

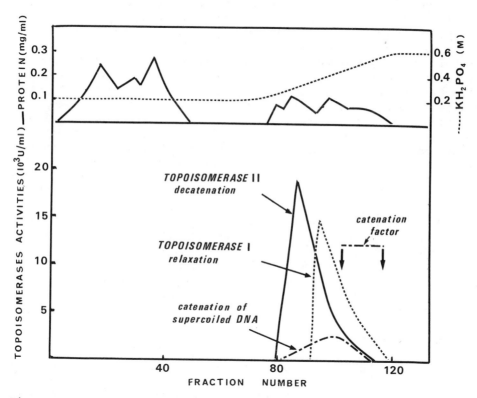

Figure 2: Phosphocellulose chromatography. Upper part: protein concentrations are indicated by the solid line. Phosphate concentration (dashed line) was measured with a Heito conductimeter. Lower part: Eluted fractions were tested in various topoisomerases assays. ······topoisomerase I, relaxation; ———— topoisomerase II, decatenation; ·—·—·—Catenation of supercoiled DNA. Finally, topoisomerase II peak fraction (fraction 85) was mixed with other fractions in the catenation reaction: Fractions 105 to 120, refered as "Catenation factor", considerably improved the catenation of supercoiled DNA by topoisomerase II.

Moreover, very little catenation was obtained by using the peak fraction of topoisomerase II (fraction 85) or more purified fractions
(IV or V) as shown in Figure 4, lane 4. However, when these fractions
are mixed with fractions eluted around 0.50 M potassium phosphate
(Figure 2) and designated as "catenation factor", large amounts of
catenated products are observed (Figure 4, lane 5). Surprisingly,
the "catenation factor" was not necessary when the DNA substrate was
relaxed (Figure 4, lane 10): this point is discussed below in more
details (see Catenation of supercoiled and relaxed DNA) as well as
the stimulatory effect of topoisomerase I on the catenation reaction.

Only the topoisomerase II fractions that did not overlap with
topoisomerase I were pooled for further fractionation (fraction III),
resulting in a considerable loss of activity (see Table 1). This
fraction was then purified on a hydroxylapatite column (fraction IV)
and finally on a double-stranded DNA-cellulose column (see experimental section), yielding fraction V. Table 1 summarizes the purification of topoisomerase II: the activity of the PEG supernatant
(fraction II) was overestimated by 3 to 4 fold with respect to the
following fractions, due to the presence of polyethylene glycol which
stimulates the reaction (6), so that the phosphocellulose column turns
out to be a good purification step, increasing the specific activity
by at least 5 fold and eliminating the contaminant topoisomerase I.
The overall purification, starting from fraction II is also illustrated in Figure 3 which represents densitometric profiles of SDS-polyacrylamide gel analysis of fractions II to V: it appears that the
purification yields two groups of polypeptides; one is a 145 K polypeptide (sometimes appearing as a triplet 150, 145, 138 K) selected
in the course of the purification; the other, composed of 66, 71 and
75 K polypeptides, is not present in all preparations and its amount
does not correlate with the level of enzyme activity. Therefore, it
is likely that the 145 K polypeptide bears the topoisomerase II
activity and the 66, 71, 75 K cluster represents trace amounts of
contaminating topoisomerase I (16).

Relaxation of supercoiled DNA

Purified rat liver topoisomerase II was able to relax a supercoiled DNA in a reaction which is dependent on ATP, as shown in
Figure 4, lanes 2 and 3. The incubation mixture was the same as for
catenation, except that histone Hl was omitted. Remarkably, relaxation was strongly inhibited by concentrations of histone Hl as low
as 10 µg/ml (Figure 4, lane 4). However, upon addition of 2 µg/ml
fraction 109 (peak of "catenation factor"), the inhibition was totally released, the DNA was relaxed and finally catenated, due to the
presence of histone Hl (Figure 4, lane 5). Our hypothesis, also supported by results on the catenation reaction, is that the "catenation factor" interacts with histone Hl, preventing inhibition. The
"factor" could not be replaced by the high mobility group protein
HMGl, a non-histone protein which interacts with histones (not shown).

Table 1: Purification of topoisomerase II

Fraction	Volume (ml)	Protein (mg)	Total activity[a] (units)	Specific activity[a] (units/mg)
I homogenate	—	—	—	—
II peg supernatant	14	11.2	8.0×10^5 (100%)	7.0×10^4 (1)
III phosphocellulose[b]	8	0.7	8.0×10^4 (10%)	11.4×10^4 (1.6)[d]
IV Hydroxyapatite	4	0.07	1.2×10^4 (1.5%)	1.7×10^5 (2.5)[d]
V double stranded DNA cellulose	0.45	0.0045[c]	9.0×10^3 (1.1%)	2.0×10^6 (28.6)

a one unit is defined as the amount of enzyme required to decatenate 50% of 0.4 µg k DNA in 30 minutes at 30ºC. Assay conditions are described in experimental procedures.

b another 10% of topoisomerase activity was lost by reducing topoisomerase I contamination.

c Estimate of protein concentration was based on silver stained polyacrylamide gel of the DNA cellulose pool.

d specific activity of phosphocellulose and hydroxylapatite were less than expected probably because of topoisomerase II degradation during dialysis.

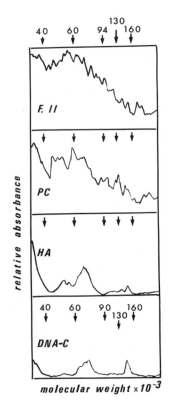

Figure 3: Densitometric profiles of active fractions in the course of rat liver topoisomerase II purification. Fraction II, PEG supernatant; fraction III, phosphocellulose (PC); fraction IV, hydroxyl-apatite (HA); fraction V, DNA cellulose (DNA-C) were run in a 7.5% acrylamide SDS-polyacrylamide gel. Migration is from right to left. The gel was stained with ultrasensitive silver staining and then scanned with a Joyce-Loebl densitometer. Molecular weight markers were respectively from right to left: RNA polymerase β subunit (160 K), β galactosidase (130 K), phosphorylase α (90 K), catalase (60 K), and aldolase (40 K). Peaks at the left side represent bromophenol blue tracking dye.

Catenation of supercoiled and relaxed DNA

As shown in Figure 4, lane 4, purified topoisomerase II could not catenate supercoiled DNA in the presence of histone H1 alone; addition of the "catenation factor" permitted the catenation, but H1 was still necessary (Figure 4, lanes 5 and 6). In contrast, relaxed DNA was readily catenated in the presence of H1 alone (Figure 4, lane 10). From this, it is clear that first, supercoiled DNA is a poor substrate for catenation (2). Second, catenation of a supercoiled

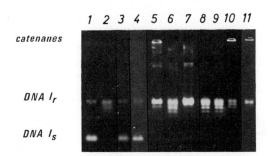

*Figure 4: Relaxation and catenation of DNA by purified topoisomerase
II. Supercoiled pBR 322 DNA (lanes 1 to 7) and relaxed pBR 322 DNA
(lanes 8 to 11) were incubated with 4 units of topoisomerase II in
the standard ATP-dependent relaxation/catenation mixture with various
additions and omissions and electrophoresed as indicated in the
experimental section. Lane 1 is pBR 322 control; lanes 2 and 3,
reactions with and without ATP; lane 4, in the presece of 10 µg/ml
histone H1; lane 5, in the presence of H1 (10 µg/ml) plus 2 µg/ml
"catenation factor"; lane 6, 2 µg/ml "catenation factor" in the
absence of histone H1; lane 7, 20 fold excess "catenation factor"
in the absence of topoisomerase II; lane 8 is relaxed pBR 322 control;
lane 9, incubation with topoisomerase II in the absence of histone
H1; lane 10, with 10 µg/ml histone H1 added; lane 11, in the presence
of both histone H1 and "catenation factor" (2 µg/ml).*

DNA by a crude fraction II alone, or by more purified fractions in
the presence of the "catenation factor" occurs in two steps: the
first is the relaxation of the DNA, and is inhibited by histone H1.
The second is the catenation of relaxed DNA which needs histone H1,
but not the "catenation factor" (Figure 4, lanes 9, 10, 11). These
findings finally explain the stimulation of the catenation by topoisomerase I which simply relaxed DNA even in the presence of histone
H1 (17).

Inhibitors of topoisomerase II

The enzyme from rat liver has been previously shown to be almost
insensitive to oxolinic and nalidixic acids, up to 250 µg/ml. In
contrast, novobiocin and coumermycin A1 are inhibitors of rat liver
topoisomerase II as of some other eukaryotic topoisomerases (1-6).
Despite the absence of precise analogy between these inhibitors and
ATP, both novobiocin and coumermycin A1 behave as competitive inhibitors of the substrate ATP (not shown). Coumermycin is the more potent
inhibitor, the enzyme activity being inhibited by 50% in the presence
of 0.3 mM novobiocin and 0.02 mM coumermycin, when ATP concentration
is 0.5 mM.

DISCUSSION

The type II topoisomerase has been purified to near homogeneity from rat liver nuclei. The activity correlated with a triplet of 150, 145 and 138 K polypeptides, a situation remarkably similar to that described in the case of the enzyme from Drosophila (3, 4).

Purified topoisomerase II was able to relax a supercoiled DNA in a reaction dependent on ATP. The reaction was strongly inhibited by histone Hl. Catenation of relaxed DNA was dependent on ATP and histone Hl; Hl could not be replaced by an additional factor. Indeed, Hl was necessary even when catenation was performed by a crude fraction II. In contrast, we have found that catenation of a supercoiled DNA substrate required Hl plus additional factor(s) which were already present in a crude fraction II, but not in more purified fractions. Our hypothesis is that these factors, namely "catenation factor" and topoisomerase I, were only required to permit the relaxation of the DNA prior to catenation.

The complex interactions that seem to take place in vitro between topoisomerase II and proteins associated with chromatin (histone Hl, topoisomerase I and perhaps the "catenation factor", a basic protein interacting with Hl) might reflect in vivo situation where topoisomerases are integral parts of chromatin structure and play a prominent role within this structure (18). Recently, a mutant of mouse cells, thermosensitive in DNA replication and affected in its chromatin structure, has been found to lack a 30 K polypeptide, presumably associated with topoisomerase II (19).

ACKNOWLEDGEMENTS

We would like to thank Dr. A.M. de Recondo (IRSC, Villejuif): a large part of this work has been made in her laboratory. We also thank Dr. J.M. Saucier for valuable discussions. This work was supported by grants from CNRS (ATP n° 6082 753).

REFERRENCES

1.   Shlomai,J. and Zadok,A. (1983) Nucleic Acids Res. 11, 4019.
2.   Goto,T. and Wang,J.C. (1982) J. Biol. Chem. 257, 5866.
3.   Hsieh,T.S. (1983) Methods in Enzymology 100, 161.
4.   Shelton,E.R., Osheroff,N. and Brutlag,D.L. (1983)
         J. Biol. Chem. 258, 9530.
5.   Benedetti,P., Baldi,M., Mattoccia,E. and Tocchini-Valentini,G.
         (1983) EMBO J. 2, 1303.
6.   Miller,K.G., Liu,L.F. and Englund,P.T. (1981) J. Biol. Chem.
         256, 9334.

7.   Duguet,M., Lavenot,C., Harper,F., Mirambeau,G. and
       De Recondo,A.M. (1983) Nucleic Acids Res.
       11, 1059.
8.   Gellert,M. (1981) Ann. Rev. Biochem. 50, 879.
9.   Liu,L.F., Nelson,E.M., Halligan,B.D. and Rowe,T.C. (1983)
       Proceedings of UCLA Symposium "Mechanisms of DNA
       Replication and Recombination", April 83, Keystone, USA.
10. Waldeck,W., Theobald,M. and Zentgraf,M. (1983) EMBO J. 2, 1255.
11. Clewell,D.B. and Helinski,D.R. (1970) Biochemistry 9, 4428.
12. Alberts,B.M. and Herrick,G. (1971) Methods in Enzymology 21, 198.
13. Laemmli,U.K. (1970) Nature 277, 680.
14. Switzer,R.C., Merril,C.R. and Shifrin,S. (1979)
       Anal. Biochem. 98, 231.
15. Bradford,M.M. (1976) Anal. Biochem. 72, 248.
16. Martin,S.R., McCoubrey Jr,W.K., McConaughy,B.L., Young,L.S.,
       Been,M.D., Brewer,B.J. and Champoux,J.J. (1983)
       Methods in Enzymology 100, 137.
17. Javaherian,K. and Liu,L.F. (1982) Nucleic Acids Res. 11, 461.
18. Mertz,J.E. and Miller,T.J. (1983) Mol. Cell. Biol. 3, 126.
19. Colwill,R.W. and Sheinin,R. (1983) Proc. Natl. Acad. Sci. USA
       80, 4644.

EXPRESSION OF SILENT GENES: POSSIBLE INTERACTION BETWEEN

DNA GYRASE AND RNA POLYMERASE

Erela Ephrati-Elizur and Becky Chronis-Anner

Department of Molecular Biology
Hebrew University - Hadassah Medical School
Jerusalem

INTRODUCTION

Numerous studies indicate that DNA gyrase is an essential component of the transcription process. Supercoiled DNA is more readily transcribed than relaxed DNA (1-4). Inhibitors of DNA gyrase inhibit the transcription of some operons in vivo (5-7) and in vitro (8-9). Transcription is also affected in a ts gyrase B mutant when grown under nonpermissive conditions (10). It was recently shown that mutants lacking topoisomerase I can grow normally only with secondary mutations that compensate for the absence of topoisomerase I activity. Several of these secondary mutations map in gyrA or gyrB. These observations suggested that DNA superhelicity is a result of a balance between topoisomerase I and gyrase activities (11-12). It was shown further that mutations in gyrA or gyrB can activate the cryptic bgl operon, however, the results obtained with various mutants do not indicate a clear cut correlation between the degree of superhelicity and gene expression (12). The differential effect of DNA supercoiling on gene expression was recently discussed by Smith (13).

The present communication shows that mutations in gyrA allow the expression of two silent genes that are normally not expressed in wild-type Escherichia coli: 1) the gene coding for adenylate synthetase, a modifying enzyme conferring resistance to streptomycin and spectinomycin, so far known to be coded only by plasmids (14) and 2) the bgl operon coding for B-glucosidase. Spontaneous mutations that activate this operon are a result of either insertions at the cryptic regulatory site (15), or by mutations in gyrA or gyrB (12). In the present study the expression of both silent genes seems to depend on the activity of both gyrase and RNA polymerase.

RESULTS

Two spontaneous mutants were isolated from a nalidixic acid resistant (Nal^r) derivative of <u>Escherichia coli</u> (strain ME3) which were resistant to both streptomycin (Sm) and spectinomycin (Sp) and at the same time lost the resistance to Nal (strains SS1, SS2). The Sm/Sp resistance was found to be due to adenylate synthetase, the only enzyme that can modify both antibiotics (14). Attempts to isolate a plasmid that carries the resistance gene failed (16), so were attempts to transfer the Sm/Sp resistance by conventional genetic methods. Strains SS1 and SS2 also express a Bgl$^+$ phenotype. The mutants' characteristics are given in Table 1.

The location of mutations responsible for Sm/Sp resistance

A Pl lysate harboring a pool of <u>Tn10</u> insertions (17) was used to map the location of the mutation causing resistance to Sm/Sp in strains SS1 and SS2. Tetracycline resistant (Tc^r) Sm/Sp^S colonies were isolated and, using a Pl lysate, the Tc^r was transferred to HfrC and HfrH. The location of the Tc^r was established, by conjugation, to be at or near <u>gyrA</u>. Transduction of strains SS1 and SS2 to Nal^r by Pl lysate of strain ME3 showed a 100% loss of Sm/Sp resistance indicating that <u>gyrA</u> is most probably the site of mutations (gyrA(1) and <u>gyrA(2)</u> respectively) responsible for the expression of the drug resistance.

Secondary mutations are required for the growth of <u>gyrA</u> mutants

To establish that <u>gyrA</u> mutation is responsible for the silent gene expression the mutation in <u>gyrA(2)</u> was transferred to the wild-type parent strain JMI. The new strain EB2 (see Table 1) expressed Sm/Sp resistance but its growth was abnormally slow on any medium and only when faster growing pappili appeared was it possible to isolate a somewhat faster growing strain (EB3). Thus the mutation <u>gyrA(2)</u> seems to require secondary mutations probably to compensate for the <u>gyrA</u> defect.

Preliminary studies have shown that the secondary mutations express a suppressor phenotype, namely, suppression of the auxotrophic requirements of the parent strain JMl and the disappearance of its ribosomal resistance to Sm (500 µg/ml). These characteristics are similar to the suppressor phenotype of the original mutant SS2 (18) and suggest that the suppressor mutation could act as a compensatory mutation. Presumably due to its presence in strain ME3 it was possible to isolate from it the <u>gyrA</u> mutants. The suppressor mutation was mapped by using the pool of <u>Tn10</u> insertions and found to be linked 90-93% to <u>leu</u>, a map position close to that of <u>mafB</u>, the gene involved in the replication of F-like plasmids (19).

Table 1: Growth characteristics of parent and mutants strains

Strain No.	Level of antibiotic resistance*					
	Streptomycin 20 µg/ml	Spectinomycin 500 µg/ml	Nalidixic acid 20 µg/ml	2 µg/ml	1 µg/ml	Salicin agar (β-glucoside)
JMI	+**	-	-	-	-	-
ME3 (Sup)	-	-	+	+	+	-
SS1 (Sup)	+	+	-	+	+	+
SS2 (Sup)	+	+	-	-	+	+
EB2	+	+	-	-	+	+

* The culture grown overnight in L-broth was diluted for single colonies and 0.02 ml drops of various dilutions plated on the selective plates.

** Ribosomal resistance

+ 100% survival

- no growth

Effect of rpoB mutations on the expression of silent genes and suppression

Twenty four independent Rif[r] mutants were isolated and all showed suppression of antibiotic resistance to varying degrees. Three mutants (rpoB4, rpoB14, rpoB15) were studied further, the rif[r] mutation was mapped and found to be about 50% linked to argH (20), therefore most probably falls in the rpoB region. Table 2 shows the suppressive effect of rpoB mutations on the expression of antibiotic resistance, Bgl[+] phenotype and the suppression of auxotrophy.[*]) These results suggest that the three factors -gyrase, RNA polymerase and an additional as yet unknown factor- could be functionally interrelated.

Table 2: The effect of rpoB mutations on the expression of silent genes and auxotrophic suppression

	Growth on*			
Strain No.	Sm/Sp 20/200/ µg/ml	Salicin agar	Minimal agar	Min-CA agar**
SS1	+	+	+	+
SS1 rpoB4	−	−	−	+
SS1 rpoB14	−	−	−	+
SS1 rpoB15	±	±	±	+

*   The culture grown overnight in L-broth was diluted for single colonies and 0.02 ml drops of various dilutions plated on the selective plates.

   +   100% survival
   ±   thin growth at 10^{-2} dilution
   −   no growth

**   Minimal agar supplemented with 0.2% casein hydrolysate

---

[*])

Due to this Rif[r] suppression it was previously erroneously thought that the auxotrophic suppression is linked to rpoB.

DISCUSSION

The results show that the expression of two silent genes not normally expressed by the wild-type culture is dependent upon a certain change in gyrase activity and suppressed by altered RNA polymerase, suggesting that in addition to an optimal state of superhelicity required for an open complex formation there may also be subtle direct or indirect interactions between the two factors -gyrase and RNA polymerase- that are abolished in Rif[r] mutants, thus suppressing gene expression. It was further shown that gyrA(2) mutation reduces drastically the culture's growth rate, however, the growth rate increases upon the appearance of secondary (compensatory) mutations, mapped at or near mafB whose function is not yet known. These secondary mutations express a suppressor phenotype which also seems to be associated with the activity of RNA polymerase since certain mutations in rpoB suppress its activity.

Figure 1 shows the observed interrelations between the phenotypes expressed by the mutations gyrA, rpoB, sup (mafB?).

In Salmonella typhimurium, topI mutants previously known as supX also exhibit non ribosomal low level antibiotic resistance (21). However, since it was shown recently that topI mutants can survive only with additional compensatory mutations in gyrA or gyrB (11-12), it is not known whether the expression of resistance is due to the alteration in gyrase or in both gyrase and topoisomerase I.

There is good evidence that supercoiled DNA is transcribed much more efficiently than relaxed DNA. A direct correlation, however, between the expression of silent genes and changes in the degree of superhelicity in gyrA or gyrB mutants was not observed (12). Altered gyrase activity may subtly change the degree of local untwisting and thus affect specific DNA configuration, such as stem-loop formation, which may act as regulatory signals for proteins involved in gene expression (22).

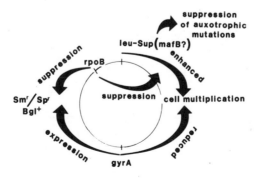

Figure 1: Interaction between gyrA-rpoB-Sup. For explanation see text.

ACKNOWLEDGEMENT

This research was supported in part by the Israel Commission for Basic Research and partly by the Leonard Wolfson Foundation for Scientific Research.

REFERENCES

1.   Botchan,P., Wang,J.C. and Echols,H. (1973) Proc. Natl. Acad. Sci. U.S.A. 70, 3077.
2.   Botchan,P. (1976) J. Mol. Biol. 105, 161.
3.   Richardson,J.P. (1975) J. Mol. Biol. 91, 477.
4.   Kano,Y., Miyashita,T., Nakamura,H., Kuroki,K., Nagata,A. and Imamoto,F. (1981) Gene 13, 173.
5.   Shuman,H. and Schwartz,M. (1975) Biochem. Biophys. Res. Commun. 64, 204.
6.   Kubo,M., Kano,Y., Nakamura,H., Nagata,A. and Imamoto,F. (1979) Gene 7, 153.
7.   Sanzey,B. (1979) J. Bact. 138, 40.
8.   Yang,H.L., Heller,K., Gellert,M. and Zubay,G. (1979) Proc. Natl. Acad. Sci. U.S.A. 76, 3304.
9.   Smith,C.L., Kubo,M. and Imamoto,F. (1978) Nature 275, 420.
10.   Mirkin,S.M. and Shmerling,Zh.G. (1982) Mol. Gen. Genet. 188, 91.
11.   Pruss,G.J., Manes,S.H. and Drlica,K. (1982) Cell 31, 35.
12.   DiNardo,S., Voelkel,K.A., Sternglanz,R., Reynolds,A.E. and Wright,A. (1982) Cell 31, 43.
13.   Smith,G.R. (1981) Cell 24, 599.
14.   Davies,J. and Smith,D.I. (1978) Ann. Rev. Microbiol. 32, 469.
15.   Reynolds,A.E., Felton,J. and Wright,A. (1981) Nature 293, 625.
16.   Birnboim,H.C. and Doly,J. (1979) Nucl. Acids. Res. 7, 1513
17.   Csonka,L.N. and Clark,A.J. (1979) Genetics 93, 321.
18.   Ephrati-Elizur,E. and Luther-Davies,S. (1981) Mol. Gen. Genet. 181, 390.
19.   Wada,C. and Yura,T. (1979) J. Bact. 140, 864.
20.   Yura,T. and Ishihama,A. (1979) Ann. Rev. Genet. 13, 59.
21.   Dubnau,E. and Margolin,P. (1972) Mol. Gen. Genet. 117, 91.
22.   Harland,R.M., Weintraub,H. and McKnight,S.L. (1983) Nature 302, 38.

POLY (ADP-RIBOSYLATION) AND DNA TOPOISOMERASE I IN DIFFERENT

CELL LINES

Ari M. Ferro, Larry H. Thompson[1] and
Baldomero M. Olivera

Department of Biology
University of Utah
Salt Lake City, UT 84112, USA

[1]Lawrence Livermore National Laboratory
Livermore, CA 94550, USA

INTRODUCTION

   With his discovery that the Escherichia coli chromosome was
circular, John Cairns first pointed out the need for a swivel in DNA
replication (1). In eukaryotes, type I DNA topoisomerase has the
enzymatic properties necessary to serve as such a swivel. DNA replication would introduce positive supertwists ahead of the replication
fork; DNA topoisomerase I has the capacity to unwind such positively
supertwisted DNA (2, 3). For this reason, a role for topoisomerase I
in replication has been suggested by many workers (2, 3). Recent
biological and phenomenological data have also led to the suggestion
that type I DNA topoisomerase may be involved in catalyzing DNA strand
exchange under certain conditions (4). The mechanisms by which the
activity of such a potentially important enzyme may be modulated are
of obvious interest.

   We recently demonstrated that calf thymus DNA topoisomerase I
can be modified by poly (ADP-ribosylation) (5). Since the addition
of each ADP-ribose moiety adds two negative charges to the protein,
poly (ADP-ribosylation) is perhaps the most drastic protein modification known (6, 7). This modification reaction is catalyzed by the
nuclear enzyme poly (ADP-ribose) synthetase. A novel feature of this
enzyme is that it absolutely requires double stranded DNA for activity (8, 9) and specifically, DNA in which the double helical structure is interrupted (10). Extensive poly (ADP-ribosylation) of DNA

topoisomerase I by the synthetase causes severe inhibition of topoisomerase activity (5). Under optimum conditions, well over 20 ADP-
ribose residues can be added to each topoisomerase I molecule, contributing over 40 negative charges to the topoisomerase protein.

All of our published work to date has been carried out using
topoisomerase and synthetase isolated from calf thymus (5, 11). Since
certain nucleic acid metabolizing enzymes (i.e., terminal deoxynucleo-
tidyl transferase; 12, 13) have only been found in the thymus and
related tissues in the immune system, it is conceivable that the
ADP-ribosylation of topoisomerase I is a specialized function of the
thymus. We report preliminary experiments to investigate this phenomenon in other cell types. We also propose a biological role for DNA
topoisomerase I modification.

EXPERIMENTAL PROCEDURES

The nuclear enzymes poly (ADPR) synthetase and Type I DNA
topoisomerase were partially purified from Chinese hamster ovary
cells. The nuclear fraction from 2.4×10^6 cells was prepared according to the procedure of Liu and Miller (14) except that 0.15% Triton
X-100 was present during the homogenization step. The enzymes were
extracted with a buffer containing a high salt concentration, and
the DNA removed from the extract, exactly as described (14). The
DNA-free extract (8.3 ml) was applied to an hydroxylapatite column
(Bio-Gel HTP, 0.5 x 8 cm) equilibrated with 50 mM Tris-Cl, pH 7.5,
1 M NaCl, 6% polyethylene glycol, 10 mM β-mercaptoethanol, 1 mM PMSF,
and the column developed according to the procedure described by
Ferro and Olivera (5) for the purification of DNA topoisomerase.

A strain of Saccharomyces cerevisiae (S288C) was grown aerobi-
cally in YEPD medium and harvested upon reaching stationary phase by
centrifugation (1000 x g) at 0° for 10 min. The cells (0.15 g) were
washed, and spheroplasts formed by the procedure of Jacobson and
Bernofsky (15) except that β-mercaptoethanol was used in place of
β-mercaptoethylamine (16). The spheroplasts were washed and lysed
following the procedure of Lange and Jacobson (17) and the suspension
of lysed spheroplasts (720 μl; referred to as the "yeast extract")
was used directly in the experiments. The product of the reaction was
analyzed by thin layer chromatography on PEI-cellulose F sheets
(E. Merck, Darmstadt) (18).

RESULTS

Poly ADP-ribosylation of DNA topoisomerase I in Chinese hamster
ovary cells

We have investigated the inhibition of DNA topoisomerase I by

poly (ADP-ribosylation) in a mutant line of Chinese hamster ovary
cells. These cells (strain EM9) are deficient in DNA repair and display elevated levels of sister chromatid exchange (19). The major
DNA topoisomerase activity and poly (ADPR) synthetase activity were
extracted from the nuclear fraction with a buffer containing high
salt, the DNA removed by precipitation with PEG, and the resulting
supernatant fraction chromatographed on hydroxyapatite (see Experimental Procedures). The fractions from the hydroxyapatite column were
then incubated in the presence or absence of NAD, the substrate for
poly (ADP-ribosylation) The incubated mixtures were assayed for DNA
topoisomerase I activity using pBR322 plasmid DNA as the substrate,
and the results shown in Figure 1. It is clear that there is a pronounced inhibition of topoisomerase I activity as a result of preincubation with NAD. These results strongly suggest that topoisomerase
I can be inhibited by poly (ADP-ribosylation) in the CHO EM9 mutant
cell line.

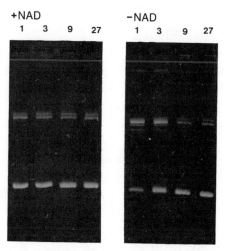

*Figure 1: Inhibition of DNA topoisomerase activity in extracts of
CHO cells by poly (ADP-ribosylation). An extract was prepared from
the CHO cell line (mutant strain EM9) and the extract chromatographed
on hydroxylapatite, as described in Experimental Procedures. An
aliquot (6 μl) of a fraction containing both poly (ADPR) synthetase
activity (0.75 units/ml; for definition of units, see ref. 5) and DNA
topoisomerase activity (2.5 x 10^5 u/ml) was incubated in a reaction
mixture (50 μl) containing 50 mM Tris pH 8.0, 10 mM MgCl$_2$, 1 mM DTT,
20 μg/ml calf thymus DNA, and where indicated, 0.2 mM (^{14}C) NAD
(15,000 cpm/nmole). The mixture was incubated 60 min at 23°, and an
aliquot withdrawn and assayed for DNA topoisomerase activity (5).
Numbers at top refer to the final dilution (x 10^{-3}) in the topoisomerase reaction mixture of the original fraction from the hydroxylapatite column. The sample which was incubated with NAD was inhibited at
least 9-fold relative to the control (incubated without NAD).*

Investigation of poly (ADP-ribosylation) in yeast

In order to determine the biological function of this extensive modification of topoisomerase I, a system where the methods of genetics and genetic engineering are highly developed would be desirable. We describe here our attempts to characterize the ADP-ribosylation phenomenon in yeast cells. In preliminary experiments (R. Sternglanz and B.M. Olivera), it was found that poly (ADP-ribose) synthetase was not assayable in yeast, and that crude yeast extracts inhibited the purified poly (ADPR) synthetase.

We have further characterized the inhibitory activity as shown in Figure 2. When a crude extract of yeast was added to a poly (ADP-ribosylation) reaction mixture containing the purified synthetase, the enzymatic activity of the enzyme decreased in a dose-dependent fashion. The inhibitory activity was not due to the enzymatic degradation of poly ADP-ribose, since upon the addition of labeled poly

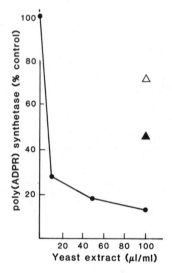

Figure 2: Inhibition of the poly (ADPR) synthetase reaction by an extract of yeast. The reaction mixture (50 μl) contained 100 mM Tris-Cl, pH 8.0, 10 mM MgCl$_2$, 1 mM DTT, 10 μg/ml calf thymus DNA, 0.2 mM or 2 mM (^{14}C) NAD (specific activity 51,000 or 5,100 cpm/nmole, respectively), poly (ADPR) synthetase (0.08 units) and yeast extract (see Methods) as indicated (μl extract/ml reaction). The mixtures were incubated at 23° for 5 min, and NAD incorporation determined (5). The data is presented ad the percent of the synthetase activity present in control reaction mixtures, which contained no yeast extract. ●——●, reaction mixtures containing 0.2 mM NAD; ▲, mixture containing 2 mM NAD; Δ, mixture containing yeast extract which had been heated at 90° for 5 min.

(ADP-ribose) to the reaction mixture, only minor breakdown was observed. However, as shown in Figure 2, the inhibition by the yeast extract could be partially overcome by increasing the concentration of NAD. This suggested that the inhibitory activity of yeast extracts might be due to an NADase.

This possibility was further explored. It was found that when the yeast extract was pre-incubated with the components of the poly (ADP-ribosylation) reaction mixture minus the synthetase, enzyme activity was completely abolished when the synthetase was later added. However, when the yeast extract was pre-incubated with the reaction mixture containing synthetase, and the NAD was added later, there was a much less severe inhibition of enzyme activity. These data support the hypothesis that the inhibitor is an NADase.

As a further test of whether poly (ADP-ribosylation) takes place, we examined whether incubation of yeast extracts with NAD would yield phosphoribosyl-AMP, the product resulting when poly (ADP-ribose) is degraded by an enzyme which attacks the pyrophosphate linkage (such as sknake venom phosphodiesterase; 20). A yeast extract was incubated with (^{32}P)-NAD and activating DNA (as in Figure 2) and the products of the reaction were analyzed by chromatography after treatment with venom phosphodiesterase. All products were degraded to AMP indicating the absence of phosphoribosyl-AMP. As a final attempt to find poly (ADP-ribosylation) in yeast extracts, we tried to assay for the enzyme using the normal conditions for the synthetase assay, but in the presence of relatively high levels of NAD. These assays were uniformly negative.

CONCLUSION AND SUMMARY

The modification of DNA topoisomerase I activity by poly (ADP-ribosylation) is a potentially important control mechanism for this ubiquitous enzyme which is widely believed to play a role in replication (2, 3). Our studies of the phenomenon so far suggest that DNA topoisomerase I activity may be regulated in response to the state of the DNA in the nucleus since the modification enzyme, poly (ADP-ribose) synthetase, is sensitive to interruptions in the normal double helical structure of DNA (10). Although this phenomenon has been most extensively characterized for the purified enzymes from calf thymus (5), preliminary evidence for poly (ADP-ribosylation) of topoisomerase I is presented in a nonthymic system, the mammalian cell line CHO EM9. Attempts to detect poly (ADP-ribosylation) in yeast cells have so far proven to be unsuccessful.

One biological role for poly (ADP-ribosylation): A speculation

The discovery of poly (ADP-ribosylation) of DNA topoisomerase I raises the question of the possible biological function of such a modification. Two observations in the literature may be relevant. At the present time, the only other enzyme known to be covalently modified and inhibited by poly (ADP-ribosylation) is a Ca^{++}-Mg^{++} dependent endonuclease found in certain eukaryotic nuclei (21). Although the biological function of this enzyme is not known, such an endonuclease, as well as DNA topoisomerase I, could introduce strand interruptions into the DNA double helix. Thus, poly (ADP-ribose) synthetase would effectively inhibit the formation of additional strand interruptions in the vicinity of a DNA structure which stimulates the synthetase (i.e., a double strand break or DNA strand interruption). A second observation in the literature is that inhibitors of poly (ADPR) synthetase stimulate sister chromatid exchange (22-26). The implication, therefore, is that normal poly (ADPR) synthetase activity is inhibitory to sister chromatid exchange.

These facts together suggest that one of the roles of poly (ADP-ribosylation) may be to depress strand exchange events in somatic cells, particularly those which might be initiated from interruptions in the double helical structure of DNA. It is well known that DNA strand breaks are highly recombinogenic. If an interruption in the double helical structure of DNA were already present, enzymes which might introduce additional interruptions (i.e., the Ca^{++}-Mg^{++} dependent endonuclease or DNA topoisomerase I) could increase the probability that such a DNA strand interruption might lead to a strand exchange event. It has been shown that in the presence of DNA termini, DNA topoisomerase I catalyzes strand exchanges in vitro. Strand exchange initiated by interruptions in the double helix would have no obvious benefit to a somatic cell and may well be detrimental: homozygosity of recessive mutations would be a consequence of such events. We suggest that a mechanism exists in somatic cells to suppress crossing over events in response to strand breaks, and that poly (ADP-ribose) synthetase is a major component in such a mechanism. The inhibition of DNA topoisomerase I might be only one of a complex series of modification reactions which are catalyzed by the synthetase to effect such a suppression.

Finally, the poly (ADP-ribosylation) of topoisomerase I may account for the inhibition of replication after DNA damage. It has been previously proposed that poly (ADP-ribose) may function in suppressing DNA synthesis (27-29), although a mechanism was not specified. Our studies indicate that the introduction of DNA strand breaks would lead to poly (ADPR) synthetase activation and as a consequence, the modification and inhibition of DNA topoisomerase I. If topoisomerase I played a role in the initiation and/or propagation of a replication fork, DNA replication would be inhibited.

ACKNOWLEDGEMENTS

       This work was supported by NIH Grant GM25654 and GM26107 from the Public Health Service.

REFERENCES

1.   Cairns,J. (1963) J. Mol. Biol. 6, 208.
2.   Wang,J.C. (1981) in: The Enzymes, pp. 331, Boyer,P.D., ed.,
         Academic Press, New York.
3.   Gellert,M. (1981) Annu. Rev. Biochem. 50, 879.
4.   Halligan,B.D., Davis,J.L., Edwards,K.A. and Liu,L.F. (1982)
         J. Biol. Chem. 257, 3995.
5.   Ferro,A.M. and Olivera,B.M. (1984) J. Biol. Chem. 259, 547.
6.   Hayaishi,O. and Ueda,K. (eds.) (1982) ADP-Ribosylation Reac-
         tions, Biology and Medicine, Academic Press, New York.
7.   Mandel,P., Okazaki,H. and Niedergang,C. (1982)
         Prog. Nucleic Acid Res. Mol. Biol. 27, 1.
8.   Yamada,M., Miwa,M. and Sugimura,T. (1971) Arch. Biochem.
         Biophys. 146, 579.
9.   Yoshihara,K. (1972) Biochem. Biophys. Res. Commun. 47, 119.
10.   Benjamin,R.C. and Gill,D.M. (1980) J. Biol. Chem. 255, 10502.
11.   Ferro,A.M., Higgins,N.P. and Olivera,B.M. (1983) J. Biol. Chem.
         258, 6000.
12.   Chang,L.M.S. (1971) Biochem. Biophys. Res. Commun. 44, 124.
13.   Bollum,F.J. (1978) Adv. Enzymol, Meister,A., ed., 47, 347.
14.   Liu,L.F. and Miller,K.G. (1981) Proc. Natl. Acad. Sci. USA
         78, 3487.
15.   Jacobson,M.K. and Bernofsky,C. (1974) Biochim. Biophys. Acta
         350, 277.
16.   Cabib,E. (1971) Methods Enzymol. 22, 120.
17.   Lange,R.A. and Jacobson,M.K. (1977) Biochem. Biophys. Res.
         Commun. 76, 424.
18.   Randerath,K. and Randerath,E. (1967) Methods Enzymol. 12A, 323.
19.   Thompson,L.H., Rubin,J.S., Cleaver,J.E., Whitmore,G.F. and
         Brookman,K. (1980) Somatic Cell Gen. 6, 391.
20.   Miwa,M. and Sugimura,T. (1982) in: ADP-Ribosylation Reactions
         Biology and Medicine, pp. 263, (Hayaishi,O. and Ueda,K.,
         eds., Academic Press, New York.
21.   Yoshihara,K., Tanigawa,Y., Burzio,L. and Koide,S.S. (1975)
         Proc. Natl. Acad. Sci. USA 72, 289.
22.   Utakoji,T., Hosoda,K., Umezawa,K., Sawamura,M., Matsushima,T.,
         Miwa,M. and Sugimura,T. (1979) Biochem. Biophys. Res.
         Commun. 90, 1147.
23.   Oikawa,A., Tohda,H., Kanai,M., Miwa,M. and Sugimura,T. (1980)
         Biochem. Biophys. Res. Commun. 97, 1311.
24.   Hori,T. (1981) Biochem. Biophys. Res. Commun. 102, 38.
25.   Watarajan,A.T., Csukas,I. and van Zeeland,A.A. (1981)
         Mutat. Res. 84, 125.
26.   Schwartz,J.L., Morgan,W.F., Kapp,L.N. and Wolff,S. (1983)
         Exp. Cell Res. 143, 377.
27.   Burzio, K. and Koide,S.S. (1970) FEBS Lett. 26, 181.
28.   Claycomb,W.B. (1976) Biochem. J. 154, 387.
29.   Berger,N.A., Petzold,S.J. and Berger,S.J. (1979) Biochim.
         Biophys. Acta 564. 90.

EFFECT OF UV INDUCED DNA LESIONS ON THE ACTIVITY OF
ESCHERICHIA COLI DNA TOPOISOMERASES: A POSSIBLE ROLE OF
THESE ENZYMES IN DNA REPAIR

Antonia M. Pedrini

Istituto di Genetica Biochimica ed Evoluzionistica
del C.N.R.
Via S. Epifanio, 14
27100 Pavia, Italy

Modifications of DNA structure by a number of DNA damaging
agents cause distortions resulting in the unwinding of the DNA helix.
We have shown that the unwinding angle due to the formation of pyrimidine dimers in a circular DNA molecule is very small and inadequate
to cause helix disruption. Although the alteration of DNA superhelicity produced by photodimers appeared very limited - at least 25 dimers were necessary in order to unwind a single topoisomer by one
superhelical turn (1) - it was enough to affect the activity of
Micrococcus luteus DNA topoisomerase I: one dimer per molecule inhibited the activity of this enzyme, by 5% (2).

In vew of the essential role of DNA topoisomerases in mantaining
a "physiological" supercoiling in vivo and the role of such super-
structure on DNA dependent functions, we thought it worthwhile to investigate whether the distortions caused by damaging agents on DNA,
might affect DNA topoisomerase activities and hence have some conse-
quences on DNA repair processes, and whether conversely, impairment
of topoisomerases activities my affect cellular recovery from UV irradiation. A role of calf thymus DNA topoisomerase I in eukaryotic
DNA repair processes has been recently suggested (3).

Activity of DNA topoisomerases on ultraviolet irradiated substrate

The activity of DNA topoisomerase I and DNA gyrase purified
from Escherichia coli has been assayed using as substrate DNA of
plasmid pAT153 irradiated at 254 nm with fluences producing a very
limited number of pyrimidine dimers per molecule in order to minimize
the effect due to the formation of minor photoproducts. The conse-

449

quence of UV light damage on the activity of DNA topoisomerase I,
when assayed in processive reaction conditions, is shown in Figure
1. Increasing amount of pyrimidine dimers caused a progresssive reduction in the amount of relaxed forms produced by the enzyme and
the simultaneous increase of the RFI band. These results indicate an
inhibition of the Escherichia coli DNA topoisomerase I activity similar to that observed for the Micrococcus luteus topoisomerase I.

The rate of DNA topoisomerase reaction has been shown by Wang
(4) to be a function of the degree of DNA supercoiling. Since photoproducts causes unwinding of the molecule, I have measured the activity of this enzyme in the presence of chloroquine, a drug which unwinds DNA. The results show that the inhibition by chloroquine is
qualitatively different from that caused by UV damage: in as much as
the drug caused not only the decrease of formation of the relaxed
species but also the progressive disappearance of the more relaxed
topoisomers. This result can be explained by the fact that at the
inhibitory concentrations the DNA is already partially relaxed as
shown in the last lane of the upper panel. DNA gyrase supercoiling
activity instead was not affected by UV damage, as already seen with
DNA gyrase purified from Bacillus subtilis (5).

At the moment remains to be explained how a very small alteration in the DNA structure consequent to the formation of a very limited number of dimers per molecule can affect the activity of the DNA
topoisomerase I unless one assumes that pyrimidine dimers, when formed
in a molecule with high superhelical density, cause a more dramatic
change in DNA structure than appreciated so far. It has been recently
shown that conformational changes such as cruciform formation and
B --- Z transition are a function of the superhelical density. It is
therefore possible envisage that at high superhelical density dimers
might facilitate local destabilisation of the helix creating single-
stranded areas to which DNA topoisomerase I will preferentially bind
in view of its high affinity for single-stranded DNA. The subsequent
passage of the opposite strand through the DNA·protein bridge will
than be hindered by the presence of the dimer. Although we have been
indirectly able to show the unlikelihood of such an effect for the
Micrococcus luteus DNA topoisomerase I, this possibility cannot be
excluded until it will be possible to demonstrate that photo-damage
modification of supercoiled DNA is not a function of DNA superhelical
density. Edenberg (6) has in fact shown an increase in sensitivity
to the single-stranded specific S_1 nuclease of UV irradiated circular
DNA directly proportional to the degree of supercoiling.

Responses of DNA topoisomerase mutants to ultraviolet irradiation

I have studied the effect of UV irradiation on the survival of
strain DM800, a derivative of Escherichia coli DM4100 containing a
deletion at 28 minutes on the chromosome. This deletion comprises
the cysB gene and the gene for DNA topoisomerase I. Figure 2 re-

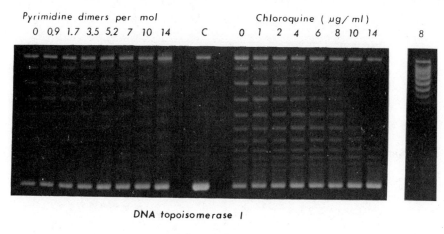

DNA topoisomerase I

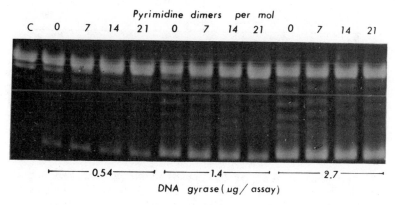

DNA gyrase (ug/assay)

Figure 1: Effect of photo-dimers or chloroquine on the activity of *Escherichia coli* DNA topoisomrases. Supercoiled or relaxed pAT153 DNA was irradiated at 254 nm at increasing doses. Calculation of pyrimidine dimer content and electrophoresis conditions were carried on as described by Ciarrocchi and Pedrini (1). Last lane above: pAT153 in presence of 8 ug/ml of chloroquine.

ports the UV survival curve of strains DM4100, DM800 and of their recA derivatives. Strain DM800 proved to be more sensitive than DM4100. The UV sensitivity was comparable with that seen for *Escherichia coli* mutants in DNA polymerase I or ligase but far less sensitive than that seen for recA mutants. Introduction of the recA mutation eliminates the component of UV sensitivity due to topoisomerase I absence indicating that in strain DM800 a recA dependent recovery from UV irradiation is partially affected.

Escherichia coli strains in which the topoisomerase I gene is deleted have a strong tendency to acquire secondary mutations that are linked to the gyrA and gyrB genes (7). This is the case also for

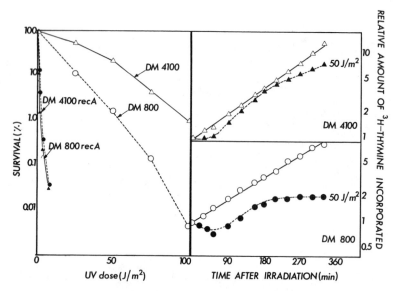

Figure 2: *(left) Colony-forming ability of irradiated cells of strains DM4100* cysB242 *(am), DM4100* recA56, *DM800* Δ *(top-cysB)204,* acrA12, gyrB225 *and DM800* recA56. *Cultures grown at 37oC at a density of about 1x10[8] cells/ml in minimal medium (Difco Bacto Davis) supplemented with 0.5% glucose and 40 μg/ml cysteine were irradiated and plated on the same minimal medium.*

Figure 3: *(right) Effect of UV irradiation on net DNA synthesis in thymine dependent derivative of DM4100 and DM800. Cells were grown for at least three generations in minimal medium containing 2 μg/ml of (*[3]*H)thymine up to a density of 1x10[8] cells/ml and then divided into two aliquots one of which received a dose of 50 J/m[2]. Incubation was continued and 0.050 ml samples were removed at intervals for determination of cold acid precipitable radioactivity.*

strain DM800 in which the compensatory mutation is linked to gyrB and results in a reduction in the average supercoiling of the chromosome (8). Since the rate of transcription from some promoters is altered by reduction in supercoiling, it is possible to surmise that the recA dependent UV sensitivity of DM800 is due to an altered induction of the recA-lexA regulatory circuit. The lack of sensitivity of the strain containing the gyrB mutation alone (data not shown), in agreement with what observed by von Wright and Bridges (9) with indipendently isolated gyrB mutants, argues against the possibility that decrease in the overall chromosome supercoiling is responsable for the UV sensitivity of the DM800 strain. It does instead suggest that the DNA topoisomerase I or an unknown gene comprised in the deletion is involved in some step of post replication repair.

   Further evidence for an impairment of recA dependent repair
pathway comes from the effect of UV light on DNA synthesis (Figure
3) and chromosome degradation (Figure 4).

   In Figure 3 we can see the effect of 50 J/m^2 at 254 nm on net
DNA synthesis in thymine derivatives of DM4100 and DM800. While
DM4100 cultures showed only a slight and transient effect in the apparent rate of thymine incorporation, DM800 cells showed a marked
effect consisting in a long delay and only partial recovery of post
irradiation DNA replication. In this type of experiment the apparent
rate of thymine incorporation reflects both DNA synthesis and degradation. In fact DM800 degraded significant amount of its DNA following
UV irradiation (Figure 4). Since control of post-irradiation DNA degradation is considered a DNA damage-inducible function, also this
result is suggestive of at least a partial impairment of post replication repair (10). Analogous indication has been obtained with a
DNA topoisomerase I mutant of Salmonella typhimurium (11). Tentatively
I will suggest that Escherichia coli DNA topoisomerase I is involved
in the recA dependent recovery from UV irradiation either indirectly
by facilitating induction of the recA-lexA regulon or directly participating in recombinational repair as suggested by the in vitro
facilitation of recA - protein mediated strand transfer (12).

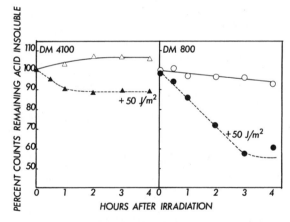

Figure 4:   Effect of UV irradiation on DNA degradation of DM4100
and DM800. Exponentially growing cultures were grown for at least
three generations with (^{3}H)thymine than washed and transferred to a
non radioactive medium. Each culture was divided in two aliquots,
one of which was irradiated with 50 J/m^2. Triplicate 0.05 ml samples
were removed at intervals for determination of cold acid precipitable
radioactivity.

ACKNOWLEDGEMENTS

I am grateful to Dr. P.C. Hanawalt, in whose laboratory part of these experiments were performed, for his encouragement and support. I thank Drs. Ann K. Ganesan and A. Falaschi for advice and valuable discussion and Drs. R. Sternglanz and A.C. Leonard for kindly providing strains. This work was supported by a research grant to the author from the European Molecular Biology Organization, by contract DE-AT03-EV700-07 with the Unites States Department of Energy and by the Progetti Finalizzati: "Medicina Preventiva e Riabilitativa" and "Controllo della crescita neoplastica".

REFERENCES

1.   Ciarrocchi,G. and Pedrini,A.M. (1982) J. Mol. Biol. 155, 177.
2.   Pedrini,A.M. and Ciarrocchi,G. (1983) Proc. Natl. Acad.
           Sci. USA 80, 1787.
3.   Ferro,A.M., Higgins,N.P. and Olivera,B.M. (1983) J. Biol.
           Chem. 258, 6000.
4.   Wang,J.C. (1971) J. Mol. Biol. 55, 523.
5.   Pedrini,A.M. and Ciarrocchi,G. (1983) in: "Proceedings of
           the National Meeting of Italian Group of DNA repair",
           Castellani,A., ed. (ENEA, Rome), p. 85.
6.   Edenberg,H.J. (1983) Cold Spring Harbor Symp. Quant. Biol.
           47, 379.
7.   Di Nardo,S., Voelkel,K.A., Sternglanz,R., Reynolds,A.E. and
           Wright,A. (1983) Cold Spring Harbor Symp. Quant. Biol.
           47, 779.
8.   Pruss,G., Manes,S.H. and Drlica,K. (1982) Cell 31, 52.
9.   von Wright,A. and Bridges,B.A. (1981) J. Bacteriol. 146, 18.
10.   Witkin,E.M. (1982) Biochimie 64, 549.
11.   Overbye,K.M., Basu,S.K. and Margolin,P. (1983) Cold Spring
           Harbor Symp. Quant. Biol. 47, 785.
12.   Cunningham,R.P., Wu,A.M., Shibata,T., Das Gupta,C. and
           Radding,C.M. (1981) Cell 24, 213.

VII
DNA Binding Proteins, Nucleases
and Poly ADP-ribose Polymerase

THE STRUCTURE OF THE BACTERIAL NUCLEOID

Kjell Kleppe, Ivar Lossius, Rein Aasland,
Knut Sjåstad, Askild Holck and Lars Haarr

Department of Biochemistry
University of Bergen
Bergen, Norway

INTRODUCTION

In prokaryotic organisms the chromosome is present in a nucleus-like body termed the nucleoid (1, 2). The nuclear membrane is, however, absent in prokaryotic cells. Most of our knowledge concerning the structure and organization of the nucleoid comes from studies with Escherichia coli. The condensed form of the chromosome can be visualized by phase contrast microscopy directly in living prokaryotic cells (3-5). The mode of division of the nucleoid can also clearly be seen. Escherichia coli cells growing in a minimal medium contain one or two separate nucleoids per cell. When this organism is grown in a rich medium the cells possess two nucleoids per cell.

The nucleoid structure has also been studied by means of the electron microscope, and conflicting results have been obtained. It is probable that these are caused by use of different fixation agents and also different ionic strength in the fixation medium. Thus, when Escherichia coli cells, grown in the presence of 0.35 M NaCl in the medium, are fixed with 0.1 % OsO_4, a markedly condensed nucleoid can be seen (6). When the salt is omitted from the growth medium the nucleoplasm appears to be more dispersed throughout the whole cell. The permeability barrier appears to be destroyed by OsO_4. Since the internal concentration of K^+ is at least 0.2 M in exponentially growing Escherichia coli cells (7), these data do suggest that in such cells the nucleoid exists in a contracted state.

Isolation and models of the "high salt" nucleoid

     Methods for the isolation of the nucleoid became available
through the work of Stonington and Pettijohn (8). The procedure involves lysing the cells with large amounts of egg white lysozyme followed
by detergent treatment and centrifugation in a neutral sucrose gradient containing 1 M NaCl (9, 10). Essentially two types of nucleoids
are obtained depending on the types of detergent used, namely cell
envelope-attached nucleoid and the cell envelope-free nucleoid. The
former contains large amounts of RNA, peptidoglycan, proteins and
lipids in addition to DNA. In the latter nucleoid there are approximately 30% RNA (w/w) and only small amounts of proteins, the major
protein being RNA polymerase. The RNA present is essentially nascent
RNA, consisting of rRNA, tRNA and mRNA (2).

     The structure of the "high salt" nucleoid is sensitive to both
RNase and DNase treatment resulting in a decrease in the sedimentation coefficient. By means of ethidium bromide titration the DNA in
the nucleoid has been shown to exist in a supercoiled state (2, 10).
Based on studies cited above, Pettijohn and coworkers (11) proposed
the clover leaf type model for the nucleoid. A similar model was also
suggested by Worcel and Burgi (12) for the Escherichia coli chromosome.
The essential features of this model, one form being shown in Figure
1, are that the DNA exists in a series of supercoiled loops, the
loops being held together by RNA chains. One DNA loop can thus be
relaxed separately without affecting the other loops. Further support
for such a model for the in vitro nucleoid came from electron micro-

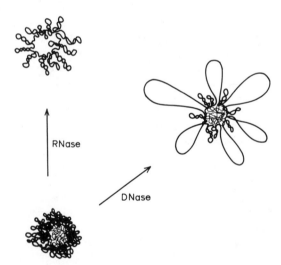

Figure 1:   *A possible model for the isolated "high salt" nucleoid.*
             From Kleppe et al. (1979), reference 1.

scopic data of Kavenoff (13). The estimate for the number of loops
has varied. Recent studies by Pettijohn, using psoralene crosslinking
and γ-irradiation, have suggested that the number of domains in the
in vitro "high salt" nucleoid is 100 $\pm$ 30. In vivo the number has been
reported to be 43 $\pm$ 30 domains (14).

Use of "high salt" nucleoid in DNA repair studies

The "high salt" method for isolation of nucleoids has proven
to be useful in studying induction of damage and repair of the chromosome in Escherichia coli. One example with methyl methanesulfonate
(MMS) (10) is shown in Figure 2A. The sedimentation coefficient of
the nucleoid decreased from 1560 S to approximately 850 S after
treatment of the bacteria for 1 hour with 25 mM MMS, suggesting that
the supercoiling has been lost. This was confirmed by titration studies
with ethidium bromide, Figure 2B. During the repair period the supercoiling was partially introduced again into the nucleoid. In the case
of mitomycin C (4, 5) large concentrations of this agent, higher than
5 M, leads to a decrease in S-value and loss of supercoiling. In the
presence of low concentrations of mitomycin-C, on the other hand,
there is virtually no loss in supercoiling but an increase in sedimentation coefficient to 2200 S. This corresponds to the mass of two
nucleoids, thus suggesting that the two nucleoids in a cell have fused
together during recombination repair. This was also verified using
phase contrast microscopy.

Models for the structure of the in vivo nucleoid

The present evidence suggest that the clover leaf model postulated from the "high salt" nucleoid data is correct. However, the
structure of the nucleoid in vivo is most probably different since
many artefacts can arise during the isolation of a nucleoid in the
presence of high salt and strong detergents. Thus, it is known that
mRNA will aggregate in high salt concentrations (15) and moreover,
rRNA and mRNA easily form complexes in high concentrations of salt
(16). A portion of the RNA molecules may also exist in a double-
stranded RNA-DNA hybrid form in the "high salt" nucleoid. Richardson
(17) has shown that such hybrids are formed when a reaction mixture
containing RNA polymerase and supercoiled DNA is treated with detergents and high concentrations of salt, i.e. conditions similar to
those used for the isolation of the "high salt" nucleoid. Stabilization
of DNA-RNA hybrids at the 3'-hydroxyl end of the RNA is thermodynami-
cally favoured under such conditions, leaving the 5'-end of the RNA
molecules essentially single-stranded. This part of the RNA could
then engage in complexes with other RNA or DNA molecules. Thus, the
separate DNA loops may represent regions on the chromosome where relatively little transcription occurs, whereas the transcriptionally
active areas are situated in the center of the isolated nucleoid.

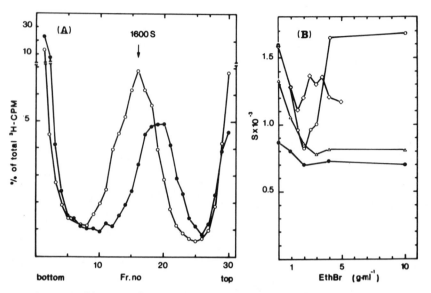

Figure 2: A Sedimentation pattern in 10-30% sucrose gradient of
          envelope-free nucleoids from MMS-treated Escherichia coli
          K12 DG75. o————o , control •————• , cells treated
          with 25 mM MMS for 1 hr at 37°C before lysis.

          B Sedimentation coefficient as a function of ethidium
          bromide concentration for envelope-free nucleoids from
          MMS-treated Escherichia coli K12 DG 75. o————o , con-
          trol •————• , cells treated with 25 mM MMS for 1 hour
          at 37°C before lysis. △————△ , cells treated for 1 hour
          with MMS, then resuspended in MMS-free medium and incu-
          bated for 1 hour or ◇————◇ 2 hours prior to lysis.
          Further details are given in reference 10.

     Autoradiographic studies of Escherichia coli cells have shown
that RNA synthesis occurs primarily in the cytoplasm and along the
membranes and not in the nucleoid centre (18). It appears, therefore,
that the location of most of the RNA chains in the centre of the isolated nucleoid has been inverted from that in the intact cell where
the RNA chains are found mainly at the outer regions of the nucleoid
or attached to the cytoplasmic membranes.

     We have earlier proposed a structure for the in vivo nucleoid
(1) which is shown in Figure 3. The basic concept of this model is
that RNA chains protrude into the cytoplasm and along the membranes
rather than being in the middle of the nucleoplasm. There are several
attachment points to the cell envelope; one is at the origin of replication and this attachment points is probably to the outer membrane
since a number of studies have shown that the proteins bound to the

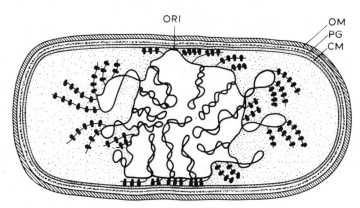

Figure 3: A schematic model of the organization of the nucleoid in vivo in Escherichia coli. Ori, origin of replication; OM, outer membrane; PG, peptidoglycan; CM, cytoplasmic membrane. From Kleppe et al., (1979) reference 1.

oriC DNA are outer membrane proteins (A). The chromosome appears to be attached to the cell-envelope throughout the whole cell cycle (20). Moreover, such an attachment point would seem to be necessary for segregation of the two daughter chromosomes. Attachment only to the inner membrane would probably be insufficient as an anchoring point as this membrane is too fluid. In addition to the origin attachment point there are probably many other binding sites to the inner membrane, but these are thought to be mediated via RNA polymerase-mRNA-ribosomes interactions to membrane and involving the synthesis of membrane and periplasmic proteins according to the signal hypothesis (1). It has been estimated that approximately 20 such contact points exist in the cell (21).

Stabilization of the in vivo nucleoid by proteins and polyamines

A central point in the model for the in vivo nucleoid is that there are still regions or domains of supercoiling. It is postulated that such domains or regions are stabilized by polyamines and proteins. All prokaryotic cells contain large amounts of polyamines and it has been shown earlier that polyamines can catalyze aggregation of particularly supercoiled DNA (22). With regard to DNA binding proteins a large number of these have recently been characterized. For a review see reference 23. Some have been given different names, but they appear to be identical species. Among the best characterized proteins are HU (identical to H2, BH2, HD, NS1, NS2, HLPII) molecular weight 9 kd, BH1 (possibly identical to HPLI, H1) and having a molecular weight of 28 kd has recently also been isolated (24). Several DNA binding proteins associated with the outer membrane have also been found to bind to an oriC DNA fragment (19). These and other proteins have also been described earlier by several other groups.

Proteins associated with DNA in vivo

There appears to be a number of proteins in the cell which can interact with the DNA. Employment of DNA-cellulose chromatography as a measure of DNA protein interaction has shown that at least 250 different intracellular proteins are bound to the DNA and can be eluted with increasing ionic strength (25). Many of these proteins could bind to the DNA-cellulose more or less in unspecific ion-exchange type manner and may not have any role in stabilizing specific DNA structure in vivo. In order to explore which proteins are bound to DNA in vivo we have recently developed a new method for isolation of the bacterial chromosome of Escherichia coli using T_4 lysozyme, Nonidet P-40 detergent and low ionic strength buffers (26). The complex isolated contains approximately equal amounts of DNA and RNA and some outer membrane fragments. The ratio of protein to nucleic acids is approximately 3:1. In analogy with eukaryotic cells we have used the term bacterial chromatin for this complex. The nucleic acids in this chromatin can easily be solubilized by micrococcal nuclease or by DNase, whereas RNase has little effect. A short sonication treatment (1-2 sec) also results in complete solubilization of the chromatin. When such solubilized chromatin is subjected to a neutral sucrose gradient centrifugation (SW 41 Ti, 38000 rpm, $5\frac{1}{2}$ hrs, 4°C) the DNA-protein complex sediments in the middle of the tube with a sedimentation coefficient of approximately 25S. Most of the proteins and all the RNA are found at the top of the gradient whereas the membrane fragments are present at the bottom (results not shown). In order to estimate the exact ratio of protein to DNA in the complex, the DNA-protein complex was subjected to metrizamide gradient centrifugation as shown in Figure 4. Essentially a single symmetrical peak was seen having a peak density of 1.16 g/cm^3 whereas the density of free Escherichia coli DNA was 1.12 g/cm^3. From the known effect of composition on the buoyant density of the nucleoprotein complexes (27) it can be estimated that the protein:DNA ratio in the present isolated complex was 0.7 (w/w).

The proteins in the DNA-protein complex were analyzed by SDS polyacrylamide gel electrophoresis and the result is shown in Figure 5A. Essentially 10 major as well as a number of minor protein bands were detected and the molecular weight of the major proteins were 160, 58, 55, 52, 49, 31, 19, 17, 16 and 10 kd. Only the protein having a molecular weight of 17 kd was acid extractable, Figure 5B, suggesting that this is the only basic protein present in the complex.

The 17 kd protein is probably identical to the BH1 or HLPI. The functions of the other proteins, except the 160 kd protein which must be β' and β subunits of RNA polymerase, are presently not known. The protein of 10 kd could correspond to HU (BH2, HLPII) but it is not acid extractable using the present conditions. At least 3 low molecular weight proteins with a molecular weight of 10 kd are present in the intact chromatin. One of them is an acidic protein, one is

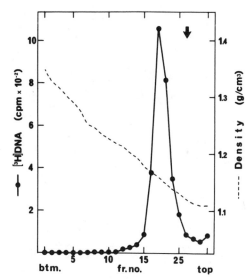

Figure 4: *Metrizamide density gradient centrifugation of the DNA-protein complex, isolated from the sucrose density gradient. The sample was mixed with metrizamide to give a final density of 1.20 g/cm³ in 10 mM Tris·HCl pH 7.5, 150 mM NaCl, 1 mM EDTA. Centrifugation was carried out at 38,000 rpm in a V65Ti-rotor for 16½ hrs at 40°C.*

neutral and one basic. These proteins are, however, found to be mostly associated with the membrane fractions and also in the RNA-protein complexes. Our results differ from those published earlier with regard to the amount of the HU protein bound to DNA in vivo (23, 28). The HU protein has, however, earlier been shown to be associated with ribosomes (29) and can become bound to membranes (30). Several of the major proteins may be associated with both membranes and DNA. This is particularly the case for the larger proteins (58 kd, 57 kd and 55 kd) (19).

Experiments are currently being carried out to determine whether defined protein-DNA subunit structures are present in the chromatin. Two possible models can be proposed for the stabilization of the nucleoid structure by proteins: A. Parts of the chromosome could possess nucleosome-like structures. B. Some proteins could serve as a scaffold for holding together the different supercoiled domains. Both models could in principle explain the stabilization of supercoiled regions by proteins. With regard to model A it should be noted that HU has been shown to be able to form nucleosomes in relaxed SV40 DNA in the presence of nicking closing enzymes (31). Such nucleosomes can, however, only form in low ionic strength buffers. Unstable

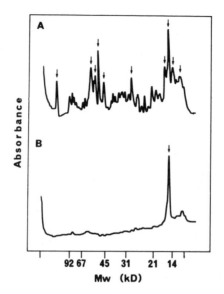

Figure 5: SDS polyacrylamide gel electrophoresis of (A) the total proteins in the DNA-protein complex from the sucrose gradient and (B) the acid extractable proteins of the complex. The acid extraction of DNA-protein complex was carried out using 0.25 N HCl. The precipitate removed by centrifugation and the supernatant was further extracted with 5 M perchloric acid. Samples were concentrated by trichloroacetic acid precipitation (25% w/v) before resuspension into sample-buffer. Electrophoresis was performed at 35 mA constant current (675 V hrs) in a 10-20% linear gradient slab-gel with the buffer-system previously described (26). Proteins were visualized with a silver staining technique and the gel-slab was scanned with a densitometer.

nucleosome-like structures have also been detected in electron microscope studies of DNA from <u>Escherichia coli</u> (32). In the case of linear DNAs, on the other hand, HU tends to catalyze the formation of condensed structures having multiple garland-like loops as suggested in model B. It remains to be seen wheter other proteins also can form such structures. Loop structures, possessing a protein core, have recently been described in the nucleoid from <u>Bacillus licheniformis</u> (33).

ACKNOWLEDGEMENTS

Parts of this work were supported by grants from the Norwegian Research Council for Science and Humanities.

REFERENCES

1.   Kleppe,K., Øvrebø,S. and Lossius,I. (1979)
       J. Gen. Microbiology 112, 1.
2.   Pettijohn,D.E. (1976) CRC Critical Rev. in Biochemistry
       4, 175.
3.   Mason,D.J. and Powelson,D.M. (1956) J. Gen. Microbiology
       71, 474.
4.   Lossius,I., Krüger,P.G., Male,R. and Kleppe,K. (1983)
       Mutation Res. 109, 13.
5.   Lossius,I. and Kleppe,K. (1981) in: Chromosome Damage and
       Repair, Plenum Press (Eds. E. Seeberg and K. Kleppe) pp. 41.
6.   Woldringh,C.L. (1973) Cytobiologie 8, 97.
7.   Schultz,S.G. and Solomon,A.K. (1961) J. Gen. Physiol. 45, 355.
8.   Stonington,D.G. and Pettijohn,D.E. (1971) Proc. Natl. Acad.
       Sci. U.S.A. 486.
9.   Worcel,A. and Burgi,E. (1974) J. Mol. Biol. 82, 91.
10. Lossius,I. Krüger,P.G. and Kleppe,K. (1981) J. Gen.
       Microbiology 124, 159.
11. Pettijohn,D.E. and Hecht,R. (1973) Cold Spring Harbor Symposia
       on Quant. Biol. 38, 31.
12. Worcel,A. and Burgi,E. (1972) J. Mol. Biology 71, 127.
13. Kavenoff,R. and Bowen,B.C. (1976) Chromosoma 59, 89.
14. Pettijohn,D.E. (1982) Cell 30, 667.
15. Boedtker,H. (1968) Methods in Enzymology 12B, 429.
16. Marcot-Queiroz,J. and Monier,R. (1965) J. Mol. Biology 14, 490.
17. Richardson,J.P. (1975) J. Mol. Biology 98, 565.
18. Ryter,A. and Chang,A. (1975) J. Mol. Biology 98, 797.
19. Hendrickson, W.G., Kusano,T., Yamaki,H., Balakrishnan,R.
       King,M., Murchie,J. and Schaechter,M. (1982)
       Cell 30, 915.
20. Korch,C., Øvrebø,S. and Kleppe,K. (1976) J. Bacteriology
       127, 904.
21. Dworsky,P. and Schaechter,M. (1973) J. Bacteriology 116, 1364.
22. Osland,A. and Kleppe,K. (1977) Nucleic Acid Res. 4, 685.
23. Geider,K. and Hoffmann-Berling,H. (1981) Annual Rev. of
       Biochemistry 50, 233.
24. Hübscher,U., Lutz,H. and Kornberg,A. (1980) Proc. Natl. Acad.
       Sci. U.S.A. 77, 5097.
25. Sjåstad,K., Haarr,L. and Kleppe,K. (1983) Biochim. Biophys.
       Acta 739, 8.
26. Sjåstad,K., Fadnes,P., Krüger,P.G., Lossius,I. and Kleppe,K.
       (1982) J. Gen. Microbiology 128, 3037.
27. Richwood,D., Birnie,G.D. and MacGillvray,A.J. (1975)
       Nucleic Acids Res. 2, 723.
28. Varshavsky,A.J., Nedospasov,S.A., Bakayev,V.V., Bakayeva,T.G.
       and Georgiev,G.P. (1977) Nucleic Acids Res. 4, 2725.
29. Mende,L., Timm,B. and Subramanian,A.R. (1978) FEBS Letters
       96, 395.

30.   Zentgraf,H., Berthold,V. and Geider,K. (1977)
           Biochim. Biophys. Acta 474, 629.
31.   Rouvieré-Yaniv,J. and Yaniv,M. (1979) Cell 17, 265.
32.   Griffith,J.D. (1976) Proc. Natl. Acad. Sci. U.S.A. 73, 563.
33.   Sloof,P., Maagdelijn,A. and Boswinkel,E. (1983)
           J. Mol. Biology 163, 277.

PROTEINS FROM THE PROKARYOTIC NUCLEOID: BIOCHEMICAL AND ^{1}H NMR STUDIES ON THREE BACTERIAL HISTONE-LIKE PROTEINS

M. Lammi[1], M. Paci, C.L. Pon, M.A. Losso[1], A. Miano[2],
R.T. Pawlik, G.L. Gianfranceschi[2] and C.O. Gualerzi[2]

Max-Planck-Institut für Molekulare Genetik
Abt. Wittmann,
Berlin, Germany

[1]Dept. of Cell Biology
University of Calabria
Calabria (CS), Italy

[2]Dept. of Cell Biology
University of Camerino
Camerino (MC), Italy

INTRODUCTION

In the past, the absence of histones and organized chromatin were regarded as characteristics of the prokaryotic cell. In reality, the approximately 4000 kb circular chromosome of bacteria is condensed in a region called the nucleoid and gentle lysis methods allowed the obtainment of chromatin-like fibers with repetitive granular structure. The granules have a diameter $\simeq$ 130 Å and are reminiscent of the eukaryotic nucleosome (1). In the structural organization of "bacterial chromatin", a role is probably played by specific DNA-binding proteins (2-5). Only recently, the characterization of the structural and functional properties of these proteins has been undertaken in various laboratories.

Here we outline the results obtained by our group on the structural and functional properties of three Escherichia coli ds-DNA-binding proteins, which we call NS1, NS2 and H-NS. The native complex containing stoichiometrically equivalent amounts of NS1 and NS2 will be referred to as NS (6), which is equivalent to the protein independently isolated by other groups and referred to us as HU (2),

467

HD (3), protein 2 (4), DBPII (5) and HLPII (7). H-NS could be equivalent to protein Hl (8), protein 1 (4) or HLPI (7). Some experiments were also carried out with BS-NS, the single protein isolated from *Bacillus stearothermophilus*, homologous to both NS1 and NS2, which has recently been crystallized, lending hope for the rapid elucidation of the three-dimensional organization of these proteins (9).

RESULTS

The main properties of NS1, NS2 and H-NS of *Escherichia coli* are listed in Table 1. NS1 and NS2 are homologous to each other and to other DNA-binding proteins isolated from a number of bacterial species. H-NS, on the other hand, is different from NS in amino acid composition and the amino terminal octapeptide: Ser-Glu-Ala-Leu-Lys-Ile-Ile....

All three proteins retain radioactive ds-DNA on nitrocellulose filters; the dose-response curve is linear for H-NS (not shown) and sigmoidal for NS (Figure 1) suggesting that the binding of this protein to DNA is cooperative.

NS is more efficient than either NS1 or NS2 alone in binding DNA (Figure 1). Unlike with NS, whose DNA binding is strongly reduced by increasing the ionic strength, binding of H-NS is stimulated by monovalent cations and by magnesium (not shown). The interaction between NS and DNA was also studied by competition with proflavin.

Table 1: Characteristics of DNA-binding proteins of *Escherichia coli* nucleoid

Properties	NS1	NS2	H-NS
Number of amino acid residues	90	90	∿135
Molecular weight	9,225	9,535	∿16,000
Lys/Arg	1.8	3.7	1.1
Basic/Acidic	1.4	1.4	1.2
Hydrophobic	45.6%	42.2%	41 %
Ala	21.1%	16.6%	8.3%
Copies/Cell	80,000 – 120,000		∿20,000
Homology	69%		–

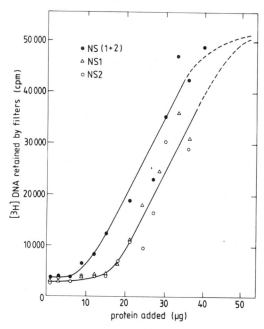

*Figure 1:   Binding of DNA by NS, NS1 and NS2. <u>Escherichia coli</u> ^{3}H-DNA
(5000 cpm/μg) and the indicated amounts of protein were incubated
15 min at 37°C in 0.1 SSC and filtered through nitrocellulose filters.*

Since the absorbance of proflavin is quenched by DNA, the competition
between proflavin and NS for DNA binding results in a decreased
quenching when increasing amounts of protein are offered (Figure 2A).
Also in this case, the dose-response curve appears sigmoidal, suggesting binding cooperativity (Figure 2B).

<u>Effect on thermal stability and transcription of DNA</u>

        NS1, NS2 and NS affect the denaturation and renaturation of
DNA by causing an increase in both Tm and extent of renaturation
without affecting the total hyperchromicity and the extent of denaturation cooperativity (11). The effect is qualitatively similar with
<u>Escherichia coli</u> and calf thymus DNA, as well as with synthetic poly-
deoxynucleotides. From the quantitative point of view, however, the
effect is greater on homologous (i.e. prokaryotic) than on calf
thymus DNA and on poly d(C,G) than on poly d(A,T) (11). Similar effects, including the above-mentioned "quantitative species-specifi-
city" were found with BS-NS. A comparison of the effects of BS-NS,
NS, NS1 and NS2 on the thermal stability of <u>Bacillus stearothermo-
philus</u> DNA is presented in Figure 3. It can be seen that both BS-
NS and NS increase the extent of DNA renaturation (Figure 3A), the
temperature at which 50% of the maximum DNA renaturation occurs
(Figure 3B), as well as the Tm of denaturation (Figure 3C). All

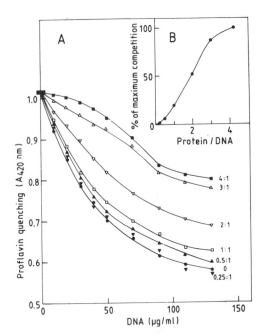

Figure 2: *NS-proflavin competition for DNA binding. The experiment was carried out in 0.1 SSC as described (10) offering increasing amounts of the indicated NS:DNA mixtures to a proflavin solution having A$_{420}$ = 1.0.*

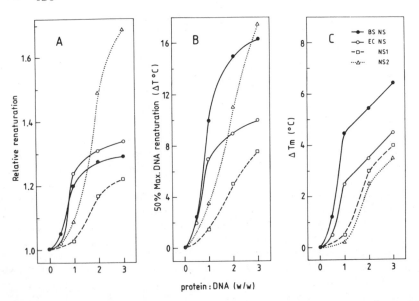

Figure 3: *Effect of NS, NS1, NS2 and BS-NS on the thermal stability of DNA. The experimental conditions are essentially those described (11).*

dose-response curves are sigmoidal with most of the effects occurring at approximately a 1:1 protein:DNA weight ratio. Isolated NS1 and NS2 are less efficient in promoting the above effects, although an excess of NS2 displays the greatest effect in restoring the original DNA hypochromicity.

Transcription is inhibited by NS and, slightly less efficiently, by NS1 and NS2. The extent of inhibition varies depending upon the type of DNA template, being two fold greater with <u>Escherichia coli</u> than with calf thymus DNA (12). The initiation is probably due to the stabilization of the double helical structure of DNA induced by the protein.

Thus, in both denaturation-renaturation and transcription experiments, NS has greater effects on prokaryotic than on calf thymus DNA. Since the two types of DNA are prepared in the same way and display, as far as we can judge, identical properties, the "quantitative species specificity" could stem from a more efficient interaction of NS with the homologous DNA. This could be explained assuming the existence, only in prokaryotic DNA, of a pattern of regularly repeated regions where the proteins bind preferentially ("phasing"); cooperative protein-protein interactions would then afford the maximum packaging effect with the lowest amount of protein. This hypothesis is illustrated in the model shown in Figure 4.

As to the effect of H-NS, also this protein protects somewhat DNA from denaturation, stimulates its renaturation and inhibits transcription (not shown).

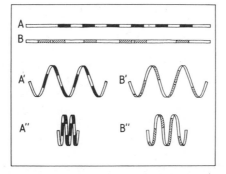

Figure 4: Hypothetical species-specific "phasing" of NS binding sites on DNA. Prokaryotic DNA - extended (A,A'), packaged (A"); eukaryotic DNA - extended (B,B'), packaged (B").

Protein-protein crosslinking

NS1 and NS2, as well as BS-NS, are normally found in aggregated forms (6, 9). Reaction with the bifunctional crosslinking reagents
DMS and DMA yields oligomers (up to hexamers), the crosslinker with
the longer arm (DMS) being more efficient (Table 2). In the electrophoretic gels, the band of dimers appears rather heterogeneous due
to the formation of more than one type of product (e.g. αβ, αα or ββ)
having different electrophoretic mobilities. The band of trimers, on
the other hand, is always very narrow indicating that geometrically
equivalent trimers are distributed and that the RCSM resonances are
more numerous in BSNS.

Protein-DNA interaction

Several lines of evidence indicated that, while protein-protein
interaction in NS is primarily hydrophobic, protein-DNA interaction
is chiefly ionic. When increasing amounts of deuterated DNA were
added to NS, the resolution of the spectrum was progressively lost
due to the broadening of all resonance lines. Aside from the Arg δCH_2
peak which was comparatively more affected, all resonances broadened
simultaneously indicating that NS binds to DNA in the aggregated form
and suggesting that the interaction involves Arg residues (Parci
et al., in preparation). Selective chemical modification of BS-NS
with 2, 3 butanedione showed that one Arg located in the positively
charged, hydrophilic region D (see Figure 5) is essential for the
interaction with DNA. Finally, involvement in DNA binding of the
single His residue of NS2 also located in the D "domain", was suggested by dye-sensitized photooxidation of this residue (Lammi et al.,
in preparation).

A model for the structural organization of NS

Hydrophobicity analysis (15) of NS1, NS2, BS-NS and NS-related
proteins from other bacteria reveals the existence of two hydrophobic
(A and C in Figure 5) and two hydrophilic regions (B and D in Figure
5) as well as two additional hydrophobic regions of variable intensity near the amino and caroxyl termini.

Combining the results of the crosslinking, chemical modifications
and ^{1}H-NMR spectroscopy, we can surmise that the NS1-NS2 dimers are
held together by a very strong hydrophobic interaction involving the
stacking formed (e.g. αβα, βαβ). The presence of DNA during the reaction depresses the formation of some of the dimers and, especially
at low protein concentrations, favors the formation of larger aggregates (Table 2). Unlike with larger aggregates, the formation of
dimers is insensitive to increases of ionic strength (>0.5 M KCl)
(13) or to drastic reductions of the NS concentration (<0.01 mg/ml)
indicating that at least one of the dimeric interactions (i.e. the
αβ one) has a strong hydrophobic character and a low dissociation
constant ($< \sim 10^{-7}$M).

Table 2: Crosslinking products of NS[*]

Product	yield (%) after 2 min reaction with			
	dimethylsuberimidate (DMS)		dimethyladipimidate (DMA)	
	-DNA	+DNA	-DNA	+DNA
Monomer	17	20	34	29
Dimer	74	42	56	51
Trimer	3	19	6	15
Tetramer	4	10	4	5
Penta and hexamer	2	9	-	-

[*] Reaction conditions will be described in a forthcoming paper (Losso et al., in preparation).

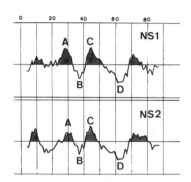

Figure 5: Hydropathic character of NS1 and NS2. The primary struc-tures of NS1 and NS2[14] were analyzed according to a computer to dis-play their hydrophobic (shaded area) and hydrophilic (non-shaded) regions.

Native H-NS appears to be aggregated and its reaction with DMS yields a substantial amount of dimers and traces of tri- and tetra-mers. The crosslinking is unaffected by the presence of DNA.

^{1}H-NMR Spectroscopy

The ^{1}H-NMR spectrum of NS is characterized by the presence of a large number of highfield perturbed Phe resonances, several shielded (RCSM) and deshielded methyl resonances and backbone NH protons quite inaccessible to the solvent (Figure 6). These features arise from the tertiary and quaternary structures of the proteins which involve strong hydrophobic interactions. Some spectral differences between NS1, NS2 and NS showed that heterologous dimers (i.e. NS1 + NS2) are formed in preference to homologous self-aggregation (13). Unlike that of the eukaryotic histones, the tertiary and quaternary structures of NS and BS-NS are totally insensitive to extensive ionic strength va-riations (13). In Figure 6 a comparison of the spectra of NS and BS-NS is also presented. Although the spectra are overall similar, it is clear that the perturbed Phe resonances are differently of two-to-four (out of six) Phe residues. Since two out of three Phe resi-dues of NS1 or NS2 are localized in the highly conserved hydrophobic region C, we suggest that this region of the molecule is involved in the major dimeric interaction. The second hydrophobic region A, which contains the largest number of conserved amino acid substitu-tions is presumably involved in the homologous dimeric interaction. Of the two hydrophilic regions, the first (B) is unlikely to be in-volved in an interaction with the DNA backbone since it contains in both NS1 and NS2, an excess of negative charges and, in the case of BS-NS, the arginine located therein was unavailable for butanedione modification (Lammi et al., in preparation). The second hydrophilic

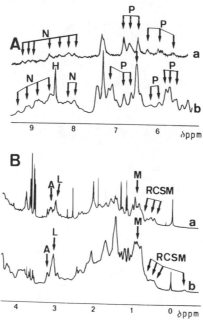

Figure 6: ^{1}*H-NMR spectra of NS and BS-NS. Aromatic (A) and alipha-tic (B) region of BS-NS (10^{-5}M, spectra a) and NS (5 x 10^{-5}M, spectra b) in 10 mM K phosphate buffer, pH 7.5 (meter reading) in 99% D$_2$O. The spectra were recorded at 20° on a Bruker WM-400 spectrometer at 400 MHz as previously described (13). N, hard-to-exchange NH protons; P, perturbed Phe ring protons; H, C-2 His proton; A, Arg δ=CH$_2$ peak; L, Lys ε=CH$_2$ peak; M, methyl protons peak; RCSM, ring current shifted methyl resonances.*

region (D), aside from being one of the most conserved of the mole-cule and overall positively charged, was shown to contain (in BS-NS) at least one arginine essential for DNA binding. Furthermore, at least part of this hydrophilic "domain" is predicted to have an α-helical structure which would orient some of the basic residues so as to favor a potential interaction with the DNA phosphate backbone in the minor groove.

Based on the available data and on the above considerations, we feel that the most likely structural arrangement of NS is that shown in Figures 7 and 8, in which the heterologous dimers are com-bined to form tetramers in such a way that the DNA binding regions are exposed with a two-fold rotational symmetry. The finding of small but significant amounts of hexamers in the crosslinking experiments suggests that additional interactions among tetramers take place probably giving rise to octamers. Whether such octamers indeed exist,

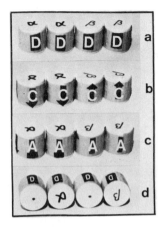

*Figure 7: Hypothetical structural organization of NS. Two NS1 mono-
mers (α) and two NS2 (β) exposing the "domains" of Figure 6. a, the
positively charged hydrophilic region D responsible for DNA interaction; b, the Phe-containing hydrophobic region C responsible for
the major heterologous dimeric interaction; c, the hydrophobic region
A responsible for the weaker homologous dimeric interaction; d, the
monomers oriented in an anti-parallel fashion to show the surfaces
probably responsible for the weaker interactions yielding octamers
from tetramers.*

and whether they are able to organize, alone or in combination with
H-NS, the prokaryotic DNA in a way similar to that of eukaryotic nucleosomes will have to await further study.

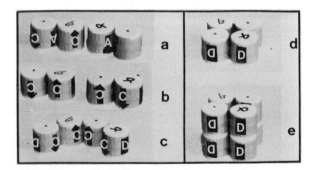

*Figure 8: Building an NS octamer. a, formation of the weak homolo-
gous dimeric interactions; b, homologous dimers exposing the strong
hydrophobic binding sites; c, formation of tetramer; d, tetramer;
e, octamer.*

REFERENCES

1.   Griffith,J.D. (1976) Proc. Natl. Acad. Sci. USA 73, 563.
2.   Rouviere-Yaniv,J. and Gros,F. (1975) Proc. Natl. Acad. Sci.
         USA 72, 3428.
3.   Berthold,V. and Geider,K. (1976) Eur. J. Biochem. 71, 443.
4.   Varshavsky,A.J., Nedospasov,S.A., Bakayeva,V.V. and
         Georgiev,G.P. (1977) Nucleic Acids Res. 4, 2725.
5.   Geider,K. and Hoffmann-Berling,H. (1981) Annu. Rev. Biochem.
         50, 233.
6.   Suryanarayana,T. and Subramanian,A.R. (1978) Biochim. Biophys.
         Acta 520, 342.
7.   Lathe,R., Buc,H., Lecocq,J-P. and Bautz,E.K.F. (1980)
         Proc. Natl. Acad. Sci. USA 77, 3548.
8.   Spassky,A. and Buc,H.C. (1977) Eur. J. Biochem. 81, 79.
9.   Dijk,J., White,S.W., Wilson,K.S. and Appelt,K. (1983)
         J. Biol. Chem. 258, 4003.
10.   Gianfranceschi,G.L., Amici,D. and Guglielmi,L. (1975)
         Biochim. Biophys. Acta 414, 9.
11.   Miano,A., Losso,M.A., Gianfranceschi,G.L. and Gualerzi,C.O.
         (1982) Biochem. Internat. 5, 415.
12.   Losso,M.A., Miano,A., Gianfranceschi,G.L. and Gualerzi,C.O.
         (1982) Biochem. Internat. 5, 423.
13.   Paci,M., Pon,C.L., Losso,M.A. and Gualerzi,C.O. (1983)
         in press.
14.   Mende,L., Timm,B. and Subramanian,A.R. (1978) FEBS Lett.
         96, 395.
15.   Kyte,J. and Doolittle,R.F. (1982) J. Mol. Biol. 157, 105.

THE ROLE OF HMG1 PROTEIN IN NUCLEOSOME ASSEMBLY AND IN

CHROMATIN REPLICATION

Catherine Bonne-Andrea, Francis Harper,
Joëlle Sobczak and   Anne-Marie De Recondo

Unité de Biologie et Génétique moléculaires
Institut de Recherches Scientifiques sur le Cancer
B.P. no 8
94802 Villejuif Cedex, France

INTRODUCTION

Much progress has been made in understanding the basic structure of chromatin, at least at the nucleosomal level. But, how the eukaryotic DNA is folded into this structure, changes its shape and unfolds is still unknown.

The process of chromatin assembly in vitro from the separated components has been widely studied. Rapid assembly in vitro at physiological ionic strength was demonstrated under the influence of different factors: nucleoplasmin (1), ribonucleic acid (2), acidic polypeptides (3).

Determination of the primary structure of HMG1 (a non histone protein of the chromatin) has revealed the extraordinary acidity of the C-terminus of the molecule, which presents a continuous stretch of 41 acidic residues (4). Such a structural feature may be of importance to the functions of HMG1. It was tempting to speculate that this high negative charge density in the C-terminus of the molecule is involved in complex formation with the very basic regions of histones. In fact, complexes between HMG1 and histone H1 have been reported (5, 6). However, at the present time, HMG1 has been reported to be inactive in nucleosome assembly in vitro (2). The methods used for the purification of this protein (exposure to extremes of pH and organic solvents (7)) could be responsible for its inactivity.

We have previously described the purification and different
properties of a single-stranded DNA binding protein from rat liver,
isolated under conditions which preserved its native structure and
which allowed us to show the different functional properties exhibited by this depending on the physiological state of the rat liver
(8, 9). This protein has recently been identified with an HMG1 protein (10).

The present work provides evidences that HMG1 interacts with
histones and is able to mediate the reconstitution of subunits very
similar to nucleosome cores at physiological ionic strength.

RESULTS

Histone-histone associations at 0.15 M NaCl in the presence of
HMG1 protein

The cross-linking reagent dimethyl-suberimidate was used to
reveal interactions among the histones in the presence or absence
of HMG1.

Figure 1 shows the results of such experiments, analyzed by
SDS-PAGE: at 0.15 M NaCl, histones were largely cross-linked into
dimers, whereas HMG1, under the same conditions, did not form oligomers. When histones and HMG1 were mixed at different HMG1-to-histone
ratios of 0.5, 1 and 2, two major additional cross-linked products
were revealed, which could correspond to tetramer and octamer of
histones on the basis of their respective electrophoretic mobility.
This result shows that HMG1 allows the association of histones at
low ionic strength.

The relative yield of tetramer and octamer appeared to be primarily proportional to the amount of histones, HMG1 did not seem to
be cross-linked with histones.

Assembly of nucleosome

We next questioned whether HMG1 was able to mediate the transfer of associated histones to DNA and to reconstitute nucleosome
subunits at a physiological ionic strength.

Supercoiling assay: Nucleosome formation was first followed
by the insertion of superhelical turns into circular relaxed DNA as
described by Germond et al. (12).

Histones preincubated with HMG1 were mixed with SV40 DNA Ir
at an ionic strength of 0.15 M NaCl. Topoisomerase I was added to
the mixture in order to relieve any extranucleosomal superhelicity.
The deproteinized DNA was then analyzed by agarose gel electrophoresis.

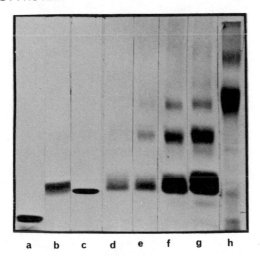

*Figure 1:   Effect of HMG1 protein on the cross-linking of histones
at 0.15 M NaCl. Concentrated histones in a 10 mM Tris-HCl, pH 8,
2 M NaCl buffer, were diluted in an HMG1 solution to 0.15 M NaCl,
pH 7.5, at different appropriate weight ratios. Samples were cross-
linked with dimethyl suberimidate, as previously reported (11),
prior to addition of SDS and analysis on a SDS 4- 20% polyacrylamide
gradient gel. Lanes a and b, histones (10 µg) untreated and treated
with dimenthyl suberimidate; lane c and d, HMG1 (5 µg) untreated or
treated with dimethyl suberimidate; lanes e, f and g, histones
(10 µg) mixed respectively to 5, 10 and 20 µg of HMG1 and treated
with dimethyl suberimidate; lane h, cross-linked histone octamer
marker.*

Figure 2 shows the distribution of superhelical turns induced
in SV40 DNA Ir molecules by a mixture of HMG1-histones. The most
efficient assembly was obtained at a histone to DNA weight ratio
of 1.5 (Figure 2B, lane 2) and an HMG1 to-histone weight ratio of
2-3 (Figure 2A, lanes 3-4). Lanes 3-4 in Figure 2B show that assembly
was inhibited at a histone-to-DNA ratio higher than 1.5, even when
HMG1 was added simultaneously with histones. When HMG1 was omitted
from the association mixture, no supercoiling was observed (lane 6).
Lanes 5 in Figure 2A and B demonstrate that there was no change in
the initial relaxed state of DNA when it was incubated with HMG1
alone at this salt concentration.

Micrococcal nuclease digestion of reconstituted complexes: To
further affirm the fidelity of core particle reconstitution in the
presence of HMG1 in vitro, we next determined the size of DNA fragment released after a micrococcal nuclease treatment from the reconstituted complex.

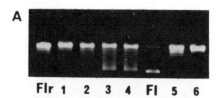

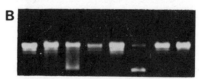

*Figure 2: Generation of superhelical turns in SV4O DNA Ir, due
to core particle assembly at O.15 M NaCl. Incubations were performed
at 37°C for 3O min. SV4O DNA Ir was at a final concentration of
1O μg/ml and was added with topoisomerase I to preincubated histone-
HMGl solution. After incubation, DNA was deproteinized and electrophoresed as previously described (13).
A) Supercoiling induced with increasing amount of HMGl at a prefixed
histone-to-DNA weight ratio of 1.5. Lanes 1 to 4, final HMGl to histone weight ratios were O.5 (1), 1.O (2), 2.O (3), 3.O (4); lane
5, histones were omitted; lane 6, HMGl was omitted.
B) Supercoiling induced at a prefixed HMGl-to-histone weight ratio
of 2 with increased amount of histones. Lanes 1 to 4, final histone-
to-DNA weight ratio of O.8 (1), 1.5 (2), 2.O (3), 3.O (4); lane 5,
histones were omitted; lane 6, HMGl was omitted. The markers were
supercoiled DNA (FI) and circular, relaxed, covalently closed DNA
(FIr).*

Furthermore, in order to prove that the extract of topoisomerase I was not required in the assembly process, complexes were reconstituted from a mixture of the four histones, with HMGl and SV40
DNA I, the latter being either relaxed (with or without topoisomerase)
or supercoiled. Reconstitutes were submitted to micrococcal nuclease
digestion. As shown in Figure 3A, only the reconstitution performed
with supercoiled DNA yielded an assembly sufficient to permit observation of a protected DNA fragment in the monomer region (lane 5);
with DNA Ir, with or without topoisomerase I, DNA fragments of 140
base pair were barely detected under these experimental conditions
(lanes 3, 4). When HMGl or histones were omitted no protected DNA
fragment appeared (not shown). These results show that the size of
about 140 base pairs released from reconstituted complex was in good
agreement with the size of the DNA fragment associated with native
nucleosome (lanes 2, 6), and that this assembly was favored by a
negative supercoiled DNA. If similar experiments were performed with
(^{3}H)-labelled DNA, and the digests acid-precipitated, percentages

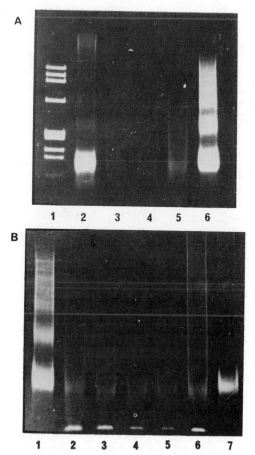

Figure 3: A) *Nuclease digestion of the products of in vitro assembly.*
SV40 DNA (1 μg), either supercoiled or relaxed, was added to histone-
HMG1 solution. Histone-to-DNA weight ratio was 1.2 and HMG1-to-histone
weight ratio was 2.0. After an incubation at 37°C for 30 min, samples
were made to 1 mM CaCl$_2$ and digested with micrococcal nuclease (1
U/μg DNA), 5 min at 37°C. The digestion was terminated by addition
of EDTA to 15 mM and the DNA fragments prepared for electrophoresis
as described in Duguet et al. (13). The sample were electrophoresed
in an 8% acrylamide slab gel in Tris/borate, EDTA buffer at 10 V/cm
for 3 h.
Lane 1, Hae III-digested φX174 DNA; lane 2, DNA isolated from native
core particles (140-200 bp); lane 3, assembly with DNA Ir supplemented
in topoisomerase I; lane 4, assembly with DNA Ir; lane 5, assembly
with DNA I; lane 6, DNA isolated from micrococcal nuclease digestion
of rat liver chromatin.
        B) Time course of assembly. Assembly and nuclease diges-
tion were performed with (^{3}H)-labelled SV40 DNA I under conditions
described above. At each time, aliquot was removed from a large-

*scale reaction: 0.1 µg for measuring the amount of DNA protected
by acid precipitation and 0.9 µg for analyzing the DNA fragments on
acrylamide gel.
Lane 1, DNA fragments isolated from micrococcal nuclease digestion
of rat liver chromatin. Lanes 2-6, the times of in vitro assembly
prior to digestion were 0,5, 30, 60 and 90 min; lane 7, DNA of native
core particle. Zero time is the time necessary to take an aliquot
for immediate digestion after the addition of DNA to histone-HMG1
solution.*

of protected DNA of 20%, 5-10% and 35-40% were obtained respectively,
for the assembly digested by micrococcal nuclease, under conditions
of Figure 3A, lanes 3, 4 and 5. Topoisomerase I thus appeared to
enhance in vitro the assembly with a relaxed DNA but the extent obtained is lower than what is obtained with a negatively supercoiled
DNA. Since SV40 DNA contained about 5000 base pairs, and 140 base
pairs were protected in each core particle, the percentage (35 to
40%) of protected DNA corresponded to an average of 10-15 nucleosomes
for the SV40 genome.

The conversion of SV40-labelled DNA into a micrococcal nuclease-
resistant DNA fragment was the most suitable technique for determining the rate of assembly in vitro (2, 14). The percentage of protected labelled DNA, was independent of the incubation time. As
shown in Figure 3B, the same amount of 140 bp DNA fragment was observed at each time, providing that the assembly was instantaneous
when histones were reconstituted with supercoiled DNA in the presence
of HMG1.

Electron microscopy of reconstituted complexes. In order to
further demonstrate that histones and supercoiled DNA are rapidly
assembled in the presence of HMG1, SV40 DNA I was added to a solution
of histones and HMG1 and directly examined by electron miscroscopy.

Electron micrographs shown in Figure 4 revealed that the histone
complexes formed on SV40 DNA I (panels B-G) are globular particles
morphologically similar to the native nucleosome (panel A) and with
the same diameter of 100 Å.

DISCUSSION

In this study, we show that HMG1 purified by its affinity for
single-stranded DNA can act as nucleosome assembly factor under physiological conditions, like nucleoplasmin described by Laskey et al.
(1). First, HMG1 interacts with the four core histones to mediate
their association in tetramer and octamer at physiological ionic
strength.

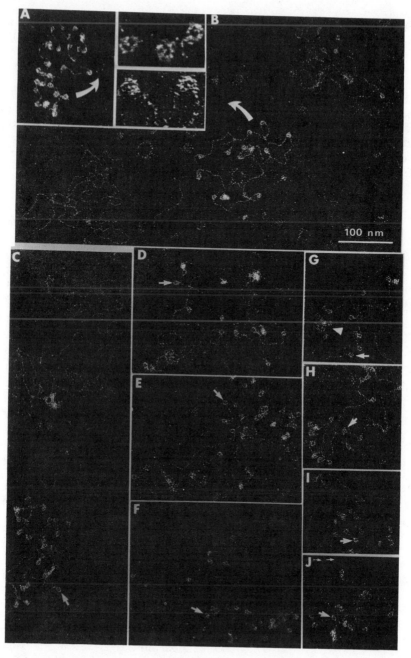

100 nm

Figure 4:   Electron microscopy study of reconstituted nucleosomes.
The complexes were prepared for dark field electron microscopy as
described in Saragosti et al. (15), (A) example of a native SV40
minichromosome. This viral chromatin displays 26 nucleosomes evenly

spaced (B, C, D, E, F, G, H, I, J). Histones, HMG1 protein and SV40
DNA I were combined as indicated in Figure 3 and the assembly product
was spread to reveal a mixture of twisted or relaxed (SV40 DNA I was
contaminated by a small amount of DNA II) naked DNA complexes with
a variable number (1 to 22) of bead particles (B, C). The perinucleosomal location of DNA is clearly visualized by uranyl acetate staining
and many particles display a central granule (small arrows) also
visible in the native core particles. The structural similarity between native and reconstituted particles is demonstrated at high
magnification (X 540.000) in the inserts indicated by curved arrows.
In the reconstituted complexes, the nucleosomes were generally evenly
spaced (D, E. F, H, I, J) but some of them were densely packed
(G, arrowhead). Bar = 100 nm.

Second, the assembly of preformed histones-HMG1 complexes and
DNA is instantaneous and forms nucleosome like structures which
satisfy most of the criteria defining native core particle: (i) Initially relaxed DNA acquires superhelical turns after complexes formation; (ii) the digestion of the complexes with micrococcal nuclease
yield a DNA fragment of approximatively 140 base pairs in length;
(iii) electron microscopy of the reconstituted complexes allows to
visualize a beaded structure with a perinucleosomal location of DNA.

Its primary structure shows that HMG1 could simultaneously interact with DNA by one part of its molecule and with histones by its
C-terminal part which presents a high negative charge density.

Our previous results (9, 16) demonstrate that HMG1 was able
both to lower the melting point of poly d(A-T) and to stimulate rat
liver DNA polymerases in vitro only when it was isolated from regenerating rat liver. This "melting activity" of HMG1, demonstrated in
vitro when it was isolated from dividing cells, may be related to
its role in facilitating the formation of, and transiently stabilizing,
single-stranded DNA whenever the need for this conformation arises
during replication or other processes such as transcription.

A nucleosome assembly factor with a specific affinity for
single-stranded DNA on domains distinct from the domain having assembly activity almost certainly plays an important role in DNA replication in eukaryotes. At the replication fork, HMG1 could in some way
stabilize single-stranded DNA regions, allowing DNA polymerase to
replicate and, in addition, mediate the rapid, correct assembly of
new histones with DNA. Moreover it was found that the nucleosomal
structure of old histones is maintained during replication in vivo
(for review see ref. 17). Different models have been proposed for
nucleosome segregation during genetic readout (transcription and
replication), in which histone octamers become transiently bound to
single-stranded DNA (18). In such processes, HMG1 would appear to be
the factor which promotes such transfer during unwinding of the DNA
helix.

Moreover, HMG1 as the histones, undergo several post-synthetic modifications in vivo. These modifications could modify the interactions of HMG1 with the DNA and histones and therefore modulate its nucleosome assembly (and deassembly) activity.

ACKNOWLEDGEMENTS

We thank Mrs Y. Florentin for expert technical assistance in electron microscopy studies, Mrs J. Grandin for typing the manuscript, M. Duguet, G. Mirambeau, M. Philippe and T. Soussi for many suggestions and discussions and for their gift of Topoisomerase I and SV40 DNA.

This work was supported by grants from Centre National de la Recherche Scientifique (60.82.756) and Ministère de l'Industrie et de la Recherche (650.79.46).

REFERENCES

1.   Laskey,R.A., Honda,B.M., Mills,A.D. and Finch,J.T. (1978) Nature 275, 416.
2.   Nelson,T., Wiegand,R. and Brutlag,D. (1981) Biochemistry 20, 2594.
3.   Stein,A., Whitlock,J.P. and Bina,M. (1979) Proc. Natl. Acad. Sci. USA 76, 5000.
4.   Walker,J.M., Gooderham,K., Hastings,J.R.B., Mayes,E. and Johns,E.W. (1980) FEBS Lett. 122, 264.
5.   Shooter,K.V., Goodwin,G.H. and Johns,E.W. (1974) Eur. J. Biochem. 47, 263.
6.   Smerdon,M.J. and Isenberg,I. (1976) Biochemistry 15, 4242.
7.   Goodwin,G.H., Sanders,C. and Johns,E.W. (1978) Eur. J. Biochem. 38, 14.
8.   Duguet,M. and De Recondo,A-M. (1978) J. Biol. Chem. 253, 1660.
9.   Bonne,C., Duguet,M. and De Recondo,A-M. (1979) FEBS Lett. 106, 292.
10. Bonne,C., Sautiere,P., Duguet,M. and De Recondo,A-M. (1982) J. Biol. Chem. 257, 2722.
11. Bonne,C., Duguet,M. and De Recondo,A-M. (1980) Nucleic Acids Res. 8, 4955.
12. Germond,J.E., Rouviere-Yaniv,J., Yaniv,M. and Brutlag,D.L. (1979) Proc. Natl. Acad. Sci. USA 76, 3779.
13. Duguet,M., Bonne,C. and De Recondo,A-M. (1981) Biochemistry 20, 3598.
14. Earnshaw,W.C., Honda,B.R. and Laskey,R.A. (1980) Cell 21, 373.
15. Saragosti, S., Moyne,G. and Yaniv,M. (1980) Cell 20, 65.
16. De Recondo,A-M., Bonne,C. and Duguet,M. (1980) in: Mechanistic Studies of DNA replication and Genetic Recombination: ICN-UCLA Symposia on Molecular and Cellular Biology (Alberts,B. and Fox,C.F. eds.) Vol. XIX, pp. 629, Academic Press, New York.

17.   De Pamphilis,M.L. and Wassarman,P.M. (1980) Ann. Rev. Biochem.
        49, 627.
18.   Palter,K.B., Foe,V.E. and Alberts,B.M. (1979) Cell 18, 451.

DNA-BINDING PROTEINS IN REPLICATING AND MITOTICALLY ARRESTED

BRAIN NEURONS

Clive C. Kuenzle

Universität Zürich-Irchel
Institut für Pharmakologie und Biochemie
Winterthurerstrasse 190
CH-8057 Zürich, Switzerland

Neurons of the cerebral cortex and the cerebellum of the rat appear to be excellently suited to study the signals mediating mitotic arrest in vivo. This is because the time course of neuronal differentiation is known exactly in both these parts of the brain and in this species. A decisive event in neuronal differentiation is the transition from proliferating precursor cells to non-dividing, terminally differentiating neurons. This takes place just about term in the cortex but only after birth in the cerebellum.

The transition from proliferating precursor cells to mitotically arrested neurons requires a switch at the gene level. DNA replication and the cell cycle must be turned off permanently, and successive sets of genes must become expressed in an orderly temporal sequence. Such events are generally thought to be mediated by non-histone chromosomal proteins, amongst which the DNA-binding proteins are the best candidates.

In order to search for such putative regulatory proteins of the cell nucleus we have undertaken a survey of the non-histone chromosomal proteins in cerebral cortex and cerebellar neurons of the rat during brain development (1, 2). We find that many nuclear proteins fluctuate in exact temporal coincidence with the arrest of cell division and the initiation of terminal differentiation. An example of this is shown in Fig. 1. It compares the low-mobility group non-histone proteins of cerebral cortex neurons at two different stages, namely 1 day before birth as opposed to postnatal day 14. Since in the development of cortex neurons birth marks the loss of proliferative capacity, these two time points are representative of the dividing and non-dividing

489

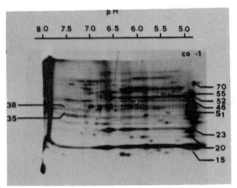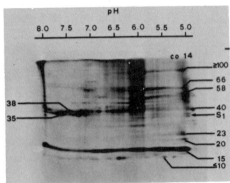

Figure 1: Low mobility group non-histone chromosomal proteins of rat cerebral cortex neurons 1 day before birth (co-1) and 14 days postnatal age (co 14). Neuronal nuclei were isolated from the cerebral cortex of rats, the chromatin was solubilized by mild treatment with micrococcal nuclease, extracted with 0.35 M NaCl and the low mobility group non-histone fraction precipitated with 2% trichloroacetic acid. The precipitated proteins were ^{14}C-labeled by reductive methylation, separated by two-dimensional gel electrophoresis and visualized by fluorography. For full experimental details, see (1). Proteins are identified by pI (scale on top) and M_r (values in kilodalton in the margins). Note the appearance around birth of two very basic proteins of M_r 35 000 and 38 000.

states, respectively. A very prominent change affects two trains of very basic proteins of M_r 35 000 and 38 000. These two proteins are almost absent 1 day before birth but very conspicuous at 14 days postnatal age. An analogous behavior was found in cerebellar neurons (granule cells); only, there the proteins appeared markedly later, in agreement with the delayed and protracted time course of neuronal differentiation in this location.

We found that both proteins bind to single-stranded DNA but not to double-stranded DNA (2). Thus, these two proteins were characterized as single-stranded DNA-binding proteins. This facilitated their isolation in preparative quantities (3) but for reasons of yield only the protein of M_r 35 000 (designated 35K protein) was amenable to biochemical characterization. Figure 2 shows a two-dimensional gel electrophoresis of the isolated 35K protein. The charge microheterogeneity, already seen in Figure 1, is clearly visible. There is evidence to indicate that this derives from posttranslational modification. In agreement with its single-stranded DNA-binding property the protein protects single-stranded DNA from digestion by S_1 nuclease in a concentration-dependent manner (Figure 3). This is brought about by mechanical shielding of the DNA rather than by DNA reannealing.

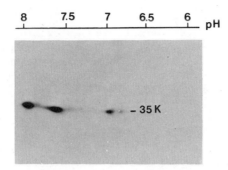

Figure 2: Two-dimensional gel electrophoresis of the isolated 35K protein from calf cerebral cortex. Isolation of the protein was by affinity chromatography on tandemly coupled columns of double-stranded and single-stranded DNA-cellulose as described (3). Electrophoretic separation was by pI (horizontal) and M_r (vertical). Note the basic character and typical charge microheterogeneity, which probably derives from posttranslational modification.

This can be inferred from the observation that the protein also protects single-stranded Ml3 DNA, which inherently is incapable of reannealing. Also, the protein does not alter the melting temperature of poly(dA-dT)·poly(dA-dT).

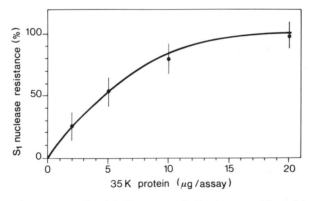

Figure 3: Resistance of single-stranded DNA to digestion by S_1 nuclease as a function of added 35K protein. The experiments were performed as described (4) except for the following modifications: (a) incubation of single-stranded DNA with 35K protein was at 37°, (b) Heat-denatured Escherichia coli (^{3}H)DNA was used as a probe, and (c) the composition of the S_1 nuclease buffer was changed to 40mM NaCl, 50mM Na acetate, pH 4.6, and 4mM $ZnSO_4$.

    An early idea was that the 35K protein might function as a
suppressor of DNA replication because, as elaborated above, it appears
during brain development exactly at the time when neurons stop dividing. However, the protein was found neither to inhibit the DNA
polymerases from various prokaryotic and eukaryotic sources nor to
block replication in an in vivo-like replication assay using permeabilized Chinese Hamster Ovary cells (5). Further evidence against
an involvement of the 35K protein in DNA replication comes from the
observation that homologous proteins present in other cell types
either do not fluctuate at all or change their levels in opposite
direction in response to artificial manipulations of proliferative
activity. For instance, neuroblastoma cells in culture, whether left
to divide or induced to differentiate and arrest proliferation by
treatment with dibutyryl cyclic AMP, retained an even level of a
homologous 35K protein, and the same was true for liver during regeneration following partial hepatectomy. On the other hand, contrary
to what is observed in cortex and cerebellar neurons, a protein with
properties very similar to the 35K protein abounds in dividing rat
phaeochromocytoma PC 12 cells but disappears upon induction of mitotic
arrest by nerve growth factor (6). It is therefore highly improbable
that the 35K protein acts to suppress DNA replication in vivo.

    Very recently, an involvement of the 35K protein in RNA tran-
scription could also be excluded. When the protein was injected into
Xenopus oocytes together with cloned genes for rRNA, histones or tRNA
(transcribed by RNA polymerases I, II and III, respectively) no effect
on transcriptional activity was detected. Thus, no specific function
could be ascribed to the protein. The relatively high concentration
in the tissues investigated together with its single-stranded DNA
binding properties and its characterization as a low-mobility group
non-histone chromosomal protein perhaps allows the conclusion that
the 35K protein is a structural, rather than a catalytic, component
of the chromatin.

    This demonstration of a developmentally fluctuating protein
should only serve to exemplify one of the approaches we are taking
to arrive at functionally important proteins acting at the chromatin
level. In the following, an alternative approach is presented, which
promises to lead to gene regulatory proteins more directly.

    Recently, it has been shown that DNA can occur in a left-handed
conformation, termed Z-DNA (for reviews, see 7,8). The term Z-DNA has
been coined to highlight the fact that in this conformation the
sugar-phosphate backbone assumes an irregular zig-zag path (9). This
is in contrast to the more common right-handed B-conformation where
the backbone ascends smoothly in a manner reminiscent of a staircase.
It is well conceivable that in vivo short tracts of Z-DNA interspersed
with longer sequences in the B-conformation could represent a regula-

tory signal acting at the gene level. Thus, we have set out to search
for Z-DNA in neurons by immunohistochemistry (10). This is possible
because, in contrast to B-DNA, Z-DNA is highly immunogenic and specific antibodies can be raised to it. Using such antibodies we have observed Z-DNA immunoreactivity in the cerebellum among other tissues.
Z-DNA immunoreactivity was clearly restricted to the cell nuclei.
However, not all nuclei were stained. Immunoreactivity associated with
cerebellar neurons was detected in the molecular and granular layer,
whereas Purkinje neurons were not stained significantly. Thus, at
least some cerebellar neurons harbor Z-DNA in their nuclei.

The case for the natural occurrence of Z-DNA in neurons could
be considerably strengthened if it were possible to show that these
cells also contain Z-DNA-binding proteins. Such proteins might serve
to induce and/or stabilize the Z-conformation. We have searched for
Z-DNA-binding proteins in cerebral cortex neurons by affinity chromatography. The nuclear proteins were fractionated on tandemly arranged columns of B-DNA (first column) and Z-DNA (second column)
covalently coupled to Sepharose. This arrangement ensured that proteins binding to all kinds of DNA would be retained on the B-DNA
column before coming in contact with Z-DNA, whereas proteins binding
to Z-DNA specifically would flow through and only be captured by the
Z-DNA column. The proteins retained on the two columns were then
eluted with salt and analyzed by two-dimensional gel electrophoresis.
As shown in Figure 4, three proteins (labeled 1, 2 and 3) of pI ca.
6.5 and M_r 60 000 - 70 000 were detected predominantly in the eluate
from the Z-DNA column. The same proteins were only weakly discernible
in the eluate from the B-DNA column. Thus, they can be regarded as
Z-DNA-binding proteins. A fourth protein (labeled 4) was present in
both eluates with perhaps a slightly stronger representation on the
Z-DNA column. One trivial interpretation could be that protein 4 lacks
the capacity to discriminate between the B- and the Z-form of DNA.
A more attractive, though for the moment purely speculative, hypothesis is that protein 4 might be a Z-DNA-inducing protein. Since the
B-DNA of the first column was represented by poly(dG-dC)·poly(dG-dC),
which has the capacity to convert from the B- to the Z-conformation,
protein 4 might have induced this conformational transition and might
then have become bound to the newly generated Z-DNA partially. The
remainder would have flown through and would have been retained on
the second column. Whether this hypothesis is true will have to be
answered by further tests. However, for the time being it might
suffice to recognize that Z-DNA-binding proteins are likely to exist
in the nuclei of cerebral cortex neurons.

In conclusion, this report has demonstrated the approaches we
are taking to arrive at functionally important proteins of the cell
nucleus but the proteins so far recovered are still awaiting their
functional assignments.

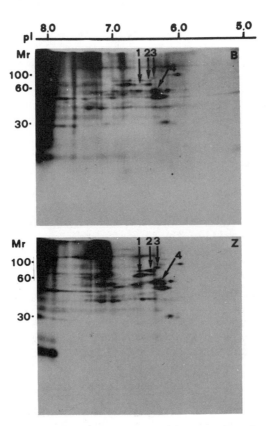

*Figure 4: Z-DNA-binding proteins of rat cerebral cortex neurons.
Neuronal nuclei were isolated from the cerebral cortex of 22-days-old
rats, the nuclear proteins were extracted with O.6 M KCl ^{14}C-labeled
by reductive methylation and desalted. The proteins were then fractionated by affinity chromatography on tandemly arranged columns of
B-DNA (first column) and Z-DNA (second column) covalently coupled to
Sepharose. The B-DNA was represented by poly(dG-dC)·poly(dG-dC) and
the Z-DNA by poly(dG-Br^{5}dC)·poly(dG-Br^{5}dC). The proteins retained on
the two columns were eluted with 3 M NaCl and analyzed by two-dimensional gel electrophoresis followed by fluorography. The top panel
(B) shows the protein pattern obtained from the B-DNA column and the
bottom panel (Z) that from the Z-DNA column. Note that proteins labelled 1, 2 and 3 travel through the B-DNA column with almost no binding,
but are retained on subsequent passage through the Z-DNA column.*

ACKNOWLEDGEMENTS

        I thank W. Blank and R. Schindler for assaying in vivo-like
replication on permeabilized CHO cells and J. Mouse and M. Birnstiel
for oocyte injection. This work was supported by Swiss National
Science Foundation, Grant 3.004-0.81.

REFERENCES

1.   Heizmann,C.W., Arnold,E.M. and Kuenzle,C.C. (1980)
         J. Biol. Chem. 255, 11504.
2.   Heizmann,C.W., Arnold,E.M. and Kuenzle,C.C. (1982)
         Eur. J. Biochem. 127, 57.
3.   Winkler,G.C. (1981) Thesis, School of Veterinary Medicine,
         University of Zürich.
4.   Cox,M.M. and Lehman,I.R. (1981) Nucleic Acids Res. 9, 389.
5.   Reinhard,P., Maillart,P., Schluchter,M., Gautschi,J.R. and
         Schindler,R. (1979) Biochim. Biophys. Acta 564, 141.
6.   Biocca,S., Cattaneo,A. and Calissano,P. (1983) Embo J. 2, 643.
7.   Zimmerman,S.B. (1982) Annu. Rev. Biochem. 51, 395.
8.   Neidle,S. (1983) Nature 302, 574.
9.   Wang,A.H.J., Quigley,G.J., Kolpak,F.J., Crawford,J.L.,
         van Boom,J.H., van der Marel,G. and Rich,A. (1979)
         Nature 282, 680.
10.   Morgenegg,G., Celio,M.R., Malfoy,B., Leng,M. and Kuenzle,C.C.
         (1983) Nature 303, 540.

SEARCHING FOR PROTEINS AND SEQUENCES OF DNA REPLICATION IN MAMMALIAN CELLS

A. Falaschi, S. Riva, G. Della Valle[1], F. Cobianchi,
G. Biamonti[1] and O. Valentini[1]

Istituto di Genetica Biochimica ed Evoluzionistica C.N.R.
Via S. Epifanio 14
I-27100 Pavia, Italy

[1]Dipartimento di Genetica e Microbiologia
Università di Pavia
Via S. Epifanio 14
I-27100 Pavia, Italy

We have approached the problem of describing the process of DNA replication of mammalian cells at the molecular level from two sides:

1)   Purification and description of the properties of proteins (presumably) involved in the advancement of the growing fork.

2)   Attempts at cloning sequences of human DNA allowing autonomous replication in mammalian cells of plasmids bearing them.

We will describe here briefly the progress we have made in these two directions.

PROTEINS OF THE GROWING FORK

DNA binding proteins

We have purified the SS-DBPs* of calf thymus with a procedure analogous to that of Herrick and Alberts (1), but assaying at each

---

*
DBP: DNA binding proteins. Kd: kilodalton. Kb: kilobases. DS- or SS-DNA: double-stranded or single-stranded DNA. PEG: polyethylene glycol. SDS-PAGE: sodium dodecylsulfate-polyacrylamide gel electrophoresis. ARS: Autonomously Replicating Sequence. TK: thymidine kinase. HSV: Herpes Simplex Virus. PMSF: phenylmethyl sulfonylfluoride.

step a functional property, namely the ability to stimulate DNA polymerase α (2), in analoqy with the known ability of the prokaryote
SS-DBPs to specifically stimulate the homologous replicative polymerases. We chose alternating poly(dAT) as the template with which
to challenge the DNA polymerase, prompted by the consideration that,
being a molecule with some features of a double-stranded structure,
it would set the α-polymerase in a situation mimicking a growing
fork. The purification procedure has already been described in detail
(2); essentially the thymus extract was fractionated with PEG-2M
NaCl to remove the nucleic acids, then applied to DS and SS-DNA cellulose (two columns in series), eluted from the SS-DNA cellulose with
NaCl and further purified with hydroxyapatite and G-75 chromatography.
We obtained in this way 3 pools (pools 1, 2 and 3) all highly specific for binding to SS-DNA. As shown in Figure 1, the pools contain
a variety of molecular forms, ranging between 24 and 30 Kd. Such
heterogenous pattern is confirmed also in isoelectric focusing. The
three pools show a variety of bands with pIs ranging between 6 and
8 (see Figure 2). Two dimension electrophoresis experiments (not
shown) indicate that some bands with the same Mr have different pI
values. Pool 3 protein seem to contain bands with relatively higher
pIs.

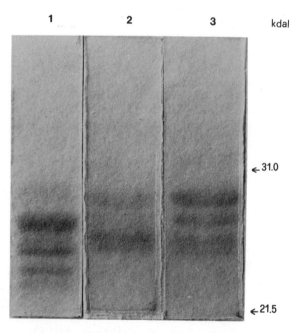

*Figure 1: SDS-polyacrylamide gel electrophoresis (12% polyacrylamide,
3.5% stacking) of different DBP pools. Lane 1: pool 1. Lane 2: pool 2.
Lane 3: pool 3.*

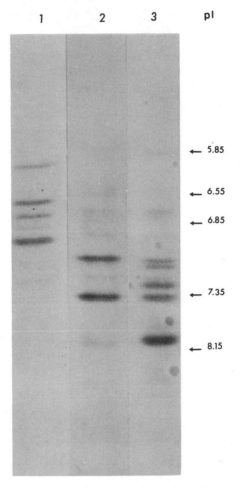

Figure 2: Separation by isoelectric focusing in ampholine gradient-polyacrylamide of the different DBP. Lane 1: pool 1. Lane 2: pool 2. Lane 3: pool 3.

This variety of molecular forms corresponds to a variety of functional properties. As shown in Figure 3, all pools stimulate α-polymerase (as expected in view of the assay used for the purification) but to variable extents. The pool giving the highest stimulatory effect is pool 3, which contains the proteins having the highest affinity for SS-DNA, being the last to be eluted from the SS-DNA cellulose. The degree of stimulation is as high as 20 fold (vs. a maximum of 4 fold for the UP1 of Herrick and Alberts in the same assay) and is highly specific for α-polymerase (no effect on β, γ and pol I). The maximum stimulation is obtained at high protein/DNA ratio (of the order of 10) in agreement with a stoichiometric inter-

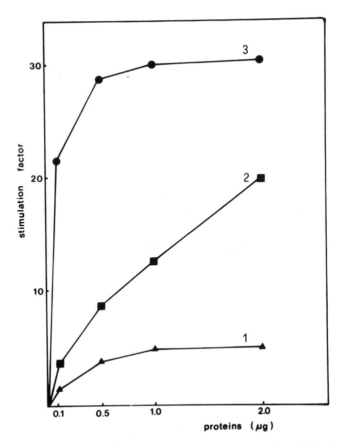

Figure 3: Stimulation of DNA polymerase α on poly(dAT) template by DBP˙(1): pool 1, (2): pool 2, (3): pool 3.
0.25 µg of poly(dAT) per assay were used (2).

action of the DBPs with template. Pool 3 is also the one showing the strongest effect on the melting temperature of poly(dAT) which is lowered from 58⁰ to 30⁰ at a 10:1 protein/DNA ratio. Thus pool 3 is composed of proteins with the highest affinity for SS-DNA, with the highest molecular weight and pI, with the strongest effect on melting and with the strongest stimulatory effect on α-polymerase.

In spite of the great physical and functional variability, the different SS-DBPs of the three pools are all antigenically related. A western blot analysis with antibodies against pool 1 shows that all the bands in the other two pools and also in fractions at different degree of purification are cross-reactive; furthermore, a band at 31 Kb, barely visible with Coomassie staining, becomes apparent in several less purified fractions (not shown).

In conclusion, we propose that we are dealing with a variety of molecular species, probably deriving from post-translational modifications of a common precursor (perhaps the 31 Kd form). The post-translational modifications are unlikely to occur in vitro during the purification, since we have added three protease inhibitors (PMSF, Pepstatin A and Na-metabisulfite) throughout the whole purification and since prolonged incubation of fractions with freshly prepared crude extracts did not change the size of the DBP. It is worthwile to mention that mild proteolytic digestion of pool 3 gives smaller polypeptides, that are still able to bind SS-DNA and to stimulate α-polymerase. This is in analogy with the persistence of such properties after mild proteolytic digestion of the gene 32 protein of T_4 (3).

DNA-dependent ATPase

We have already described the properties of the main SS-DNA dependent ATPase purified from HeLa cells (4, 5). The partially purified protein, devoid of any contaminating nuclease, polymerase or topoisomerase activities, had a Mr (as native molecule) of 110 Kb, showed a limited helicase activity with a 3'→5' polarity of the unwound strand (like the helicases associated at the growing fork of T_4 and of T_7 DNAs) and could stimulate α-polymerase up to 30 fold on fork-like templates, including poly(dAT). We could purify further the ATPase on SS-DNA cellulose, taking advantage of the fact that it is not eluted by a K-phosphate gradient; so, with two subsequent gradients, (a K-phosphate one, followed by a NaCl gradient) the ATPase elutes in the second gradient, whereas the α-polymerase stimulatory activity elutes early in the first gradient (5). Both activities were purified to homogeneity and shown to reside on two proteins totally pure as judged from the SDS-PAGE analysis, corresponding single polypeptides of 68 Kd (ATPase) and 28 Kd (α-polymerase stimulation) respectively. Interestingly enough, the sum of the two Mrs is close to the one of the unfractionated ATPase. The helicase activity is not recovered, nor can it be reconstituted by mixing the two pure proteins.

The α-stimulating activity resides in a protein which binds to SS- and DS-DNA with equal affinity. This protein is distinct from the SS-DBPs described in the previous paragraph because it binds equally well to SS and DS-DNA, is antigenically unrelated to them, and gives high stimulation of α-polymerase on poly(dAT) at low protein/DNA ratio (0.2 vs. 10 for the SS-DBP), pointing to an effect on the catalytic activity of the enzyme rather than on a simple physical effect on the template. We are trying to reconstitute the helicase activity with these and possibly other molecules that might have been fractionated away in the process.

ATTEMPTS AT CLONING DNA REPLICATION ORIGINS

We have approached the task of isolating origins of human DNA
replication with a strategy inspired by the isolation of autonomously
replicating sequences (ARS) in yeast (6). To this purpose we have
constructed plasmid pOE3 (see Figure 4) by inserting into pAT-153
the TK gene of HSV, containing the Tn5 neo^r gene placed under the
TK promoter (7, 8). This plasmid expresses in Escherichia coli the
resistance to Tetracyclin, Ampicillin and Neomicin, but can also
transform mammalian cells to the ability to grow in the presence of
G-418, an analog of Gentamycin which inhibits protein synthesis of
mammalian cells. This plasmid has a single Bam HI site in the tet^r
gene. We have thus prepared a partial Sau 3A digest of human DNA
extracted from the HL60 cell line; we have isolated the molecules
ranging between 7 and 10 Kb and have cloned them in the Bam HI site.
After transformation of Escherichia coli we have selected 20 independent plasmids spanning approximately 180 Kb of human DNA and we have
assayed each of them for the ability to transform mouse L cells and
human KB cells to G-418 resistence by the Ca-phosphate mediated gene
transfer. In view of the higher susceptibility to transformation of
murine cells compared to human cells, we have used 2 and 10 µg of
DNA per plate respectively. As Table 1 shows, no dramatic difference
in transformation efficiency was observed (considering that in yeast
one gets up to 1000 fold greater transformation efficiency with cloned
ARSs). The G-418 resistant colonies in each plate were pooled and
grown up to 2 x 10^7 cells; low molecular weight DNA was extracted by
the Hirt procedure and assayed for : (i) properties and structure of
the recovered plasmids; (ii) presence of sequences hybridizable to
pOE3; (iii) resistance of the recovered plasmids to Mbol, an enzyme
that cannot digest A-methylated DNA and which therefore reveals the
presence of input, unreplicated DNA.

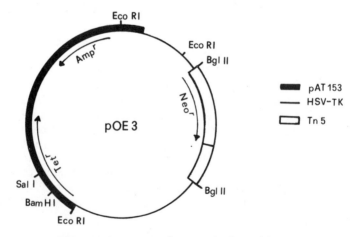

Figure 4:   Structure of plasmid pOE3.

Table 1:   Efficiency of transformation to G-418 resistance of
           plasmids containing human DNA inserts[a]

Plasmid[b]	DNA insert size (Kb)	Transformation frequency[c]	
		Mouse L cells	Human KB cells
pOE3	0	2.4×10^{-4}	6.6×10^{-4}
pOE6-1	9.1	12.0×10^{-4}	15.0×10^{-4}
pOE6-2	13.9	7.2×10^{-4}	3.4×10^{-4}
pOE6-8	8.0	3.4×10^{-4}	0.5×10^{-4}
pOE6-12	7.2	2.6×10^{-4}	1.2×10^{-4}
pOE6-15	8.2	0.7×10^{-4}	0.8×10^{-4}

a) The efficiency of transformation of the other 15 plasmids (see
   text) are not shown since all fell within the same range as those
   reported here.

b) The amount of plasmid DNA used for transformation was 2 μg and
   10 μg per 100 mm plate of mouse L cells and human KB cells respectively.

c) Cells (1×10^6 per 100 mm plate) were transformed with plasmid DNA
   using the calcium phosphate precipitation technique without car-
   rier DNA, and colonies were counted at 20 to 25 days after growing
   in selective medium.

     The results show that, even after 20 generations (required to
arrive to 2×10^7 cells from the initially transformed single colonies)
molecules related to the input are still present in the Hirt supernatant. These molecules are essentially all resistant to MboI, indicating that they are passively diluted and not replicated autonomously. This is true also for pOE3 that does not contain human sequences.

     In order to rule out the possibility that the TK sequence or
the selective pressure applied to the introduced plasmids may increase
the ability of the exogenous circles to persist in the cells, we measured such ability on pAT-153, a plasmid lacking both TK gene and the
G-418 resistance selectable genes. To this aim we performed two co-
transformation experiments. In the first one, L cells were cotransformed with pAT-153 and pTK at a 10:1 ratio, selecting for TK+ and
measuring the persistence of pAT-153 in the Hirt supernatant. In the
second experiment KB cells were cotransformed with pAT-153 and pOE8
at a 1:1 ratio (pOE8 is an analog of pOE3 missing most of the TK
gene) selecting for G-418 resistance and characterizing the Hirt

Table 2: Rearrangements in recombinant plasmids extracted from transformed cells

Plasmid	Plasmid size (Kb)	Murine L cells			Human KB cells		
		Number of bact.col. examined	Rearranged (b) n°	Rearranged %	Number of bact.col. examined	Rearranged n°	Rearranged %
pAT-153	3.6	156	0	<0.70	300	0	<0.3
pOE3	8.9	18	11	61	10	1	10
pOE6-1	18.0	36	36	100	150	0	<0.6
pOE6-2	22.8	28	28	100	-	-	-
pOE6-8	16.9	22	18	81	4	1	25
pOE6-12	16.1	22	16	73	10	0	<10
pOE6-1 HA (a)	14.6	12	5	42	-	-	-
pOE6-1 HB (a)	10.2	12	0	<9	-	-	-

a) These two plasmids are two rearranged DNA molecules rescued from the Hirt supernatant of the first transformation with pOE6-1.

b) Only massive rearrangements resulting in drastic size reductions or loss of antibiotic resistance were considered.

supernatant as above. The results of these two experiments obtained both by analysis of bacterial transformation, and by Southern blotting, show that the non selectable pAT-153 persist for over 20 generations, without replicating.

Thus, the results are consistent with the contention that, in Ca-phosphate mediated DNA transformation, a high number of molecules enters the cells (up to 10^5 plasmids per cell, in agreement with the data of Loyter et al.(9)); these plasmids are then passively diluted away, but they persist for as long as 20 generations without undergoing extensive degradation. Since at longer times no plasmids are detectable in transformed cells, it is reasonable to assume that transformation to TK[+] and G-418 is sustained, at least at later times, by integration into the chromosome.

Thus, more exacting selective procedure will have to be applied in order to give adequate advantage to the potentially replicating molecules. It is also possible that greater superstructural specificities are required in mammalian cells for allowing the recognition of chromosomal origins. In any case it will be necessary to cope with the tendency of exogenous molecules to integrate into the chromosome, thereby reducing the selective advantage of the (possible) freely replicating circles.

Finally, we have observed that the majority of the plasmids recovered from L cells have undergone drastic rearrangements (see Table 2), always accompanied by reductions in size; the rearranged molecules are more stable at this level, as judged from the structure of the molecules recovered after a second cycle of transformation in L cells. Instead, in KB cells, no such rearrangements are observed (see Table 2) and the plasmids are recovered intact by the criteria just mentioned.

Thus, murine and human cells seem to differ drastically in the tendency to rearrange exogenous DNA molecules, an observation that may be related to the markedly lower transformability of the latter with respect to the former. In any case the lack of rearrangements in KB cells is potentially a very useful property in all experiments which require a passage of recombinant plasmids in mammalian cells.

ACKNOWLEDGEMENTS

The experiments described in this paper were performed with the support of the Progetti Finalizzati "Controllo della Crescita Neoplastica" and "Ingegneria Genetica e Basi Molecolari delle Malattie Ereditarie" Consiglio Nazzionale delle Ricerche, Rome. G. Biamonti was supported by a fellowship from Sorin Corp. (Saluggia) and O. Valentini was a fellow of the Fondazione A. Villa Rusconi.

REFERENCES

1. Herrick,G. and Alberts,B. (1976) J. Biol. Chem. 251, 2124.
2. Riva,S., Clivio,A., Valentini,O and Cobianchi,F. (1980)
        Biochem. Biophys. Res. Commun. 96, 1053.
3. Moise,H. and Hosoda,J. (1976) Nature 259, 455.
4. Cobianchi,F., Riva,S., Mastromei,G., Spadari,S., Pedrali-Noy,G.
        and Falaschi,A. (1979) Cold Spring Harbor Symp. Quant.
        Biol. 43, 639.
5. Biamonti,G., Cobianchi,F., Falaschi,A. and Riva,S. (1983)
        EMBO Journal 2, 161.
6. Stinchomb,D.T., Thomas,M., Kelly,J., Selker,E. and Davis,R.W.
        (1980) Proc. Natl. Acad. Sci. USA 77, 4559.
7. Colbère-Garapin,F., Haroduicenau,F., Kourilsky,D. and Garapin,
        A.C. (1981) J. Mol. Biol. 150, 1.
8. Southern,P.J. and Berg,P. (1982) Jour. Molec. and Applied
        Genetics 1, 327.
9. Loyter,A., Scangos,G.A. and Ruddle,F.H. (1982) Proc. Natl.
        Acad. Sci. USA 79, 412.

DIADENOSINE TETRAPHOSPHATE AND DIADENOSINE TETRAPHOSPHATE-BINDING

PROTEINS IN DEVELOPING EMBRYOS OF ARTEMIA

Alexander G. McLennan and Mark Prescott

Department of Biochemistry
University of Liverpool
P.O. Box 147
Liverpool, L69 3BX, United Kingdom

INTRODUCTION

Diadenosine $5',5'''-P^1, P^4$-tetraphosphate (Ap_4A) has been detected in a number of eukaryotic cells and tissues and its intracellular concentration found to vary between 0.05 to 0.1 pmol/10^6 cells
in resting cells to 5-10 pmol/10^6 cells in rapidly dividing populations (1). It has recently been found to bind tightly to one subunit
of a high molecular weight DNA polymerase-α complex from calf thymus
(2) and HeLa cells (3). These facts have contributed to the idea that
Ap_4A may be an important regulatory nucleotide required for the
initiation of DNA replication (4).

In order to gather more information on this nucleotide, we
have studied it during the development of the brine shrimp Artemia.
When the ametabolic encysted gastrulae of Artemia are placed in
aerated seawater, development resumes and free-swimming larvae hatch
some 12-16h later. During this period there is extensive RNA and
protein synthesis, but DNA replication does not begin until after
hatching (5). Hence Artemia may offer a useful system for the study
of Ap_4A function. Here we report on the detection and developmental
changes of Ap_4A in Artemia and on proteins which display an affinity
for it.

RESULTS

Ap$_4$A has been positively identified by reverse phase HPLC of TCA-extracts of 24h Artemia larvae to which a "spike" of (^{3}H)Ap$_4$A (52,000 dpm/pmol) was added. The Ap$_4$A elutes from the column with a specific activity of 655 dpm/pmol (Figure 1). The specific activity of the Ap$_4$A is constant across the peak which strongly suggests that the compound responsible for the reduction in specific activity of the "spike" is Ap$_4$A. The apparent increase in specific activity at the leading edge of the peak is due to a small amount of (^{3}H)Ap$_5$A in the "spike".

In agreement with previous results (7), ATP reaches a maximum of 3 nmol/10^6 cells between 8 and 16h and then begins to increase again in marked contrast to Ap4A (Figure 2, bottom panel). These results clearly indicate that the size of the Ap$_4$A pool is regulated independently of the ATP pool and that the maximum Ap$_4$A coincides with the onset of DNA replication.

Binding of (^{3}H)Ap$_4$A to protein in extracts of Artemia can be detected by a nitrocellulose filter binding assay. When protein extracts from 18h larvae are fractionated on Matrex blue gel, four peaks of binding activity are seen (Figure 3).

Further evidence of the idenity of Ap4A comes from t.l.c. of pooled fractions from the C$_{18}$ column on PEI-cellulose. The specific

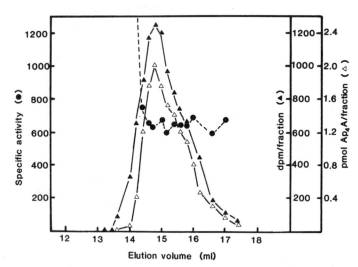

Figure 1: HPLC of TCA-extract of 24h Artemia larvae. 200 μl of a TCA extract containing 0.5 μCi (^{3}H)Ap4A were injected on a Waters RP 18 Radial-Pak cartridge and run in 0.1 M potassium phosphate, pH 7.2, 12% methanol, 5 mM tetrabutylammonium hydroxide. Fractions were assayed for radioactivity and Ap4A (6).

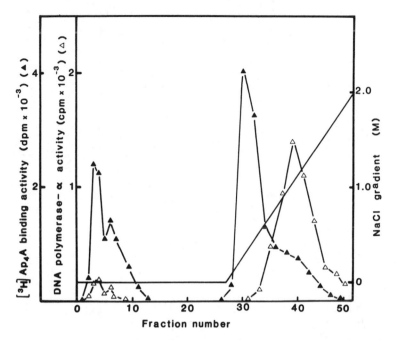

Figure 2: Variation of Ap₄A and adenine mononucleotide levels during development. Timed samples were analysed by HPLC as described for Figure 1 and the specific activity of the eluted Ap₄A measured. Values represent the means of 3 or 4 determinations (± one S.D.). ATP, ADP and AMP were also measured by bioluminescence assays and (³H)thymidine incorporation into hatched larvae by standard procedures.

activity of the label is unaltered by this separation while treatment with snake venom phosphodiesterase yields ATP and AMP only.

The intracellular concentration of Ap_4A in 24h larvae is calculated to be 3.5 pmol/10^6 cells. Figure 2 (top panel) shows the change in Ap_4A content during redevelopment and its relation to the onset of DNA replication. Desiccated cysts retain a significant quantity of Ap_4A: 0.47±0.17 pmol/10^6 cells. On rehydration the level rises to a maximum of 5.70±0.35 pmol/10^6 cells around the time of emergence (E). After hatching (H), when DNA replication commences, there is a steady decline up to 48h.

Heterogeneity is also seen on DEAE-cellulose. None of the binding activities are associated with Ap_4A-degrading activity; diadenosine tetraphosphatase elutes from Matrex blue gel at 1.5 M NaCl. Of the activity which co-chromatographs with DNA polymerase-α, only the unbound fraction has not yet been separated from the polymerase. It

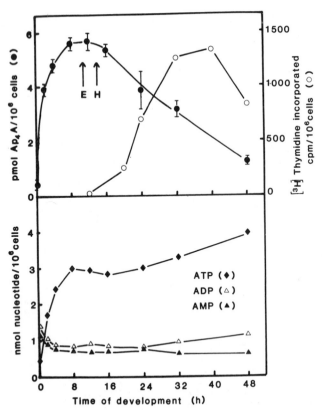

Figure 3: *Fractionation of Ap₄A-binding protein on Matrex blue gel.*
An extract of 18h larvae was applied to a column of Matrex blue gel
in 10 mM potassium phosphate, pH 7.5, 1 mM EDTA, 10% glycerol, 5 mM
mercaptoethanol. The flow-through fraction was collected and applied
to a second column of Matrex blue gel to ensure that any fractiona-
tion observed was not due to overloading. The second column was
eluted with a NaCl gradient and fractions assayed for binding of
(³H)Ap₄A and DNA polymerase-α.

remains to be determined whether further association between DNA
polymerase-α and Ap₄A-binding proteins occurs at other developmental
times and what function these proteins have during embryogenesis.

REFERENCES

1.   Rapaport,E. and Zamecnik,P.C. (1976) Proc. Natl. Acad. Sci.
         USA, 73, 3984.
2.   Grummt,F., Waltl,G., Jantzen,H-M., Hamprecht,K. Hübscher,U.
         and Kuenzle,C.C. (1979) Proc. Natl. Acad. Sci. USA
         76, 6081.

3.   Rapaport,E., Zamecnik,P.C. and Baril,E.F. (1981)
          J. Biol. Chem. 256, 12148.
4.   Grummt,F. (1978) Proc. Natl. Acad. Sci. USA 78, 371.
5.   Hentschel,C. and Tata,J.R. (1976) Trends Biochem. Sci. 1, 97.
6.   Ogilvie,A. (1981) Anal. Biochem. 115, 302.
7.   Warner,A.H. and Finamore,F.J. (1967) J. Biol. Chem. 242, 1933.

A RIBONUCLEASE H FROM YEAST STIMULATES DNA POLYMERASE IN VITRO

Robert Karwan, Hans Blutsch[1] and Ulrike Wintersberger

Institute for Tumorbiology/Cancer Research
University of Vienna
Borschkegasse 8a
A-1090 Vienna, Austria

[1]Institute for Molecularbiology, University of Vienna

Recently we have purified a ribonuclease-H (RNase-H) activity from the yeast, <u>Saccharomyces cerevisiae</u> (1) which differs in many of its properties (see Table 1) from ribonucleases-H described earlier for this organism by others(2, 3). The enzyme specifically degrades the RNA part of a DNA-RNA hybrid in an endonucleolytic mode and is completely inactive with single-stranded RNA or native DNA (Table 2).

Table 1: Properties of ribonuclease H from Yeast[a]

Molecular weight	70.000	
pH - Optimum	7.1	
Mg^{++} - Optimum	10	mM
Activity in presence of Mn^{++} (between 0.5 and 20 mM)	<6	%
Inhibition by N-ethylmaleimide 1 µM	46	%
5 µM	25	%
20 µM	<1	%

[a] Molecular weight was determined by glycerol gradient centrifugation, gel filtration and SDS polyacrylamid gel electrophoresis. Enzyme activity was determined as described (1).

513

Table 2: Substrate specificity of ribonuclease H from yeast

Substrate (1 μg/100 μl)	Amount of enzyme in 125 μl assay[a] mixture	Acid soluble nucleotides after incubation for	
		30 min	120 min
DNA-RNA hybrid	0.04 μg	14.5	49.5
(native)	0.16 μg	54.5	74.5
DNA-RNA hybrid (denatured)	0.16 μg	0.6	3.1
DNA (native)	0.16 μg	0.0	0.0

[a] Assay of enzyme activity were carried out as described earlier (1).

During the isolation procedure, which includes a chromatography on a DNA cellulose column, we have observed the co-purification of an activity highly and specifically stimulating in vitro DNA synthesis by yeast DNA polymerase A. The enhancement of the DNA polymerisation reaction by our RNase-H preparation (Table 3) is strongly dependent on the proportion between template, DNA polymerase and RNase-H in the reaction mixture. It is most pronounced if very small amounts of DNA polymerase react with the gapped template-primer poly(dA)oligo $(dT)_{8-12}$ (5:1) but can also be observed with "activated DNA" as template. Interestingly, Koerner and Meyer (4) recently reported the purification of a DNA binding protein from the Novikoff hepatoma which stimulates DNA synthesis by mammalian DNA polymerase β in a similar fashion. This protein, however, has not been tested for RNase-H activity.

As the DNA polymerase stimulating ability accompanies also our most purified preparations of RNase-H and as the activity coincides exactly with the peaks of RNase-H during glycerol gradient centrifugation as well as during gelfiltration we have suggested that it might be an integral part of the same protein molecule (1). Proteins combining DNA synthesis stimulating and RNase-H activity in a single polypeptide chain have not been described for any other organism until now (5). As such enzymes might be of considerable interest and perhaps play a role during eukaryotic DNA replication we have looked for another criterion proving that the two activities reside in one protein. It was shown by Dirksen and Crouch (6) that hydrolysis of poly(rA) from the synthetic poly(rA)poly(dT) hybrid by RNase-H from Escherichia coli could be blocked by addition of the polysaccharide dextran to the reaction mixture. We tested our enzyme preparation and found the same effect of dextran on the yeast RNase-H indicating a possible similarity of the mechanism of action between the enzymes

Table 3: Stimulation of yeast DNA polymerase A in vitro
by ribonuclease H

Enzymes in 125 µl assay mixture		Factor of stimulation[a]
DNA polymerase A (units)	RNase H (µg)	
0	3.2	no polymerization
0.003	0.2	8.2
	3.2	54.0
0.006	0.2	7.9
	3.2	33.0
0.012	0.2	5.7
	3.2	14.4

[a] Template-primer: poly(dA)(dT)$_{8-12}$ (5:1), 0.25 µg/100µl, assay
conditions as described earlier (1).

from the two different species. Then we investigated the in vitro
DNA polymerisation reaction with yeast DNA polymerase A alone as
well as in presence of RNase-H. Remarkably, DNA polymerase was
totally unimpaired by the presence of dextran at concentrations up
to 0.1 mM. The stimulation effect of the RNase-H preparation, however,
was abolished by the same amounts of dextran which also inhibited
the RNase-H activity (Figure 1). This result further supports, although not directly proves, the hypothesis that it is the RNase-H
molecule itself which enhances the DNA synthesis reaction and not
any undetected contamination of our enzyme preparation.

Thus the question arises by what mechanism a protein, which
possibly might be involved in the initiation of DNA replication by
generating a proper RNA primer (6, 8, 9), could influence the comparatively simple DNA polymerase reaction in vitro and whether this
finding reflexes an additional physiological role of the yeast
RNase-H. As the acceleration of DNA synthesis was found more pronounced in assays to which DNA polymerase was added after short
preincubation of the template with RNase-H than in those of simultaneous addition of the two enzymes (Karwan and Wintersberger,
unpublished observation) we suspect that RNase-H does not directly
interact with the DNA polymerase molecule but rather with the DNA.
The ability of RNase-H to retain single-stranded M-13 DNA to nitrocellulose filters which we have observed in preliminary studies

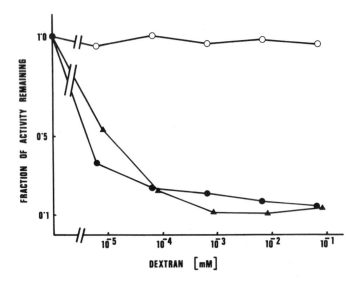

Figure 1: Inhibition of RNase-H and DNA polymerase A stimulating
activity by dextran: RNase-H was reacted with poly(rA)poly(dT) as
described by Dirksen and Crouch (6) in presence of increasing amounts
of dextran T 70 (Pharmacia) employing 0.2 µg of enzyme per 125 µl
assay (▲). Activity of DNA polymerase alone (o) was measured under
standard assay conditions (0.05 units/125 µl assay) (7), the RNase-H
stimulated polymerisation reaction (●) was carried out with 0.01
units of DNA polymerase A and 2 µg RNase-H per 125 µl assay; dextran
was added in amounts given in the figure.

agrees with this conception. On the other hand, we have found that
ongoing DNA synthesis with DNA polymerase A using the gapped template
can still be accelerated to a certain degree by later addition of
RNase-H (Figure 2). This effect was observed even with such small
amounts of DNA polymerase which themselves did no longer exhibit
measurable DNA polymerisation (data not shown). Probably, RNase-H
purified by us, which by the way is a relatively abundant enzyme of
yeast cells, is a component of the protein complex responsible for
DNA replication. Besides its possible involvement in the initiation
of DNA synthesis (for the generation of a proper RNA primer or/and
for its removal) it might facilitate the polymerisation reaction per
se by binding to the template and thereby influencing its conformation. Perhaps, DNA polymerase A thus becomes more efficient and
processive. To test these ideas experimentally will be the aim of
our further investigations.

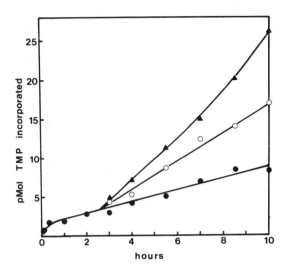

*Figure 2:   Stimulation of ongoing DNA synthesis by addition of
RNase-H: After DNA polymerase A (O.2 units per ml of incubation
mixture (7)) had reacted with poly(dA)(dT)$_{8-12}$(12.5 µg/ml) for 2.5
hours the standard incubation mixture was divided into three parts,
two of which were supplied with RNase-H (O-O 1.7 µg/ml,▲ - ▲ 8 µg/ml,
●-● no addition). The increase in volume of the reaction mixture
by the addition of the RNase-H solution was compensated by adding
appropriate amounts of buffer, Mg^{++} and deoxyribonucleosidetriphos-
phates, the lower concentrations of template and DNA polymerase were
taken into consideration by acid precipitation of correspondingly
larger aliquots (50 µl versus lO µl) starting with the 3-hour time
point. Polymerisation is given in TMP incorporated per aliquot.*

ACKNOWLEDGEMENTS

        We are grateful to Erhard Wintersberger for many helpful discus-
sions and to Anneliese Karwan for cheerfully contributing to the
completion of the manuscript. The work was supported by grant 4326
to Ulrike Wintersberger from the Fonds zur Förderung der wissen-
schaftlichen Forschung in Oesterreich.

REFERENCES

1.   Karwan,R., Blutsch,H. and Wintersberger,U. (1983)
             Biochemistry 22, 5500.
2.   Wyers,F., Sentenac,A. and Fromageot,P. (1976) Eur. J. Biochem.
             69, 377.
3.   Wyers,F., Huet,J., Sentenac,A. and Fromageot,P. (1976)
             Eur. J. Biochem. 69, 385.

4.   Koerner,T.J. and Meyer,R.R. (1983) J. Biol. Chem. 258, 3126.
5.   Crouch,R.J. and Dirksen,M.-L. (1982) in: "Nucleases" S.M. Linn
         and R.J. Roberts, eds., Cold Spring Harbor Laboratory,
         Cold Spring Harbor.
6.   Dirksen,M.-L. and Crouch,R.J. (1981) J. Biol. Chem. 256, 11569.
7.   Wintersberger,U. and Wintersberger,E. (1970) Eur. J. Biochem.
         13, 11.
8.   Selzer,G. and Tomizawa,J.-I. (1982) Proc. Natl. Acad. Sci. USA
         79, 7082.
9.   Hillenbrand,G. and Staudenbauer,W.L. (1982) Nucl. Acids Res.
         10, 833.

CATALYTIC ACTIVITIES OF HUMAN POLY(ADP-RIBOSE)POLYMERASE

AFTER SDS-PAGE

A. Ivana Scovassi, Miria Stefanini, Fabrizio Bonelli
and Umberto Bertazzoni

Istituto C.N.R. di Genetica, Biochimica ed Evoluzionistica
Via S. Epifanio 14
27100 Pavia, Italy

Poly(ADP-ribose)polymerase is an eukaryotic enzyme, localized
in the nucleus, which uses NAD as substrate to catalyze the ADP-
ribosylation of the polymerase itself and of a variety of acceptor
proteins (see Figure 1). Enzyme activity is dependent on the presence
of DNA strand breaks (1) and is stimulated in vivo by DNA-damaging
agents. Several lines of evidence indicate that poly(ADP-ribose) is
implicated in different processes, such as DNA repair (2), cell differentiation (3) and sister chromatid exchange (4), all of which
imply modifications of chromatin structure. The precise function of
this enzyme is still much debated. Lehmann and coworkers postulated
(5) that the polymerase reduces the steady-state level of breaks in
the cellular DNA by stimulating the activity of DNA ligase II (6) or
by altering the structure of chromatin so that DNA ligase can more
easily join strand breaks (7). Olivera suggested that the interaction
with DNA lesions results in poly(ADP-autoribosylation) of the enzyme,
until repulsion between the activating DNA and the negatively charged
polymer causes the dissociation of the enzyme from the chromatin (8).
Recently, this group has demonstrated that DNA topoisomerase I from
calf thymus is modified and inactivated by the poly(ADP-ribose)poly-
merase (9, 10).

The i situ detection of the functional activities of enzymes
after SDS-polyacrylamide gel electrophoresis (PAGE) has proven to be
particularly valuable for DNA polymerases (11-13). We have devised
an activity gel technique for the human poly(ADP-ribose)polymerase.
The assay includes the separation of crude extracts and purified
fractions of the enzyme on SDS-PAGE, renaturation of proteins whithin
the gel, incubation of the intact gel with (^{32}P)NAD, removal of non-

519

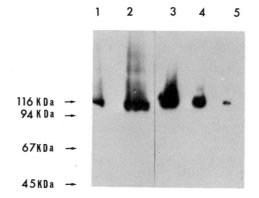

Figure 1: Reaction catalysed by poly(ADP-ribose)polymerase.

incorporated NAD by TCA and autoradiography. The method was developed by using extracts and partially purified fractions of human poly(ADP-ribose)polymerase. Samples for electrophoresis were prepared essentially as described by Spanos et al. (11) and run on a slab gel containing 7.5% polyacrylamide and 100 µg/ml of activated DNA. Renaturation of the catalytic activities of the enzyme was achieved by removing SDS and washing extensively the gel at 4°C with Tris buffer. The assay was carried out by incubating the intact gel in the enzyme reaction conditions (8) containing (^{32}P)NAD. Optimal incorporation was obtained after 8-12 h at 37°C. When DNA was omitted from the gel, the activity bands were still detected, though at much lower intensity; on the contrary, the presence of the specific inhibitor 3-aminobenzamide during the incubation reaction, completely abolished the appearance of the bands (results not shown).

Activity gel analysis of HeLa cell extracts revealed a major activity band of poly(ADP-ribose)polymerase with a molecular mass of 116-120 kDa (Figure 2, lane 1), corresponding to the same band obtained with the partially purified enzyme from the same cells (Figure 2, lanes 3-5). Three other high mol. wt. bands were also distinctly visible in the autoradiograms with sizes of about 125 kDa and 145 kDa, respectively. In extracts of cells treated with dimethyl sulfate (DMS), a 7-10 fold stimulation was evident, with the strongest increase occurring for the 116-120 kDa doublet (Figure 2, lane 2). When partially purified preparations of the enzyme were run in the activity gel it appeared that the intensity of the active bands tended to decrease with the extent of purification (Figure 2, lanes 3, 4 and 5).

In conclusion, the activity gel for poly(ADP-ribose)polymerase appears to be particularly useful for studying the structure of the

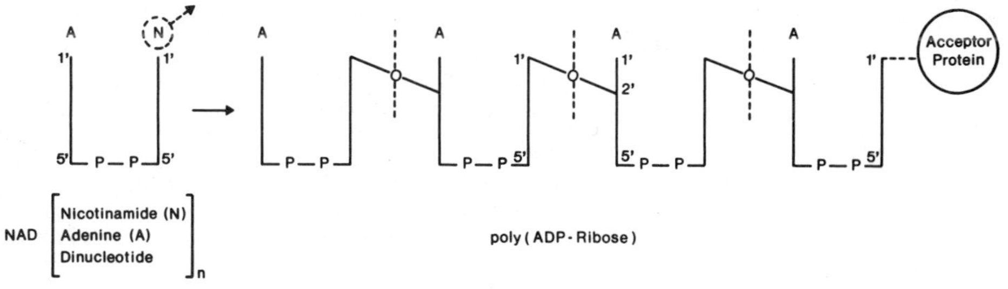

ACCEPTORS : Poly (ADP-R) Polymerase ; Histones ; HMG Proteins ; Ca-Mg Endonuclease ; DNA Ligase II ; RNA Polymerase II ; DNA Topoisomerase ; T-antigen ; Polyoma Virus protein.

INHIBITORS : 3-Aminobenzamide ; Nicotinamide.

Figure 2: Activity gel analysis of poly(ADP-ribose)polymerase catalytic activities in crude extracts and purified preparations from HeLa cells.
Lanes: 1, extract from control HeLa cells; 2, extract from HeLa cells treated with 1 mM DMS for 1 h; 3, nuclear extract (fraction I, 10 units of polymerase); 4, ammonium sulfate fraction (fraction II, 10 units); 5, phosphocellulose fraction (fraction III, 10 units).
M_r markers were β-galactosidase (116 kDa), phosphorylase B (95 kDa), bovine serum albumin (67 kDa) and egg white ovalbumin (43 kDa).

enzyme and its response to treatment of cells with DNA-damaging agents. The presence of several activity bands raises the question of the heterogeneity of the enzyme as already discussed for DNA polymerases (11-13). The observation that the catalytic activities are less visible in the activity gel as more extensively purified preparations of the poly(ADP-ribose)polymerase are electrophoresed, indicates that the purified proteins have a reduced renaturation capacity in comparison to "native" peptides present in crude extracts.

ACKNOWLEDGEMENTS

This work was supported in part by contract BIO-428-81-I of the Radiation Protection Programme of the Commission of the European Communities (contribution n° 2071). U.B. is a scientific official of C.E.C.

REFERENCES

1. Benjamin,R.C. and Gill,D.M. (1980) J. Biol. Chem. 255, 10502.
2. Berger,N.A., Sikorski,G.W., Petzold,S.J. and Kurohara,K.K. (1979) J. Clin. Invest. 63, 1164.

3.   Mandel,P., Okazaki,H. and Niedergang,C. (1982) in: "Progress
         in Nucl. Acids Res. Mol. Biol.", Academic Press, 27, 1.
4.   Hori,I. (1981) Bioch. Biophys. Res. Commun. 102, 38.
5.   James,M.R. and Lehmann,A.R. (1982) Biochemistry 21, 4007.
6.   Creissen,D. and Shall,S. (1982) Nature 296, 271.
7.   Ohaishi,Y., Ueda,K., Kawaichi,M. and Hayaishi,O. (1983)
         Proc. Natl. Acad. Sci. USA 80, 360.
8.   Ferro,A.M. and Olivera,B.M. (1982) J. Biol. Chem. 257, 7808.
9.   Ferro,A.M., Higgins,N.P. and Olivera,B.M. (1983) J. Biol.
         Chem. 258, 6000.
10. Ferro,A.M. and Olivera,B.M. (1984) J. Biol. Chem. 259, 547.
11. Spanos,A. Sedgwick,S.G., Yarranton,G.T., Hübscher,U. and
         Banks,G.R. (1981) Nucleic Acids Res. 9, 1825.
12. Hübscher,U., Spanos,A., Albert,W., Grummt,F. and Banks,G.R.
         (1981) Proc. Natl. Acad. Sci. USA 78, 6771.
13. Scovassi,A.I., Torsello,S., Plevani,P., Badaracco,G.F.
         and Bertazzoni,U. (1982) EMBO J. 1, 1161.

VIII
Fidelity of DNA Replication

FIDELITY OF DNA REPLICATION IN VITRO

Alan R. Fersht

Department of Chemistry
Imperial College of Science and Technology
London, SW7 2AY, England

INTRODUCTION

DNA is replicated with a fidelity far higher than that expected from the specificity of Watson-Crick base pairing. The overall error rate in the replication of the chromosome of Escherichia coli is only one mistake in 10^8-10^{10} nucleotides polymerised (1) yet theoretical and experimental work suggests that non-complementary base pairing occurs at a frequency of up to 10^{-4} - 10^{-5} (2). There is strong evidence that the fidelity of DNA replication is boosted by editing or proofreading mechanisms that excise errors of misincorporation (3). For example, all known prokaryotic DNA polymerases have a 3'→5'-exonuclease activity that is specific for the removal of mismatched nucleotides from the growing 3'-terminus during replication (4). There are also post-replicative mismatch repair systems that patrol newly replicated DNA, repairing mismatches in the daughter strand (5). This paper concerns the contribution of the polymerase to accuracy and addresses the following problems: measurement of the overall accuracy of the major replicative DNA polymerase of Escherichia coli in vitro and its relationship to the accuracy in vivo; the accuracy of the other DNA polymerases; measurement of the frequencies of non-complementary base pairings catalysed by a DNA polymerase; role of the 3'→5'-exonuclease activity; the specificity and magnitude of proofreading; what are the limits on the contribution of proofreading to fidelity and why the need for post-replicative mismatch repair?

RESULTS AND DISCUSSION

Systems for measuring accuracy of DNA replication in vitro

      The best current method of measuring the accuracy of a DNA poly-
merase in vitro is to copy under controlled conditions the DNA of an
amber mutant of a bacteriophage and express the DNA in vivo to produce progeny. The accuracy of replication may be simply measured by
scoring the proportion of revertants to amber stock in the progeny
(6-8). This procedure both avoids the artifacts inherent in the use
of synthetic templates and allows that accuracy in vitro to be compared with that in vivo measured by classical genetic studies.

      The bacteriophage mainly used to date is φX174. Various enzyme
systems have been used so far for measuring the accuracy in vitro,
each having its particular virtues and drawbacks. The first three
systems mimic the three stages of φX174 replication in vivo. The first,
the method of Weymouth and Loeb (6) mimics the SS→RF stage. A restriction fragment primer is annealed upstream to an amber codon in ssDNA
and the primer extended past the codon using Pol I from Escherichia
coli. The mixture of synthetic DNA and parental template is then
transfected into Escherichia coli spheroplasts. The outstanding merit
of this system is its simplicity, generality and ability to replicate
under different conditions. For example, replication will take place
in the presence of Mn^{2+} or other mutagens and so their effects may
be quantified (9) The contributions to fidelity of other proteins,
such as SSB, may be noted (10). Further, a variety of other prokaryotic
polymerases may be assayed (11). A drawback for very precise kinetic
studies is that the DNA used for the transfection contains >50% parental DNA and the expression of synthetic DNA is somewhat variable, the
results tending to be reproducible only within a factor of two or
three (12, 13). A more recent variation of this procedure is to use
the self priming activity of purified polymerase-α (14) or polymerase-
α in extracts of Xenopus laevis eggs (15) to initiate DNA replication.
We have also introduced the following important controls and improve-
ments in the oligodeoxynucleotide-primed procedure that eradicate the
variable results (14, 16). Rather than use a restriction fragment
primer, we make a synthetic primer; for example, the 12-mer that covers
the G and A of the amber codon of φX174aml6:

5'TTTTCCCAGCT   +dGTP
PHAGE----AAAAGGGTCGATTTAGA----PHAGE   (a)

We then synthesize for a control a second primer that covers the whole
of the amber codon but contains a mismatch at the relevant position
where a mismatch is being forced to occur (e.g. dGTP opposite the T
of the amber codon in (a)). Shown next:

5'CCCAGCCTGAATCT
PHAGE----AAAAGGGTCGGATTTAGA----PHAGE   (b)

Both the fidelity assay and the control are replicated in parallel and ligated. They are both subjected to S_1 nuclease digest to
remove residual ssDNA and transfected at the same time into the same
batch of spheroplasts. The proportion of amber mutants and wild type
phage produced from the control replication are measured in parallel
with the fidelity assay. The control thus acts to measure the frequency of expression of the mismatch in the synthetic (-) strand DNA
in the heteroduplex given that it can vary for the following reasons:
the efficiency of mismatch repair in the spheroplasts; the viability
of the phage produced from the DNA under these conditions; artifacts
caused by (variable) residual parental ssDNA in the assay; artifacts
caused by non-specific priming from contaminating fragments of the
Escherichia coli chromosome (this is minimized by using a synthetic
primer at high concentration rather than a restriction fragment; loss
of 5'region of primer by strand displacement. We find that the control gives, in general, from 40% to 100% expression of the minus
strand, depending on the age and batch of spheroplasts. It appears
that the mismatch repair systems can become defective with increasing
age of the spheroplasts, so allowing the minus strand expression to
dominate. When there is poor replication, the frequency of minus
strand expression in the control can drop to 10% or less. In all
cases, however, the measurements of accuracy determined from the
fidelity assay are highly reproducible when normalized relative to
the control.

The second method, that of Liu et al (7) performs, the RF→RF
stage using the T_4 enzyme system. This has the merit of producing
much progeny phage from the parental template (17, 18) and also
permits different components of the enzyme system to be omitted to
probe their importance in fidelity. A drawback is that both strands
of the template are copied thus complicating the analysis of the
effects of substrate bias kinetics.

The third method, that of Fersht (8), uses the RF→SS replication.
The virtue of this system is that it is thought to be the one used
in vivo and uses the important DNA polymerase III holoenzyme. Other
advantageous features are: many progeny single strands are produced
from one template strand so that the phage produced from the transfection of total DNA are phenotypically progeny; only one strand
of DNA is copied so that kinetic analysis of pool bias studies is
facilitated; single-stranded DNA is produced so that there are no
complications from mismatch repair mechanisms now known to remove
errors of mismatches from duplexes. Although this method does not
have the wide range of general applications of that of Weymouth and
Loeb (6), it is ideally set up for measuring the accuracy of DNA
replication in vitro and comparing it with the accuracy in vivo in
order to study the kinetic laws of spontaneous mutation arising from
base substitution.

Using this method, we were able to show for the first time that
phage DNA may be replicated in vitro with a fidelity similar to that

reported in vivo. Base substitution mutations, however, were found
to be induced by biasing the concentrations of dNTPs according to well
defined rate laws. In particular, it was shown that the concentration
of the next nucleotide to be incorporated may be crucial (8, 19).

Choice of mutant of ϕX174

One particular mutant aml6 (AB5276G→T), proved most useful in
these studies (19, 20). Seven of the eight possible revertants produced in vitro via a variety of mispairings have been isolated and
their DNA sequenced. The revertants may be grouped into classes according to their temperature sensitivity of growth, facilitating subsequent analysis and enabling the relative frequencies of misincorporations arising from purine:pyrimidine, purine:purine and pyrimidine:
pyrimidine mismatches to be calculated.

The cost-selectivity equation - A simple format for the analysis of fidelity

Most replication and proofreading mechanisms may be reduced to
the basic form of eq. (c). The DNA polymerase-DNA complex first binds
the deoxynucleoside triphosphate and adds the nucleotide to the primer
terminus (insertion). There is then the choice between elongation or
excision. The overall selectivity in incorporating correctly matched
and incorrectly matched nucleotides thus has two factors representing
(i) discrimination in the initial insertion and (ii) discrimination
in editing.

$$\text{E.DNA} + \text{dNTP} \underset{f}{\rightleftharpoons} \text{E.DNA.dNTP} \rightarrow \text{E.DNA-N} \underset{\underline{f}'' \text{ excision}}{\overset{\underline{f}' \text{ elongation}}{<}} \quad (c)$$

The discrimination in editing may be further subdivided into
discrimination in elongation and in excision. The three discrimination
factors are defined by

$$\underline{f} = \frac{\text{rate of insertion of correctly matched nucleotide}}{\text{rate of insertion of incorrectly matched nucleotide}} \quad (d)$$

($\underline{f}$ is the reciprocal of the misinsertion frequency)

$$\underline{f}' = \frac{\text{rate of elongation from correct match}}{\text{rate of elongation from mismatch}} \quad (e)$$

$\underline{f}''$ = $\dfrac{\text{rate of excision of mismatch}}{\text{rate of excision of correct match}}$ (f)

The selectivity, $\underline{S}$, (which is equivalent to the overall specificity of the polymerase) is defined by the ratio of the overall rates $(\underline{v})$ of incorporation of the correct $(\underline{v}_c)$ and the particular incorrect $(\underline{v}_i)$ nucleotide allowing for the imbalances of the concentrations of dNTPs. That is:

$$\underline{v}_c/\underline{v}_i = \underline{S}(dNTP)_c/(dNTP)_i \qquad\qquad (g)$$

The misincorporation frequency, v, of the polymerase is given by $v = \underline{v}_i(\underline{v}_c + \underline{v}_i)$, i.e. by $v = \underline{v}_i/\underline{v}_c$ since $\underline{v}_i \ll \underline{v}_c$. Thus v is inversely proportional to $\underline{S}$ and is given by:

$$v = (dNTP)_i/(dNTP)_c\underline{S} \qquad\qquad (h)$$

Another important phenomenon is the cost of proofreading $\underline{C}$. This results from the excision of correctly incorporated dNMPs by the $3' \rightarrow 5'$-exonuclease activity of the polymerase and is defined by the ratio of the rate of production of dNMP in solution to that of total dNTP consumption (i.e. to the sum of the rates of dNMP incorporation and excision reactions). $\underline{S}$ depends upon $\underline{f}$, $\underline{f}'$, $\underline{f}''$ and $\underline{C}$ in a simple manner, the cost-selectivity equation (2):

$$\underline{S} = \underline{f}(1 + (\underline{f}'\underline{f}'' - 1)\underline{C}) \qquad\qquad (i)$$

We have measured $\underline{S}$, $\underline{f}$ and $\underline{C}$ for a combination of systems, as follows, and have calculated the proofreading specificity, $\underline{f}'\underline{f}''$.

Measurement of misinsertion frequencies (1/f)

The misinsertion frequencies catalysed by Pol I from Escherichia coli have been measured using kinetics methods that separated insertion from overall misincorporation (2, 22) and from direct measurements of misincorporation by pol-α where the misincorporation frequency is equal to the misinsertion frequency (Table 1). The various combinations of purine:purymidine and purine:purine mispairs are seen to arise at a frequency of about $10^{-5} - 10^{-4}$ while the pyrimidine: pyrimidine mispair is some 200 times less frequent.

Overall accuracy of Pol III holoenzyme in ϕX174 RF$\rightarrow$SS in vitro

The rate laws for misincorporation listed in Table 2 (20, 21) show that in two cases, the G:T and C:T mediated misincorporations, there is a dependence on the concentration of the following nucleotide to be incorporated. This is diagnostic of the presence of proof-

Table 1: Misinsertion frequencies

Mispair (monomer:template)	Pol I	Pol-α
G:T	8 $\times 10^{-5}$	7.6$\times 10^{-5}$
C:A	2.5$\times 10^{-5}$	<2 $\times 10^{-6}$
G:A	1.1$\times 10^{-5}$	<4 $\times 10^{-6}$
A:G	-	3 $\times 10^{-5}$
G:G	-	4.4$\times 10^{-5}$
C:T	5 $\times 10^{-7}$	-

reading by the 3'→5'-exonuclease activity of the polymerase (8). The
lack of the "following nucleotide effect" for the G:A mediated misincorporation is not necessarily inconsistent with the presence of
proofreading (21), but it is likely that little proofreading of the
purine:purine mediated misinsertion is occuring in practice here
(see below).

The fidelity of the enzyme in vivo can be estimated from these
rate laws by insertion of the values of the concentrations of dNTPs
in vivo into the rate laws in vitro (Table 2). The overall agreement
between calculated and observed gross mutation frequencies is excellent, but a finer analysis reveals discrepancies between the rates
of mutation in the RF→SS and the SS→RF stages (20).

Dissection of the specificity of DNA polymerase III holoenzyme

Insertion of the misinsertion frequencies, overall misincorporation frequencies and cost into the cost-selectivity equation gives
the data listed in Table 3. It is seen that the proofreading step is
directed primarily to the removal of purine:pyrimidine mediated misinsertions (2, 20, 21).

Limits imposed upon the fidelity of polymerases by the cost-selec-
tivity equation

It is seen from equation i that the selectivity of a polymerase
tends to a maximum as C tends to 1. The upper limit is $S = ff'f''$
(for $f'f'' \gg 1$). Since it is most unlikely that the specificity of
proofreading, $f'f''$, is greater than the specificity in the original
checking of pairing in the insertion reaction, f (19), and is now
seen in Table 3 to be less, S must be less than f^2. Thus, for most

Table 2: Rate laws for reversion in vitro and estimated reversion
          frequencies in vivo

Change (TAG→)	Base-pairing (monomer:template)	Rate law in vitro	Estimated ν in vivo*
GAG	G:A	1×10^{-6} (dGTP)/(dTTP)	5×10^{-7}
AAG	A:A	3×10^{-7} (dATP)/(dTTP)	4×10^{-7}
TGG	G:T	1×10^{-2} (dGTP)^2/(dATP)	$(1-5) \times 10^{-7}$
TCG	C:T	1×10^{-4} (dCTP)(dGTP)/(dTTP)	$(2-10) \times 10^{-9}$

*Calculated by substitution of the concentrations of dNTPs in vivo
into the rate law in vitro (19).

of the mispairs listed in Table 1, $\underline{S} < 10^9$. Since $\underline{C}$ must be less than
about 0.1 - 0.2 in practice in order not to consume too much of the
dNTPs wastefully, $\underline{S} < 10^8$. Hence a DNA polymerase with a simple 3'→
5'-exonuclease proofreading activity is unable to give the required
selectivity of 10^{10} observed in vivo in Escherichia coli (1). There
is thus the need for post-replicative mismatch repair in vivo to
obtain higher selectivity (2).

     The cost selectivity equation may be used to analyse the accu-
racy of mutant DNA polymerases. For example, Echols et al (23) have
measured the cost of editing by Pol III mutants from mutD and dnaQ
mutator strains of Escherichia coli and found values of 20% and 50%
respectively of the cost of wild type enzyme. Insertion of these
values into equation i shows that if the only effect of these muta-
tions is to decrease the 3'→5'-exonuclease activity (i.e. $\underline{f}$ and $\underline{f'f''}$
are unaffected), then the selectivities of the mutants would be lower
by only factors of 5 and 2 respectively! The mutator phenotype must,
if those measurements are correct, result from the mutant polymerases
being either deficient in specificity of insertion or specificity of
proofreading.

Table 3: Dissection of specificity of DNA polymerase III

Base pairing monomer:template	Specificity (S)	Initial discrimination[b] (f)	Contribution of editing to specificity (S/f)	Discrimination in editing (f'f")	Cost (C)
G:A	1×10^6	9×10^4	11	100	0.1
C:T[a]	2×10^8	5.5×10^6	36	300	0.13
G:T[a]	2×10^6	1.2×10^4	170	1.3×10^3	0.13

Notes: [a]The selectivity is a function of (dGTP) for these (see Table 1). The results are calculated for (dGTP) = 50 µM, the approximate value in vivo.

[b]Data from pol l.

REFERENCES

1. Fowler,R.G., Degnen,G.E. and Cox,E.C. (1974) Mol. Gen. Genet. 133,179.
2. Fersht,A.R., Knill-Jones,J.W. and Tsui,W.-C. (1982) J. Mol. Biol. 156,37.
3. Fersht,A.R. (1981) Proc. Roy. Soc. ser.B, 212,351
4. Brutlag,D. and Kornberg,A. (1972) J. Biol. Chem. 247,241.
5. Glickman,B.W. and Radman,M. (1980) Proc. Nat. Acad. Sci. U.S.A. 77,1063.
6. Weymouth,L.A. and Loeb,L.A. (1978) Proc. Nat. Acad. Sci. U.S.A. 75,1924.
7. Liu,C.C., Burke,R.L., Hibner,U., Barry,J. and Alberts,B.(1979) Cold Spring Harbor Symp. Quant. Biol. 43,469.
8. Fersht,A.R. (1979) Proc. Nat. Acad. Sci. U.S.A. 76,4946.
9. Tkeshelashvili,L.K., Shearman,C.W., Zakour,R.A., Koplitz,R.M. and Loeb,L.A. (1980) Cancer Res. 40,2455.
10. Kunkel,T.A., Meyer,R.R. and Loeb,L.A. (1979) Proc. Nat. Acad. Sci. U.S.A. 76,6331.
11. Kunkel,T.A., Eckstein,F., Mildvan,A.S., Koplitz,R.M. and Loeb,L.A. (1981) Proc. Nat. Acad. Sci. U.S.A. 78,6734.
12. Kunkel,T.A. and Loeb,L.A. (1980) J. Biol. Chem. 255,9961.
13. Kunkel,T.A., Schaaper,R.M., Beckman,R.A. and Loeb,L.A. (1981) J. Biol. Chem. 256,9883.
14. Grosse,F., Krauss,G., Knill-Jones,J.W. and Fersht,A.R. (1983) EMBO J. 2,1515.
15. Cotes,A., Méchali,M. and Fersht,A.R., in preparation.
16. Shi,J.P. and Fersht,A.R. (1983) in preparation.
17. Hibner,U. and Alberts,B.M. (1980) Nature 285,300.
18. Sinha,N.K. and Haimes,M.D. (1981) J. Biol. Chem. 256,10671.
19. Fersht,A.R. and Knill-Jones,J.W. (1981) Proc. Nat. Acad. Sci. U.S.A. 78,4251.
20. Fersht,A.R. and Knill-Jones,J.W. (1983) J. Mol. Biol. 165,633.
21. Fersht,A.R. and Knill-Jones,J.W. (1983) J. Mol. Biol. 165,669.
22. Fersht,A.R., Shi,J.-P. and Tsui,W.-C. (1983) J. Mol. Biol. 165,655.
23. Echols,H., Lu,C. and Burgers,J. (1983) Proc. Natl. Acad. Sci. U.S.A. 80,2189.

REPLICATION OF φX174 DNA BY CALF THYMUS DNA POLYMERASE-α :

MEASUREMENT OF ERROR RATES AT THE AMBER-16 CODON

F. Grosse, G. Krauss, J.W. Knill-Jones[1]
and A.R. Fersht[1]

Zentrum Biochemie, Abt. Biophysikalische Chemie
Medizinische Hochschule Hannover
Konstanty-Gutschow-Strasse 8
D-3000 Hannover, FRG

[1]Department of Chemistry
Imperial College of Science and Technology
London SW7 2AY, UK

INTRODUCTION

It is now generally accepted that in higher eukaryotes DNA
polymerase-α is the replicative enzyme (1). Recently we have purified
to near homogeneity a subspecies of DNA polymerase-α (9S-enzyme),
which contains a catalytically active core peptide (148 kd) and
several smaller subunits (59 kd, 55 kd, and 48 kd). We have reported
elsewhere (2) that the 9S polymerase-α is able to copy long stretches
of single-stranded phage DNA. Furthermore, it contains a powerful
primase activity. Thus, it can be regarded as a true replicative
entity. In the present study we have replicated in vitro the single-
stranded DNA of the bacteriophage φX174 (amber 16). The occurence
of revertants to several temperature-sensitive mutants have been
evaluated (3-5) in order to gain insight in the accuracy of the eukaryotic replicase.

RESULTS

In vitro replication of φX174 DNA

Single-stranded φX174 DNA was copied nearly completely within
90 minutes by using the self-priming reaction of DNA polymerase-α
for the initiation of DNA synthesis. With different oligonucleotides

535

as primers (the restriction fragment H15 (6), the synthetic 12mer
and 14mer oligonucleotides described below), an elongation by about
1000 nucleotides was observed within 90 minutes. No remaining purely
single-stranded DNA was detectable, indicating that more than 95% of
the amber sites have been copied. Increasing the concentration of one
dNTP up to 5 mM and decreasing that of an other dNTP to 0.05 mM
(nucleotide pool bias) did not change the rate and extent of replication significantly.

Fidelity assay

After replication of phage DNA, either by using the primase
activity or oligonucleotide primers, the DNA was transfected to
spheroplasts of the permissive host Escherichia coli C600 (supE).
Within 10 minutes, enough phage particles of unchanged or mutated
phages were produced, which then were used to infect intact Escherichia coli cells. Plating of infected spheroplasts on Escherichia
coli CQ2 (supF, permissive for ϕX174am16) scored the number of unchanged, i.e. correctly replicated phage DNA. Plating of infected
speroplasts on Escherichia coli C (a wild-type strain, nonpermissive
for ϕX174am16) scored the number of mutations at the amber-site. From
the size of plaques on Escherichia coli C at different temperatures
it is possible to distinguish seven different phenotypes. Since all
these phenotypes can be correlated to distinct genotypes at the amber
codon, a direct evaluation of the frequencies of base:base mispairs
is possible (Table 1).

ϕX174am16 DNA, replicated in vitro with and without added primers, exhibited an overall reversion frequency of $1.2 \cdot 10^{-4}$ (revertants
over unchanged progenies), i.e. an error rate of 1/8300 for the sum
of the seven possible mutations. Evaluation of the temperature de-
pendencies of revertant plaque sizes showed that wild-type and wild-
type-related strains (ψwt, ts$_{42}$, and ts$_{43}$, representing the sum of
C:T, G:T, G:A, and T:T mispairs) were present at a frequency of
$4.4 \cdot 10^{-5}$ (1/22700). For ts$_{38}$ (corresponding to an A:G mispair) we
have determined a reversion frequency of $4.8 \cdot 10^{-5}$ (1/20700) and for
ts$_{34/35}$ (sum of C:A and G:G mispairs) the reversion frequency was
$2.3 \cdot 10^{-5}$ (1/42600). All the values for accuracy data were corrected
for mismatch repair and minus-strand expression by a factor of 0.38,
see below.

The data from unbiased experiments are not very accurate, because the number of plaques scored and the number of representatives
(of each plaque size) replated were rather low. To improve the measurement of reversion frequencies, it is necessary to put pressure
on each position of the amber codon by biasing the nucleotide pool.

Table 1: Nucleotide substitutions possible at aml6 codon and resulting reversion frequencies

Base incurred in mispair at aml6 T . A . G			Pool bias	Resulting phenotype	Frequency of misinsertion (unbiased)	Error rate
G	-	-	G vs A	ψwt	$7.6 \cdot 10^{-5}$	1/13,500
C	-	-	C vs A	wt	not determined	-
T	-	-	T vs A	ts43	$< 2 \cdot 10^{-6}$	$< 1/500,000$
-	A	-	A vs T	unknown	not determined	-
-	G	-	G vs T	ts42	$< 4 \cdot 10^{-6}$	$< 1/250,000$
-	C	-	C vs T	ts34	$< 2 \cdot 10^{-6}$	$< 1/500,000$
-	-	A	A vs C	ts38	$2.8 \cdot 10^{-5}$	1/35,700
-	-	G	G vs C	ts35	$4.4 \cdot 10^{-5}$	1/23,000
-	-	T	T vs C	ochre	not determined	-

Pool bias studies

Figure 1 shows the dependence of the reversion frequency to a distinct phenotype on the bias of one dNTP over a second one. In all experiments, we observed a strictly linear relationship between the reversion frequency and the nucleotide pool bias. From the slope of the curves of six different bias experiments, we have evaluated the reversion frequencies for the unbiased case of the six corresponding base:base mismatches. The data of these frequencies are given in Table 1. G:T, G:G, and A:G mispairs were predominant with frequencies of $7.6 \cdot 10^{-5}$ (1/13500), $4.4 \cdot 10^{-5}$ (1/23000), and $2.8 \cdot 10^{-5}$ (1/35700), respectively, whereas C:A, G:A, and T:T mispairs were below the limit of detection ($<5 \cdot 10^{-6}$).

Two of the nine possible permutations at the amber codon could not be evaluated, because the A:A mispair corresponds to an unknown phenotype, and the T:G mispair leads to the ochre nonsense codon. The bias of dCTP over dATP (C:T mispair, leading to true wild-type) gave no viable progenies for unknown reasons.

Incorporation of the first nucleotide following a 3'-end

The synthetic dodecamer primer 5'TTTTCCCAGCCT3' covers the A and G position of the am16 codon, allowing only the incorporation of a base opposite the T. The misinsertion of G opposite T expressed a pseudo-wild-type phenotype with a frequency of $1.5 \cdot 10^{-4}$, which is slightly higher than that from the pool bias studies (Table 1). However, there was no increase of mutation frequency with increasing the nucleotide pool bias (5).

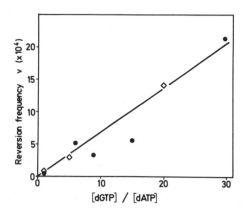

Figure 1: The effect of nucleotide pool bias on the reversion frequency (ν) to a distinct phenotype (ψwt).
● Priming with the primase activity of pol-α.
◇ Priming with a restriction fragment of φX174, located 37 nucleotides upstream from the amber codon.

Control for expression of mutations in the minus-strand

Mismatch repair mechanisms of the Escherichia coli spheroplasts correct mismatches in either of the two strands to a varying degree (7). To take account for these effects, φX174 DNA was replicated by using a synthetic 14mer primer 5'CCCAGCCTGAATCT3'. This covers the amber codon completely in forming a G:T mispair, which corresponds to the formation of a ψwt-phenotype. The expression of the minus-strand could thus be determined to be 38%.

DISCUSSION

Three main conclusions can be drawn from our work:
1. The eukaryotic replicase, DNA polymerase-α, is surprisingly in-accurate. This cannot be ascribed to contamination with terminal transferase, because addition of up to five units of TdT to the re-plication assay did not increase the error rate significantly. The low fidelity of DNA polymerase-α is certainly due to the fact that - unlike the prokaryotic and most viral replicases - it does not possess an integrated proofreading activity.
2. A proofreading exonuclease, which acts separately from the poly-merase activity, is in higher eukaryotes rather unlikely. The misin-sertion frequencies reported here, clearly represent a lower limit of those occurring. Misinsertions that are not elongated by polyme-rase-α will have a high chance of being corrected by the 3'->5' exo-nucleases of the bacterial replication and repair system. Thus, our data represent stably incorporated, i.e. misinserted and elongated errors. To explain the eukaryotic mutation rate of $10^{-9} - 10^{-11}$ in vivo (8), powerful postreplicative mismatch repair systems must be invoked.
3. The error rates of polymerase-α confirm the Topal-Fresco model of base:base mispairing, which is based on the estimated frequencies of enol- and imino- tautomers and on syn/anti conformations of the bases (9).
The linear relationship of reversion frequencies on pool biases and the very similar misinsertion frequency for the first nucleotide following a primer are not in agreement with Hopfield's energy relay model (10).

REFERRENCES

1.   Kornberg,A. (1980) in: DNA-Replication, published by Freeman, San Franciso.
2.   Grosse,F. and Krauss,G., submitted for publication
3.   Fersht,A.R. and Knill-Jones,J.W. (1981) Proc. Natl. Acad. Sci. USA 78, 4251.
4.   Fersht,A.R. and Knill-Jones,J.W. (1983) J. Mol. Biol. 165, 633.
5.   Grosse,F., Krauss,G., Knill-Jones,J.W. and Fersht,A.R. (1983) EMBO J. 2, 1515.

6. Brown,P.D. and Smith,M. (1977) J. Mol. Biol. 116, 1.
7. Baas,P.D. and Jansz,H.S. (1972) J. Mol. Biol. 63, 557.
8. Drake,J.W. (1969) Nature, 221, 1132.
9. Topal,M.D. and Fresco,J.A. (1976) Nature 263, 285.
10. Hopfield,J.J. (1980) Proc. Natl. Acad. Sci. USA 77, 5248.

IX
DNA Methylation and DNA Methylases

STUDIES ON THE ROLE OF dam METHYLATION AT THE ESCHERICHIA COLI

CHROMOSOME REPLICATION ORIGIN (oriC)

Patrick Forterre, Fatima-Zahra Squali, Patrick Hughes
and Masamichi Kohiyama

Institut Jacques Monod
Université Paris VII
2, place Jussieu
75251 Paris Cedex 05, France

INTRODUCTION

The 245 bp replication origin of Escherichia coli (oriC) contains 11 GATC whereas only one is expected at random (1). 8 of these
GATC are conserved in the ori consensus sequence of the Enterobacteriacae (2). The regions around oriC are also GATC rich. This observation raises the question, why so much GATC in and around oriC ? A
current view relates this phenomenon to the systematic adenine methylation of the GATC in Escherichia coli (dam methylation) (3). One
hypothesis is that dam methylation is required for the functionning
of oriC. For instance, dam methylation could help melting of DNA at
oriC since A(CH₃)-T base pairs are less stable than A-T ones (4).
Indeed dam methylation lowers the calculated DNA duplex stability
profile of oriC (4). In contradiction with the above hypothesis, a
dam⁻ mutant without residual methylation at oriC grows well (5).
Nevertheless, since a dam⁻ polA⁻ double mutant is lethal (6), one can
imagine that new DNA initiation mechanism dependent on DNA polymerase I occurs in dam⁻ mutants. A second hypothesis to explain the
abundance of GATC in and around oriC is that GATC tend to concentrate
mismatch repair enzymes in that region, enhancing conservation of
their DNA sequence (1). Indeed, a current hypothesis is that dam
methylation directs the mismatch repair enzymes towards correction
of the new DNA strand (7).

In vitro replication of a GATC unmethylated oriC containing plasmid

To determine if dam methylation was required for oriC dependent DNA replication in vitro, we compared replication of a methylated (Dam⁺) and a GATC unmethylated (Dam⁻)oriC containing plasmid (pOC42) using an enzyme system prepared according to Fuller et al. (8). pOC42 is a chimera of pBR322 and the oriC containing PstI fragment of pCM959 (9). To avoid replication through any polI dependent mechanism at oriC or pBR322 origin we prepared the enzyme system from a polA⁻ strain. As expected, this extract was able to replicate pOC42 but not pBR322. The absence of dam methylation in pOC42 isolated from a dam⁻ strain was confirmed by the use of the isoschizomers DpnI, MboI and Sau3A, which respectively cut only methylated GATC sequences, unmethylated ones, and both (10, 11). Using the same protocol, we found that Dam⁻ plasmids were not methylated during their incubation with the enzyme system.

When in vitro replication was performed according to Fuller et al., i.e., in the absence of added KCl, dNTP incorporation with pOC42 Dam⁻ as template was only 40 to 60% of that obtained with Dam⁺ pOC42 (Figure 1). However, the template capacity of both plasmids was the same in the presence of KCl (50 to 80 mM), a representative experiment is illustrated in Table 1. If dam methylation helps DNA melting at oriC, one should expect that Dam⁻ plasmid replication is more sensitive to relaxation by gyrase inhibitors. In fact, Dam⁺ and Dam⁻ pOC42 replications were inhibited by novobiocine in the same concentration range. In agreement with Tabata et al. (12), we observed that oriC dependent dNTP incorporation surprisingly did not start at

Table 1: Effect of added KCl on Dam⁺ and Dam⁻ pOC42 replication in vitro

Added KCl (mM)	dNTP incorporated (pmoles)		
	Dam⁺	Dam⁻	Dam⁺/Dam⁻
0	215	83	(0,38)
40	323	248	(0,76)
60	288	260	(0,90)
75	252	270	(1,07)
100	277	157	(0,56)

The reaction was performed according to the legend of Fig. 1 except for the addition of KCl in the reaction mixture. Incubation was for 60 min at 30°C.

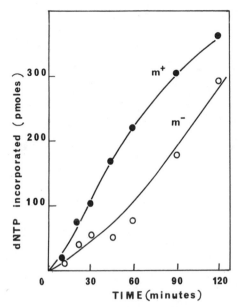

Figure 1: In vitro DNA replication kinetics of Dam⁺ and Dam⁻ pOC42.
*The plasmids (960 pmoles in nucleotides) have been incubated at 30°C
with an enzymatic extract (360 µg of proteins) prepared from the
strain 1002 (F⁻, str^R, rha⁻, polAl) in a reaction mixture for DNA
replication (50 µl total). At the time indicated, the reaction was
arrested by addition of TCA (5%) and the radioactivity in the TCA
insoluble material was determined. Preparation of the enzymatic fraction and composition of the incubation mixture were according to
Fuller et al. (5). Black circle: Dam⁺ pOC42; white circle: Dam⁻ pOC42.*

oriC but anyway on the oriC plasmid; specific initiation at oriC
nevertheless occured upon addition of partially purified dnaA protein
(data not shown). We found that, in any cases, the location of DNA
initiation was the same for Dam⁻ and Dam⁺ plasmids.

We concluded that dam methylation is dispensable for the functionning of oriC in vitro. Because the salt concentrations (added KCl
plus salt in the enzyme extract) which abolish the difference between
Dam⁺ and Dam⁻ plasmids replication were in the physiological range,
dam methylation may also be, dispensable for initiation of DNA replication in vivo. Nevertheless, it remains to be seen if DNA replication
starts at oriC in a dam⁻ strain or at a secondary origin. Our results
also demonstrate that DNA polymerase I is not required for DNA initiation in vitro even when GATC is unmethylated. Therefore, lethality
of dam⁻ polA⁻ double mutants should not be provoked by a defective
initiation of DNA replication.

     The difference between template efficiency of Dam$^+$ and Dam$^-$
plasmids at low or high KCl concentration (Table 1) supports the idea
that some components of the initiation mechanism interact with GATC
sequences and their methyl groups. In the absence of methylation,
these interactions could be abnormal outside a sharp range of KCl
concentration.

Dam methylation in other bacterial replicons

     Certain regions in the sequence of oriC are highly conserved
in all enterobacteria and in Vibrio harveyi; all these bacteria have
their GATC methylated (Dam$^+$) and their oriC sequences are GATC rich
(1-2). These obersevations led to the hypothesis that dam methylation
helps conservation of the DNA sequence through the action of mismatch
repair at oriC (1). Nevertheless, if we consider the whole eubacterial
urkingdom, the enterobacteria and Vibrio harveyi are closely related
(Figure 2). The selective pressure for conservation of strategic sequences at oriC may have been sufficient to explain the observed
similarity between their oriC sequences without specific intervention
of mismatch repair. Furthermore, some doubts exists if mismatch repair
is really instructed by dam methylation; for instance, hemi-methylated
DNA cannot be detected at the replication fork (5) and mismatch repair
has been demonstrated in Streptococcus which lacks dam methylation
(20). This last point prompted us to determine the phylogenetic distribution of dam methylation among bacterial species. For this purpose, we used the phylogenetic tree for eubacteria constructed by
Fox et al. (18) and the data in the literature on the occurence of
dam methylation among bacteria. Figure 2 indicates that dam methylation exists in one Cyanobacteria, one Moraxella, and in the subgroup
which includes the Enterobacteriacae and the species Haemophilus. We
think that within this subgroup, dam methylation is a recently acquired character which appeared after the divergence between the
Escherichia coli lineage and those of Chromatium minutissimum. In the
alternative hypothesis, i.e., an ancient trait, it should have been
lost independently several time (see Fig. 2). The restricted distribution and recent appearance of dam methylation argue against a role
of this phenomenon in some universal process such as mismatch repair.
In alternative to the methyl-instructed hypothesis, it has been sug-
gestet that mismatch repair is instructed by gaps or nicks in the
newly synthesized DNA (20). If the methyl-instructed hypothesis is
false, then the numerous GATC sites in and around oriC could not be
there to improve the sequence conservation of that region.

DISCUSSION

     Our enzymatic and phylogenetic studies suggest that over rep-
resentation of GATC sequences in and around oriC may be unrelated to
their methylation. Alternatively, palindromic GATC sequences could
help the formation of DNA secondary structure, or else, GATC could

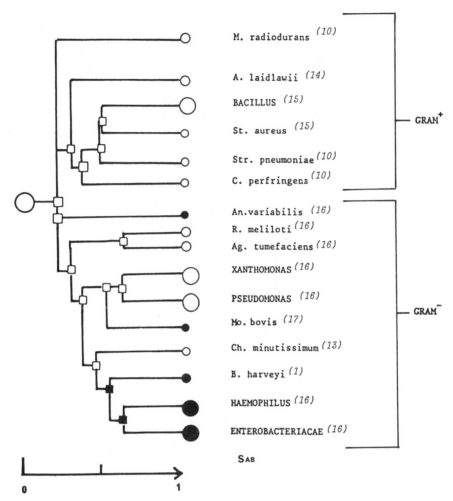

Figure 2: Schematic representation of the phylogenetic relationship between Dam⁺ and Dam⁻ eubacteria: This representation has been drawn from the evolutionary tree constructed by Fox et al. after comparative analysis of ribonuclease T1 16sRNA oligonucleotide sequences (18), and from the Bergey's manual data (19). We have approached the genus Haemophilus and the family Enterobacteriacae because a DNA sequence analogous to the Escherichia coli dam gene has been found in all bacteria tested of this genus and that family (16). Presence or absence of dam methylation has been deduced from published data (see reference on the figure). Genetic abreviation used are the following: M, Micrococcus; A, Acheloplasma; St, Staphylococcus; C, Clostridium; An, Anabaena; R, Rhizobium; Ag, Agrobacterium; Mo, Moraxella; Ch, Chromatium; B, Benecka. (White circle): Dam⁻ specie, genus or family; (black circle) : Dam⁺ specie, genus or family.

be part of a redundant recognition site for some initiation factor.
In that way, they could be methylated to protect them against restriction endonucleases produced by foreign bacteria. This idea is supported by the last compilation of restriction endonucleases target
sequences which indicates that GATC and sequences including GATC
are among the most frequent (21). Indeed, the phylogenetic data, which
indicate that dam methylation is characteristic of a bacterial subgroup, together with the high number of restriction enzymes which
recognize GATC sequences strongly suggest that dam methylation is
related to the phenomenon of modification-restriction. Finally, it
should be emphasized that, up to now, the only known bacterial replication origin sequences belong to the Enterobacteriacae and to a
related organism, Vibrio harveyi. These bacteria are all Dam$^+$ and
their ori sequences are all GATC rich (1). It would be very interesting to know if the ori sequence of Dam$^-$ bacteria, not too far remote
phylogenetically from the Dam$^+$ ones (for instance a Pseudomonas) is
GATC rich or not.

REFERENCES

1. Zyskind,J.W. and Smith,D.W. (1980) Proc. Natl. Acad. Sci. USA,
        77, 2460.
2. Zyskind,J.W., Cleary,J.M.- Brusilow,W.S.A., Harding,N. and
        Shmith,D.W. (1983) Proc. Natl. Acad. Sci. USA, 80, 1164.
3. Razin,A. and Friedman,J. (1982) in: Prog. Nucleic Acids Res.
        and Mol. Biol. 25, 33.
4. Kohiyama,M., Jacq,A. and Reiss,C. (1982) in: New approaches
        in eukaryotic DNA replication, Plenum press ed., New York,
        London, 235.
5. Szyf,M., Gruenbaum,Y., Ureli-Shoval,S. and Razin,A. (1982)
        Nucl. Acids Res. 10, 7247.
6. Marinus,M.G. and Morris,N.R. (1974) J. Mol. Biol. 85, 309.
7. Wagner,R.J. and Meselson,M. (1976) Proc. Natl. Acad. Sci. USA,
        73, 4135.
8. Fuller,R.S., Kaguni,J.M. and Kornberg,A. (1981) Proc. Natl.
        Acad. Sci. USA, 78, 7370.
9. Messer,W., Heiman,B., Meijer,M. and Hall,S. (1980) Symp. Molec.
        Cell Biol. 19, 161.
10. Lacks,S. and Greenberg,B. (1977) J. Mol. Biol. 114, 153.
11. Geier,G.E. and Modrick,P. (1979) J. Biol. Chem. 254, 1408.
12. Tabata,S., Oka,A., Sugimoto,K., Takanami,M., Yasuda,S. and
        Hirota,Y. (1983) Nucl. Acids Res. 11, 2617.
13. Hall,R.S. (1971) in: The modified nucleosides in nucleic acids,
        Columbia university press ed., New-York, London, 281.
14. Dybvig,K., Swinton,D., Maniloff,J. and Hattman,S. (1982)
        J. Bact. 151, 1420.
15. Dreiseikelmann,B. and Wackernagel,W. (1981) J. Bact. 147, 259.
16. Brooks,J.E., Blumenthal,R.M. and Gingeras,T.R. (1983)
        Nucl. Acids Res. 11, 837.

17.   Gelinas,R.E., Myers,P.A. and Roberts,R.J. (1977) J. Mol. Biol.
       114, 169.
18.   Fox,G.E., Stackebrandt,E., Hespell,R.B., Gibson,J., Maniloff,J.,
       Dyer,T.A., Wolf,R.S., Balch,W.E., Tanner,R.S., Magrum,L.J.,
       Zablen,L.B., Blakemore,R., Gupta,R., Bonem,L., Lewis,B.J.,
       Stalh,D.A., Lvehrsen,K.R., Chen,K.N. and Woese,C.R.
       (1980) Science, 209, 457.
19.   Buchanan,R.E. and Gibbons,N.E. (1974) Bergey's manual of
       determinative bacteriology, Williams and Wiltems eds.,
       Baltimore, ed 8.
20.   Lacks,S.A., Dunn,J.J. and Greenberg,B. (1982) Cell 31, 327.
21.   Roberts,R.J. (1983) Nucl. Acids Res. 11, r135.

DNA METHYLATION AND DNA STRUCTURE

S. Spadari, G. Pedrali-Noy, M. Ciomei, A. Rebuzzini,
U. Hübscher[1] and G. Ciarrocchi

Istituto di Genetica Biochimica ed Evoluzionistica
C.N.R.
I-27100 Pavia, Italy

[1]Institut für Pharmakologie und Biochemie
Universität Zürich-Irchel
CH-8057 Zürich, Switzerland

Novel DNA substrates, such as hemimethylated DNA, supercoiled, partially or fully relaxed DNA, either intact or UV-irradiated, have been used to study either some aspects of the mechanism of DNA methylation or some aspects of the effect of pathological and physiological DNA methylation on DNA conformation.

Detection of DNA methylase activity after SDS-PAGE

The results are shown in Figure 1. High M.W. unmethylated DNA of M. lysodeicticus was polymerized into a minigel of 0.6 ml total volume to serve as substrate for HpaII methylase (left) or for DNA methylase activity present in cytoplasmic (middle) and nuclear (right) calf tymus extracts. S-adenosyl-L-(methyl-^{3}H)methionine was used as label cofactor for DNA methylases.

Isolation from HeLa cells of a DNA methylase that methylates hemimethylated DNA more efficiently than unmethylated DNA

We have isolated a DNA methylase from the nuclei of HeLa cells. The final preparation, purified approximately 1,000 fold through high-speed centrifugation, DEAE-filtration, DEAE-chromatography,P-11 chromatography and native DNA cellulose chromatography, showed activity with native, denatured DNA or hemimethylated DNA as substrates. However the relative rate of methylation of hemimethylated DNA

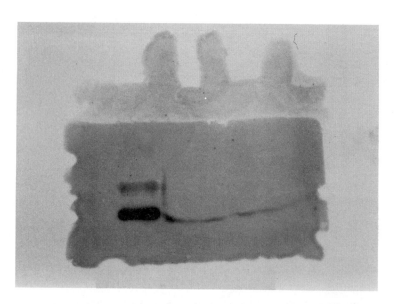

Figure 1: Autoradiogram of DNA methylase activities following SDS-polyacrylamide gel electrophoresis and in situ renaturation of the enzymes. Preparation of the enzymes for electrophoresis, electrophoretic separation in a 9% polyacrylamide gel and renaturation of the activities were performed as described in a minigel system (1,2). DNA methylase activity was measured by incubating the minigel in 2.5 ml mixture containing 50 mM Tris-HCl (pH 7.5), 10 mM EDTA, 1 mM dithiothreitol and 1 μM S-adenosyl-L-(methyl- H)methionine (10,000 cpm/pmol). After incubation at 37°C for 10 hr, the gel was treated with proteinase K (1 mg/ml) containing 1% (w/v) SDS and further incubated for 2 hr at 37°C. Finally the DNA in the gel was precipitated, the unincorporated label washed out and the gel prepared for autoradiography as described (1).

(prepared from RNA-primed single stranded DNA using DNA polymerase I in the presence of deoxyribotriphosphates including 5-methyl dCTP) was approximately 6 times higher than with native DNA.

These results suggest that we have isolated a DNA methylase acting preferentially on half-methylated sites and thus able to lead to the maintenance of the methylation pattern on the DNA during replication.

Effect of DNA structure on the mechanism of action of DNA methylase

The results of methylation kinetics of different forms of pAT153 plasmid DNA by HpaII DNA methylase are shown in Figure 2.

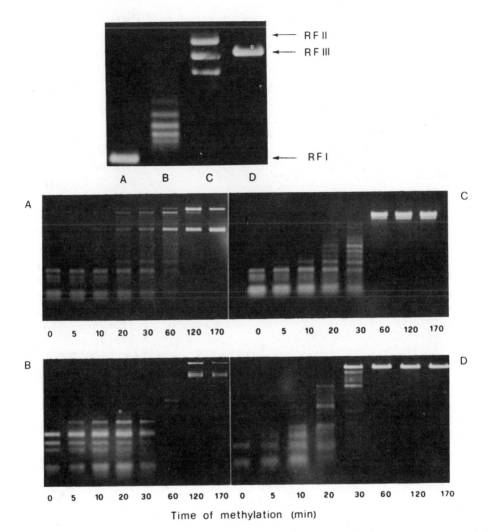

Figure 2: Methylation kinetics of different forms of pAT153 plasmid DNA by the HpaII methylase. The top part indicates the electrophoretic mobilities of different forms of pAT153 DNA used for the methylation reaction; A: negatively supercoiled pAT153 (RFI); B and C: topoisomers partially relaxed with the M.Luteus topoisomerase I; D: linear form (RF III) obtained by treating pAT153 RFI with Eco Rl. Each DNA form was then used separately for methylation kinetics by HpaII DNA methylase. Methylation was then followed by time-increasing resistance to HpaII endonuclease.

The appearance of final products of DNA methylation at early times (Figure 2A) suggests that HpaII methylase methylates supercoiled DNA in a processive way, while the accumulation of intermediate pro-

ducts together with the appearance of final products only at late
times suggest a non-processive mode of action on relaxed DNA (Figure
2D). Intermediate patterns of methylation are observed with partially
relaxed DNAs (Figure 2B and C).

        This mode of action by HpaII methylase on DNA with different
topological states might favour the idea that in vivo a DNA methylase
might methylate the newly synthesized DNA in a distributive way. Thus
the fully methylated parental strand is still distinguishable from
the newly synthesized, undermethylated DNA to allow post-replication
DNA repair to take place on the daughter strand. Full methylation of
the new strand could eventually be achieved by processive methylation
after the DNA has returned into a supercoiled form.

Structural alterations of pathologically or physiologically methylated DNA

        Many different physical and chemical events such as UV-irradia-
tion (3, 4), depurination or alkylation (5) cause DNA unwinding detectable as a reduction in the degree of supercoiling of partially
relaxed DNA topoisomers. But does a physiological modification of
DNA components alter the tertiary structure of the DNA?

        In the attempt to answer this question we have compared the
electrophoretic mobility of partially relaxed DNA topoisomers following methylation with methyl methanesulfonate (MMS mainly methylates
the N-3 and N-7 positions of purines - pathological methylation -)
and following methylation of the C_5 position of cytosine mediated
by the HpaII DNA methylase in the sequence C^{m}CGG (physiological methylation).

        As a substrate we have used partially relaxed DNA topoisomers
of the plasmid pAT153. The results are shown in Figure 3.

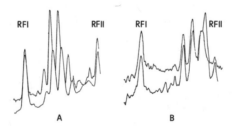

*Figure 3: Alteration of DNA electrophoretic mobility on agarose gel
of pathologically or physiologically methylated DNA. Partially relaxed
pAT153 DNA was methylated in A with MMS (30 methylated bases/molecules)
and in B with HpaII DNA methylase (44 methylated bases/molecules). Upper lanes are the unmethylated control DNAs. Part of the fully relaxed
molecules in B are still sensitive to the action of HpaII endonuclease
utilized to control the degree of methylation.*

Pathological methylation with MMS (Figure 3A) alters the tertiary structure of DNA as revealed by the lower electrophoretic mobility in agarose gels of single DNA topoisomers of pAT153. The value of DNA unwinding for pathological methylation, calculated from the average number of modified sites at which one topoisomers band would reduce the number of superhelical turns by one, was -3.4º for N-7, N-3 methylpurine. Physiological methylation (all 44 CCGG sequences per molecule were methylated as revealed by the total acquired resistance to HpaII DNA methylase) causes no detectable alteration in electrophoretic mobility of single pAT153 topoisomers (Figure 3B). These results might suggest a possible mechanism of recognition of damage sites by repair mechanisms that are not single-damage specific.

<u>DNA methylation is probably not inhibited by modifications of DNA tertiary structure but is inhibited by DNA modifications that interfere with the processive scanning function of DNA methylase</u>

The results of Figure 4 show that UV-irradiation (the pAT153 RFI contained an average of 5.5 pyrimidine dimers per 1 Kb), known to alter the DNA tertiary structure (3, 4), only slightly inhibits the rate of DNA methylation of pAT153 DNA by HpaII methylase.

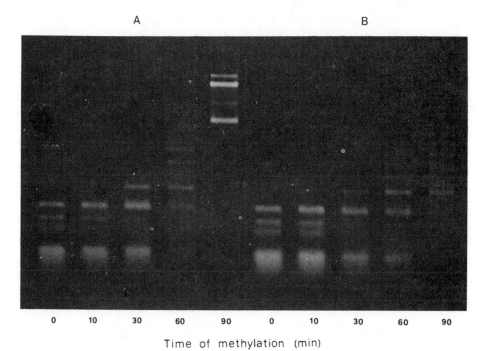

Time of methylation (min)

Figure 4:   Methylation kinetics of intact (A) or UV-irradiated (B) pAT153 plasmid DNA by HpaII methylase. pAT153 RFI DNA contained an average of 20 pyrimidine dimers per molecule corresponding to 5.5 pyrimidine dimer per 1 Kb. DNA methylation was followed by time-increasing resistance to HpaII endonuclease.

In contrast, other DNA damaging events that also alter the DNA tertiary structure (such as depurination or alkylation), strongly inhibit DNA methylation. Approximately 45% inhibition of DNA methylation was observed when 1 Kb of DNA contained an average of 0.2 apurinic sites (6). This is probably ascribable to the interference of modified sites with the processive scanning function of DNA methylase.

ACKNOWLEDGEMENTS

We thank EMBO for a short term fellowship to S.S. This work was partially supported by the Programma Finalizzato "Controllo Malattie infettive", "Oncologia" and "Medicina preventiva e riabilitativa" and by the Swiss National Science Foundation, Grant 3.006-0.81.

REFERENCES

1.   Spanos,A. and Hübscher,U. (1983) in: Methods in Enzymology
              (Hirs,C.W.S. and Timasheff,S.N., eds.) Academic Press,
              New York 91, 265.
2.   Hübscher,U. (1983) EMBO J. 2, 133.
3.   Ciarrocchi,G. and Pedrini,A.M. (1982) J. Mol. Biol. 155, 177.
4.   Ciarrocchi,G. and Sutherland,B.M. (1983) Photochem. Photobiol.
              38, 259.
5.   Ciomei,M., Spadari,S., Pedrali-Noy,G. and Ciarrocchi,G. (1984)
              Nucl. Acids Res. 12, 1977.
6.   Wilson,V.L. and Jones,P.A. (1983) Cell 32, 239.

CONTRIBUTORS

Higgins,P.	63,69	Mirambeau,G.	423
Hoffmann-Berling,H.	385	Modjtahedi,N.	113
Hohn,T.	121	Müller,B.	331
Holck,A.	457	Munn,M.	17
Horiuchi,K.	185	Muzyczka,N.	151
Huber,B.	241	Olivera,B.M.	63,441
Hübscher,U.	321,551	Orr,E.	395
Hughes,P.	543	Ottiger,H.-P.	321
Jansz,H.S.	221,231	Otto,A.M.	337
Johanson,K.O.	315	Overbeeke,N.	107
Jongeneel,C.V.	17	Paci,M.	467
Karawya,E.	343	Pato,M.	69
Karwan,R.	513	Pawlik,R.T.	467
Kleppe,K.	457	Pedrali-Noy,G.	169,551
Knill-Jones,J.W.	535	Pedrini,A.M.	449
Knippers,R.	127	Pfeiffer,P.	121
Koerner,T.J.	355	Philippe,M.	295
Kohiyama,M.	543	Plevani,P.	281
Kool,A.J.	107	Pon,C.L.	467
Kornberg,A.	3	Prescott,M.	507
Krauss,G.	307,535	Prieto,I.	35
Kuenzle,C.C.	489	Pülm,W.	127
Kwant,M.M.	93	Rebuzzini,A.	551
Lacatena,R.M.	215	Reimann,A.	45
Lanka,E.	265	Rein,D.C.	355
Lammi,M.	467	Resibois,A.	69
Laquel,P.	121	Reygers,U.	385
Lavenot,C.	423	Riva,S.	497
Lazaro,J.M.	35,193	Rommelaere,J.	143
LeBowitz,J.H.	77	Sala,F.	169
Linder,P.	209	Salas,M.	35,193
Litvak,S.	249	Samulski,R.J.	151
Lossius,I.	457	Schindler,R.	331
Losso,M.A.	467	Schnabel,A.	373
Lother,H.	395	Schneider,E.	331
Lurz,R.	395	Scovassi,A.I.	519
Mangi,G.	281	Sedgwick,S.G.	363
Manlapos-Ramos,P.	63	Seiter,A.	385
van Mansfeld,A.D.M.	221	Sheinin,R.	295
Marilley,M.	163	Shlomai,J.	409
Markau,U.	241	Sjastad,K.	457
McHenry,C.S.	315	Sobczak,J.	479
McLennan,A.G.	507	Solari,A.	249
McMacken,R.	77	Spadari,S.	169,551
Mellado,R.P.	35	Spanos,A.	363
Meyer,R.R.	355	Sperka,S.	241
Meyer,T.F.	45	Squali,F.-Z.	543
Miano,A.	467	Srivastava,A.	151